To Tony

It was a pleasure working with you at Hersh Park.

Lou Samuels.

6/29/84

METALLOGRAPHIC POLISHING BY MECHANICAL METHODS

Third Edition

LEONARD E. SAMUELS
Chief Superintendent
Department of Defence
Materials Research Laboratories
Australia

American Society for Metals
Metals Park, Ohio 44073

Copyright © 1982
by the
AMERICAN SOCIETY FOR METALS
All rights reserved

No part of this book may be reproduced, stored in a retrieval system, or transmitted, in any form or by any means, electronic, mechanical, photocopying, recording, or otherwise, without the prior written permission of the publisher.

Nothing contained in this book is to be construed as a grant of any right of manufacture, sale, or use in connection with any method, process, apparatus, product, or composition, whether or not covered by letters patent or registered trademark, nor as a defense against liability for the infringement of letters patent or registered trademark.

Library of Congress Catalog Card No.: 82-71790

ISBN: 0-87170-135-9

SAN: 204-7586

PRINTED IN THE UNITED STATES OF AMERICA

In memory of
PAUL M. UNTERWEISER
1916–1981

Preface

IN ITS BROADEST SENSE, metallography is the study of the internal structure of metals and alloys, and of the relation of structure to composition and to physical, chemical and mechanical properties. Many methods have been devised to determine internal structure, but microscopical examinations have always been among the more important. For most of the history of metallography they have been carried out by means of the optical microscope. The optical microscope has been joined in more recent years by the transmission and the scanning electron microscope, both of which now play significant roles. Nevertheless, there still is, and seemingly always will be, a place for optical microscopy in both industry and research, just as there is still a place for the visual examination of hand specimens and macro-examinations at low magnifications.

Any examination to reveal the structure of metals by optical microscopy involves three distinct processes: the preparation of a sectioned surface; the development of the structure on this prepared surface by a suitable etching process; and the actual microscopical examination of the surface. The three stages form an integrated whole, and the achievements of the over-all process are inevitably limited by the lowest standard attained by any one of the three. No one stage can be overlooked, and arguments as to their relative importance are pointless.

This book is concerned with the first of the three stages — namely, surface preparation; even at that, it is concerned only with mechanical methods of surface preparation. The approach is based on the assumption that optical metallography is a sufficiently important laboratory tool to warrant serious attention and that it is a tool that will find its full usefulness only when it is given this serious attention. The over-all objective is to provide an understanding of underlying principles, so that each new problem met with in the laboratory can be solved intelligently rather than by relying on intuition or traditional recipes.

The present book is based on one with the same title previously published by Sir Isaac Pitman and Sons Ltd, editions being published in 1967 and 1971. It is, however, a greatly expanded and revised version, incorporating much new and previously unpublished information.

Acknowledgments

THE ASSISTANCE RECEIVED over many years from my colleagues at Materials Research Laboratories is gratefully acknowledged. They have assisted immeasurably with many discussions, experiments and ideas. The names of these helpers are too numerous to include here, but special mention must be made of Dr. P. Dunn, who provided much of the basic information on plastics used in Chapter 2; Dr. R. W. Johnson, who provided the basic information on embedded abrasive in Chapter 4; and Dr. C. W. Weaver, who originated many of the ideas incorporated in the section on edge retention in Chapter 9. The quantitative data on abrasion and polishing rates owe much to the meticulous and thorough experimentation of Mr. B. Wallace, ably assisted by Mrs. A. Bonassin.

The following also kindly granted permission to reproduce illustrations:

- A. R. Entwistle, University of Sheffield (Fig. 1.2)
- D. Caplan, National Research Council of Canada (Fig. 2.2)
- A. Calabra, Dow Chemical Corp. (Fig. 2.7)
- M. Hatherly, University of New South Wales (Fig. 4.1b, 4.1d and 4.2f)
- B. Lawn, University of New South Wales (Fig. 7.3 and 7.4)
- B. J. Hockey, U.S. National Bureau of Standards (Fig. 7.8)
- Leco Corporation (Fig. 8.27)
- Buehler Ltd. (Fig. 8.30)
- J. M. Dickenson, Los Alamos Scientific Laboratory (Fig. 9.5)
- N. F. Kennon, University of New South Wales (Fig. 9.22)
- T. Piotrowski, Engelhard Industries Division (Fig. 9.40, 11.1b and 11.1c).

Contents

1 Introduction *1*

    Appendix 1-A: Sorby: The Founder of Metallography *3*
    References *7*

2 Specimen Mounting *9*

    Mounting by Means of Mechanical Clamps *9*
    Mounting in Plastic Cylinders *11*
        Requirements of Mounting Plastics *11*
        Properties of Mounting Plastics *12*
        General Fields of Usefulness of Various Plastics *20*
    Molding Methods for Thermoplastics and Thermosetting Plastics *21*
    Molding Methods for Casting Plastics *23*
    Vacuum Impregnation With Casting Plastics *25*
    Mount Dimensions *26*
    Mounting of Thin Specimens *27*
    Marking for Identification *28*
    Removal of Specimens From Plastic Mounts *28*
    Appendix 2-A: Method of Preparing a Conducting Plastic *28*
    Appendix 2-B: Method of Manufacturing a Mold for Epoxy Resins From a Polyvinyl Chloride Dipping Compound *29*
    Appendix 2-C: Method of Manufacturing a Mold for Epoxy Resins From a Cold-Cure Silicone Rubber *29*
    References *30*

3 Abrasive Machining: Principles of Material Removal *31*

    A Model of Abrasive Machining *32*
        The Basis of the Model *32*
            Characteristics of Coated Abrasive Papers *40*
            Mathematical Development of the Model *46*
            Influence of Specimen Material on Abrasion Rate *48*

Hardness *48*
Critical Rake Angle *50*
Efficiency of the Cutting and Plowing Processes *51*
Influence of Abrasive on Abrasion Rate *52*
Influence of Abrasion Fluids *54*
Usefulness of the Model *54*
Abrasion Rates Achieved in Practice *55*
Measurement of Abrasion Rates *55*
Characteristics of Waterproof Silicon Carbide and Aluminum Oxide Papers *55*
Materials That Cause Little Deterioration in the Abrasion Rate (Groups 1 and 2) *57*
Materials That Cause Severe Deterioration in the Abrasion Rate (Group 3) *62*
Summary *69*
Characteristics of Some Nonwaterproof Types of Aluminum Oxide Papers *70*
Characteristics of Diamond Laps *71*
Fine Fixed-Abrasive Laps *76*
Externally Charged Laps *77*
Internally Charged Laps *79*
Vitreous-Bonded Abrasive Wheels and Laps *79*
Appendix 3-A: Abrasive Grain Size *80*
Appendix 3-B: Methods of Determining Abrasion Rates *80*
Appendix 3-C: Plastic-Diamond Laps *82*
Appendix 3-D: An Abrasive-Wax Lap for Fine Abrasion *83*
References *84*

4 Machining With Abrasives: Surface Deformation *87*

Detection of Plastic Deformation by Optical Microscopy *87*
Cubic Metals Having High Stacking-Fault Energies *87*
Cubic Metals Having Low Stacking-Fault Energies *89*
Metastable Alloys Susceptible to Strain-Induced Transformations *91*
Massive Twinning and Recrystallization *94*
Deformation Bands *94*
Lamellar Structures *95*
Origin of the Deformed Surface Layer on Abraded Surfaces *98*
Structure of the Plastically Deformed Layer Produced During Abrasive Machining *99*
General Pattern of Deformation *99*
The Fragmented Layer *103*
Translation Parallel to the Surface *107*
Depth of the Deformed Layer *110*
Metallographic Characteristics of the Fragmented Layer *114*
Modifications of the Structure of the Deformed Layer Due to Strain-Induced Transformations *121*

Modifications of the Structure of the Deformed Layer Due to Massive Twinning *123*
Modifications of the Structure of the Deformed Layer Due to Recrystallization *126*
Modifications of the Surface Structure Resulting From Heating of the Surface *126*
Embedded Abrasive *131*
Appendix 4-A: Etching Methods *134*
Appendix 4-B: Taper Sectioning *137*
References *139*

5 Polishing With Abrasives: Principles *141*

Mechanisms of Polishing: Existence of the Beilby Layer *142*
    The Beilby Theory of Polishing *142*
    Beilby's Original Microscopical Evidence *143*
    Evidence of the Attainment of High Transient Temperatures *145*
    Continuity of Overgrowth of Deposits *149*
    Reflection Electron Diffraction *150*
    Removal of Material During Polishing *154*
    Direct Observations *154*
    Assessment of the Beilby Theory *154*
Mechanical Cutting Mechanisms of Polishing *154*
Chemical-Mechanical Mechanisms of Polishing *159*
Chemical and Electrochemical Mechanisms of Polishing *161*
Variables in Metallographic Polishing *161*
    Type of Abrasive *162*
    Type of Polishing Cloth *167*
Polishing Rates Achieved in Practice *169*
    Determination of Polishing Rates *169*
    Method of Adding the Abrasive *171*
    Grade of Abrasive *172*
    Type of Diamond *175*
    Quantity of Abrasive *175*
    Type of Polishing Cloth *179*
    Polishing Fluid *184*
    Load Applied to the Specimen *187*
    Specimen Material *187*
    Summary of Optimum Polishing Conditions *190*
Appendix 5-A: Method of Grading Aluminum Oxide Polishing Abrasive by Levigation *191*
Appendix 5-B: Method of Preparing a Carrier Paste for Diamond Abrasives *192*
References *193*

6 Polishing With Abrasives: Surface Deformation *195*

General Characteristics of the Plastically Deformed Layer *195*
    Structure of the Layer *195*

Translations Parallel to the Surface 198
Depths of the Deformed Layers 199
Modification of the Structure of the Damaged Layer Due to Formation of Deformation Twins 199
Modification of the Structure of the Damaged Layer Due to Recrystallization and Strain Relief 200
Embedded Abrasive 201
References 201

7 Brittle Materials: Principles 203

Mechanisms of Abrasion and Polishing 203
Surface Damage 209
References 213

8 Principles of the Design of Preparation Systems 215

Basic Concepts 215
Abrasion Artifacts 216
  Artifacts Originating in the Fragmented Layer 216
  Artifacts Originating in the General Deformed Layer 220
Polishing Artifacts 233
  Enhancement of Polishing Scratches by Etching 234
  Degradation of Grain Color Contrast 235
  Development of Scratch Traces by Etching 237
Practical Preparation Procedures 241
  Basic Preparation Sequences 241
  Final Polishing 243
Standard Procedures 248
  Cutting the Section 248
  Preliminary Machining 250
  Metallographic Abrasion 251
  Polishing 253
  Interstage Cleaning and Drying 260
Appendix 8-A: Operating an Electromagnetic Vibratory Polisher 262
References 265

9 Advanced and Special Preparation Methods 267

Advanced Methods of Final Polishing 267
  Skidding Techniques 267
  Chemical-Mechanical Techniques 270
    Conventional Etch-Attack Techniques 272
    Electromechanical Polishing 273
    Vibratory Polishing 279
  Alternate Etch-and-Polish Technique 285
  Chemical Polishing 288

Retention of Edges   *288*
   Effects of the Abrasion Stages   *289*
   Effects of the Polishing Stages   *294*
   Special Mounting Techniques   *298*
   Summary   *303*
Correct Representation of Cavities   *303*
Retention of Nonmetallic Inclusions   *305*
Retention of Graphite in Ferrous Alloys   *307*
Very Soft Materials   *313*
Very Hard Materials   *316*
Brittle Materials   *319*
Surface Oxides and Scales   *322*
Precious Metals and Refractory Metals   *326*
Polarization Contrast   *330*
Appendix 9-A: A Water – Propylene Glycol Polishing Fluid   *333*
Appendix 9-B: Electroplating Methods for Edge Protection   *333*
Appendix 9-C: Brashear Process for Silvering Prior to Electroplating   *335*
References   *336*

10 Nonabrasive Techniques   *339*

Etch Cutting and Machining   *339*
Spark Cutting and Machining   *340*
Microtome Cutting and Machining   *344*
References   *349*

11 Procedures for Some Common Metals and Alloys   *351*

Aluminum   *351*
   Artifacts   *352*
   Final Polishing   *352*
      Soft Alloys   *352*
      Alloys of Moderate Hardness   *353*
      Alloys Containing Intermetallic Phases   *353*
Beryllium   *354*
   Artifacts   *354*
   Final Polishing   *356*
   Note on Toxicity   *357*
Chromium   *357*
   Artifacts   *357*
   Final Polishing   *357*
Copper   *357*
   Artifacts   *358*
   Final Polishing   *359*
Gold   *359*
   Artifacts   *359*
   Final Polishing   *360*

Indium *360*
Iron-Base Alloys: Austenitic *361*
   Artifacts *361*
   Final Polishing *361*
Iron-Base Alloys: Ferritic and Martensitic *362*
   Artifacts *362*
   Final Polishing *362*
Iron-Base Alloys: Graphite-Containing *362*
Lead *362*
Magnesium *362*
   Artifacts *362*
   Final Polishing *362*
Molybdenum *364*
   Artifacts *364*
   Final Polishing *364*
Nickel *364*
   Artifacts *364*
   Final Polishing *364*
Platinum *364*
   Artifacts *364*
   Final Polishing *364*
Silver *365*
   Artifacts *365*
   Final Polishing *365*
Tin *365*
   Artifacts *365*
   Final Polishing *365*
Titanium *366*
   Artifacts *366*
   Final Polishing *367*
Tungsten *369*
   Artifacts *369*
   Final Polishing *369*
Uranium *369*
   Artifacts *369*
   Final Polishing *369*
   Note on Safe Handling *369*
Zinc *370*
   Artifacts *370*
   Final Polishing *370*
Zirconium *370*
   Artifacts *370*
   Final Polishing *370*
References *371*

APPENDIX: Glossary of Terms Used in Metallography *373*

INDEX *379*

CHAPTER 1

Introduction

OF THE THREE STAGES involved in the over-all process of metallographic examination (preparation, etching and microscopical examination), the preparation stage is without doubt the most neglected. It appears to involve so much skill and to be so laborious that it is often dismissed as a tedious step even by people who will go to a great deal of trouble during the microscopical examination itself. This is a disastrous attitude, because the most thorough planning of an experiment, immaculate etching, brilliant examinational methods and inspired interpretations will all be to no avail if preparation of the specimen has been so poor that not all the information available has been revealed or, worse still, that false information has been introduced.

Any experimenter who wishes to obtain reliable information by the use of optical metallography must first become proficient at specimen preparation. The history of the development of microscopical metallography provides a warning here. C. S. Smith[1] has pointed out that metallography arose two centuries later than biological microscopy because, until Sorby's pioneering work commenced in 1863, there was "nothing available for study but distorted fractures and brutally burnished or abraded surfaces". Sorby himself was well aware that his outstanding success compared with the successes of his contemporaries was largely due to the quality of the polish that he was able to produce. He advised that ". . . [the final] polish must not be one which gives bright reflection but one which may show all the irregularities of the material and is as far removed as possible from a burnished surface".[2] This is a pertinent warning to this day. It constitutes one of the main themes of this book.

A major difficulty encountered in any attempt to treat metallographic specimen preparation systematically is that the needs of metallographers differ so widely. The problems with which they are confronted differ in degree, and the standard required in the final result may vary independently. The literature consequently abounds in descriptions of techniques which differ in principle as well as in detail. Many of these techniques are perfectly satisfactory alternatives. But many others are quite unsatisfactory, being based either on unsound principles or on subjective impressions rather than controlled experimentation. Indeed, a major factor which has inhibited the development of sound preparation techniques has been the paucity of experiments yielding data of an objective kind.

It is also surprising that, a full century after Sorby's classic work, specimen preparation is still regarded by some as being largely an art. This is true only in the sense that a degree of art is involved in the preparation of a photomicrograph that is arresting in appearance. It is no longer true when it comes to preparing surfaces which reliably reveal all the information available without introducing any false information; and this, after all, is what counts. Our approach will therefore be to establish as many principles as possible on which the design and operation of satisfactory and optimum polishing sequences can be based. Ways in which these principles can be applied in practice will not be neglected, although specific "recipes" will be avoided.

An examination of the metallographic preparation process in outline can be simplified by the introduction of a number of terms,* many of them arbitrary and some even somewhat artificial. They are adopted here only because they provide a convenient basis on which to subdivide the discussion.

Initial sectioning normally is carried out by a mechanical *cutting* process which produces a rough and not very flat surface. Cutting consequently is usually followed by a process, such as filing or grinding, which flattens the surface but which still produces a rather rough finish. Such treatments will be termed *preliminary machining* processes. It is usual then to mount the specimen in a device which facilitates handling, and thereafter to subject it to treatments aimed primarily at progressively improving its surface finish. It will be convenient to distinguish between two general types of these finer finishing treatments. Treatments of the first type, which are performed immediately after preliminary machining, characteristically employ abrasive particles that are supported by a comparatively firm backing in which they are fixed rigidly. They will be termed *abrasion processes*. There is a limit to the quality of finish that can be achieved by these abrasion processes and it is necessary to turn to a rather different type of treatment when this limit has been reached. Treatments of this second type are called *polishing processes*.

The metallographer now has at his command two distinct types of polishing processes. The first type uses abrasive particles, and can be classified as *mechanical polishing*. The second involves chemical dissolution of the surface. It is called *chemical polishing* when straight chemical attack is involved, and *electrochemical* or *electrolytic polishing* when the chemical attack is assisted by the application of an external current. We shall be directly concerned in this book only with mechanical polishing processes, although some attention will be given to chemical polishing as a supplement to mechanical polishing.

The most obvious aim of the over-all process is to produce a polished, or specularly reflecting, surface. It might even be concluded from many

---

*A Glossary of Terms is presented on pp. 373-377.

writings on the subject that this is the sole aim. However, the objective of the subsequent microscopical examination is to determine the true structure of the material at the plane of sectioning. This cannot be done unless the final polished surface is fully representative of this plane as it existed prior to sectioning. All the structural features characteristic of this plane must be detectable, and no false structures must be observable. These considerations impose the following further requirements on the final surface:

1. Surface layers which might obscure structural features must not be present. Examples include plastically smeared layers and the layer known as the Beilby Layer.
2. False structures (artifacts) that might be detected during the subsequent examination must not have been introduced.
3. All desired fields of view must be coplanar within the depth-of-field limits of the system to be employed for examination.
4. The surface must be adequately free from stains and other chemical blemishes.

The full significance of these factors will become apparent as our discussion proceeds. They should, however, always be kept in mind, because much of the discussion is directed towards elucidating points connected with these basic requirements.

The discussion will be centered almost entirely around the needs of optical microscopy. However, the principles apply equally well to other methods of surface examination, such as scanning electron microscopy, x-ray diffraction topography and low-load hardness testing. With these methods, of course, allowances have to be made for differences in image-forming characteristics, such as the much greater depth of field and the somewhat greater resolution of scanning electron microscopy. Allowance also must be made for the sensitivity of the examinational method to plastic strains.

Many of the principles discussed also have applications to other, less specialized methods of surface preparation, such as those used in industrial machining and finishing, and to the effects of these processes on the characteristics and properties of the surfaces produced. Such matters will, however, not be discussed specifically.

APPENDIX 1-A
Sorby: The Founder of Metallography

Henry Clifton Sorby (Fig. 1.1) was a member of a family of Sheffield master cutlers. He took no part in the management of the family business himself (it was sold in 1844) but was wealthy enough to devote himself to a scientific career. He in fact devoted his whole life to science and was not so much an amateur as a self-financed researcher. Sorby was catholic in his scientific interests, and tended to maintain them in a particular activity for only a limited period before moving to another. If, however, there was one central theme to his work it was geology; his interest in

FIG. 1.1. Henry Clifton Sorby (1826–1908)

metals was only a passing one. The main achievement of his life undoubtedly was the foundation of the science of mineragraphy. This led to an interest in meteorites, then to an interest in metals (especially irons and steels), and thus to the foundation of the science of metallography as his second great achievement.

Sorby carried out his work on irons and steels principally during the period between July 1863 and October 1865, and even then it was only one of many activities. It was on July 28, 1863, that he recorded in his diary that he had "discovered" the structure of an iron. It was not until 1886 and 1887, however, that his results were recorded in a journal with a wide readership.[3,4] He carried out some additional work during the preparation of these papers, but did not return to metallography thereafter. It was during these brief spells that he devised techniques for preparing sections of metals which, for the first time, enabled their true structure to be seen. Moreover, by perceptive observation he identified all of the major microstructural constituents of ferrous materials (the constituents now known as *graphite, cementite, pearlite, austenite* and the *phosphide eutectic*). He recognized that iron was composed of a number of crystal grains, that pearlite resulted from the decomposition of a constituent which was homogeneous at high temperatures, that the hardness of quenched steels was attributable to the suppression of this decomposition, and that softening during subsequent tempering was due to the separation of the hard constituent into ferrite and cementite. He also realized that iron underwent an allotropic change on heating. These are awesome achievements considering that he started from scratch and that they were achieved after such a short period of investigation.

It is his preparation methods that interest us most here. He described them in one of his own publications[4] as follows:

> In the majority of cases slices about 1/10th of an inch in thickness were cut in the required direction by a circular saw, and further reduced in thickness by filing or grinding until sufficiently thin not to make the final preparation heavy and clumsy. One side was then fixed with hard Canada balsam to the glass, on which it was finally kept, and the upper surface made as flat as possible by rubbing on emery paper placed on a sheet of plate glass, using in the first case somewhat coarse, and finally the smoothest paper employed in preparing steel plates for engraving. It was, however, found that in many cases no good results could be obtained by polishing directly after using the fine emery-paper; because the surface was so much modified by the scratching. This exterior surface was, therefore, ground off by using fine-grained water-of-Ayr stone and water, until all trace of the scratches was removed. It was then polished on wet cloth stretched flat on a piece of wood, using in the first place the finest-grained crocus,* and lastly the very best and finest-washed rouge, so as to obtain a beautiful polish, almost or altogether free from even microscopic scratches, without any of such surface disturbance as is caused by a forced polish. This latter usually looks far better, because the minute cavities and other irregularities are hidden.

The preparation techniques used today, and discussed in this book, do not differ in principle. Note again that Sorby recognized that it was not the "looks" of a surface that mattered but whether or not it was free from disturbances which "hid" cavities and other irregularities or which "modified" the surface. It was, however, a leisurely age of cheap labor. It is recorded that a laborer was engaged to prepare a particular specimen. He took five weeks to do so, for a total wage of about $15.

Following his mineragraphic practices, Sorby cemented a cover glass onto the polished and etched surface (etching was carried out in a dilute acid, usually nitric acid). The initials HCS, identifying information, and the year of preparation were engraved on the microscope slide. This has the advantage that his specimens can still be identified positively, and several have been preserved at the University of Sheffield, England. Some of them remained in sufficiently good condition to be photographed in 1963 with a modern metallographic microscope. A selection of these photomicrographs, which were prepared by Dr. A. R. Entwistle, is reproduced in Fig. 1.2. It is fairly easy these days to exceed the standards of specimen preparation achieved by Sorby in these specimens, and cer-

---

*A grade of ferric oxide supplied for industrial polishing.

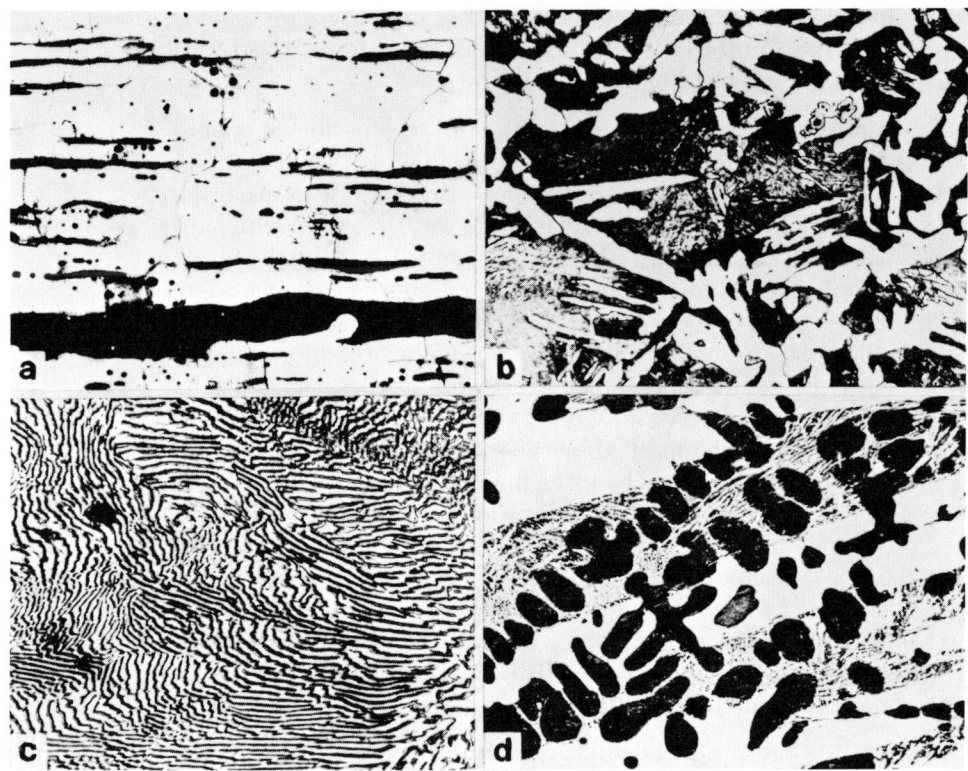

FIG. 1.2. Specimens of irons and steels prepared by Sorby in 1864, as photographed by modern optical microscopy. The specimens were etched in a 1% aqueous solution of nitric acid.

These photomicrographs were taken in 1963 by Dr. A. R. Entwistle of the University of Sheffield, using a modern optical microscope. The titles etched by Sorby on the specimen slides, and the magnifications of the modern photomicrographs, are as follows: **(a)** "00 Iron. Longitudinal. HCS. 1864." 300×. **(b)** "Hammered. Piled 5. Long. Park Gate as bloom. HCS. 1864." 150×. **(c)** "Hard 00 Iron. Reconverted. Plane of Plate. HCS. 1864." 300×. **(d)** "Stirion White Cast Iron. HCS. 1864." 50×.

tainly to do so in under five weeks' preparation time. However, it is only necessary to view a selection of published photomicrographs of, say, 20 years ago to realize how incredibly good was the standard that he achieved. More importantly, all of the structures reported by Sorby are still accepted as being correct structures. It would also still be accepted that he obtained all the information from them that was possible with the microscopes available to him.

It is probable, however, that Sorby did not really understand why his preparation methods succeeded in revealing true structures while those of his contemporaries, such as Wedding and Martins who were working contemporaneously in Germany,[5] did not. Sorby believed that it was because they did not use the final abrasion stage on the water-of-Ayr stone, and this may indeed have been significant. So too may have been Sorby's use of a rough-polishing stage with the crocus abrasive and the long time that obviously was spent at each preparation stage. The principles developed in this book would indicate that all three were desirable features of Sorby's technique that were not enjoyed by his contemporaries or many who followed him. It was perhaps not until the 1930's that the outstanding work of Vilella[6] began to point to sound explanations of the origins of artifact structures, and consequently to the development of reliable preparation methods. The concepts in this book are in fact a further development of the ideas originated by Vilella.

REFERENCES

1. C. S. Smith, "A History of Metallography", University of Chicago Press, Chicago, 1960.
2. H. C. Sorby, in H. G. C. Beale, "How to Work with the Microscope", London, 1867.
3. H. C. Sorby, *J. Iron Steel Inst.*, 1886, *28*, 140.
4. H. C. Sorby, *J. Iron Steel Inst.*, 1887, *31*, 255.
5. H. Wedding, *J. Iron Steel Inst.*, 1885, *27*, 187.
6. J. R. Vilella, "Metallographic Technique for Steels", American Society for Metals, Cleveland, 1938.

CHAPTER 2
Specimen Mounting

MOUNTING OF SPECIMENS in plastic cylinders (or, occasionally, in mechanical clamps) is desirable for many reasons. Handling is facilitated; mounts can be made in standard shapes and sizes and be designed to have no sharp corners which might damage polishing papers or cloths; friable or fragile specimens, and fragile assemblies the components of which must be maintained in their original relationships, can be mounted before sectioning, so avoiding fragmentation and distortion during this process; and retention of the edges of specimens can be greatly improved.

A specimen may be mounted before sectioning, after sectioning, after preliminary machining or even after some additional abrasion treatment. However, the stage at which mounting is carried out has little influence on either the mounting process itself or the subsequent preparation stages. Specimens are now almost universally mounted in cylinders of plastic, although there are still some instances in which mechanical clamping arrangements are useful. Specimens were once cast into blocks of sulfur or of low-melting-point alloys, but these materials cause so much trouble during both preparation and etching that they are not considered here; they have no advantages over modern plastics.

MOUNTING BY MEANS OF MECHANICAL CLAMPS

The specimen, or group of specimens, is clamped into a more massive block, the whole assembly being arranged so as to be of a convenient size for handling. Examples are given in Fig. 2.1. The material of the clamp should preferably be similar in nature to the specimen material with respect to both composition and hardness. If this is not possible, it should at least have similar abrasion and polishing characteristics when retention of the specimen edges is important (see Chapter 9). It must also have similar etching characteristics, or be inert to the etching solutions, when the specimen has to be etched after polishing; alternatively, it must be insulated electrically from the specimen. Cleaning of the assembly between preparation stages needs to be particularly thorough.

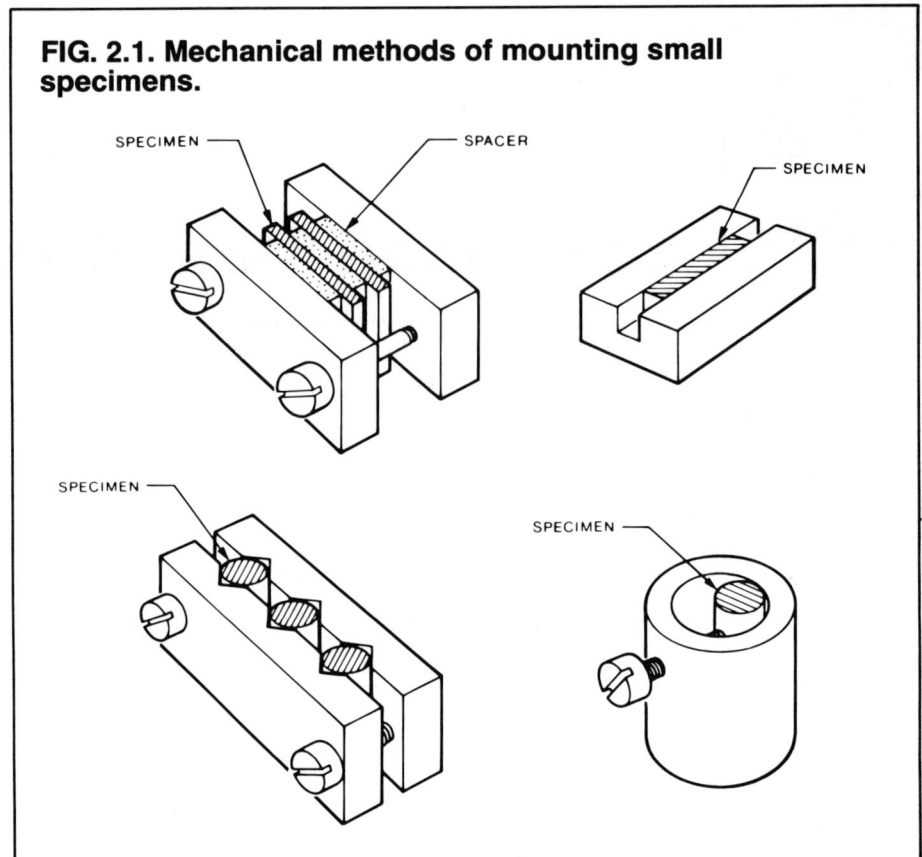

FIG. 2.1. Mechanical methods of mounting small specimens.

A further problem is encountered after etching: it is difficult to obtain close contact between the specimen and clamp, and the etchants tend to seep out of the resultant gap, causing staining along the specimen edges. This difficulty can sometimes be overcome by inserting thin spacers of a soft metal between the clamp and the specimen (Fig. 2.1, top left),[1] but the spacer material must again be of a type which does not interfere with the etching process itself. Films of a plastic material can also be used. A rather similar result is achieved if the surfaces of the specimen are precoated with a thick layer of phenolic or epoxy resin lacquer,[2] particularly if the assembly is clamped up before the resin sets.

Clamping pressure is also important. Formation of gaps between assembled components is encouraged if the pressure is too low; specimens may be damaged if it is too high.

The very nature of the clamping procedure, and the precautions that have to be associated with it, limit its application, and the method is cumbersome at best. All of this tends further to restrict application of mechanical clamping to cases where mounting in plastics is not possible for some special reason.

MOUNTING IN PLASTIC CYLINDERS

The requirements of the plastic used in a metallographic mount frequently are quite demanding. None of the many plastics available at present meets all of these demands, so that it is desirable to have several available from which the most suitable for a particular application can be selected. Proper selection requires that the needs of the particular application be carefully compared with the known properties of the available plastics. Let us first consider, therefore, the requirements of mounting plastics.

Requirements of Mounting Plastics

1. The mounting process must not physically damage the specimen by causing either distortion or structural changes which would be detectable in the subsequent microscopical examination. Likewise; it must not heat the specimen to an extent that would cause detectable structural changes. These are mandatory requirements.
2. Adequate resistance to physical distortion at elevated temperatures is desirable if the specimen must be heated during polishing, etching or washing.
3. Adequate resistance to the chemical reagents and solvents into which the mounted specimen must be immersed is required. Attack of this nature becomes significant when it causes marked deterioration of the plastic or when the specimen surface is stained by reaction or solution products.
4. It is desirable that a fissure should not form at the specimen/plastic interface. This becomes a necessary requirement when seepage of solutions from the fissure would cause staining of the prepared surface. This is also advantageous when good retention of the specimen edges is desired — particularly when thin, irregular surface films (such as oxides) are present on the specimen.
5. It may be desirable for the plastic to penetrate and fill small pores and crevices in the specimen (e.g., when the pores and crevices allow seepage of solutions during preparation and etching).
6. The abrasion and polishing rates of the plastic must be similar to those of the specimen when good retention of the edges of the specimen is desired.
7. Significant electrical conductance is desirable if, for example, the specimen is to be electrolytically polished or etched, or examined in a scanning electron microscope or an electron probe microanalyzer.
8. Sufficient transparency to permit recognition of features on the side surfaces of the specimen is an advantage in certain cases.
9. The reflectivity of the plastic may need to be such as to provide good contrast against the edges of the specimen.
10. Other factors being equal, the plastic should be simple to mold and readily available.

11. Costs may be a final consideration, although they should be kept in perspective with the total cost of the examination.

Properties of Mounting Plastics

Let us now compare these requirements one by one with the properties listed in Table 2.1* for various plastics.

Damage to Specimen. Two general types of plastics are available: namely, those which must be molded under pressure at elevated temperatures, and those which can be cast as a liquid which subsequently polymerizes at atmospheric pressure and at a temperature close to room temperature (casting types). The pressure applied during molding in the first case can cause the following types of damage to the specimen: fracture of friable materials; distortion of fragile specimens; and introduction of deformation artifacts in certain alloys (e.g., brass[7] and zirconium[8]). Pressure damage is completely avoided by the use of casting plastics. The temperature (up to 200°C) that must necessarily be attained with molding plastics may also induce structural changes in the specimen, such as tempering in quench-hardened steels and aging in precipitation-hardened alloys. Some temperature rise is also possible with epoxy casting plastics, although it can be kept small (see p. 24).

Distortion of the Plastic at Elevated Temperatures. The molding plastics can be further subdivided into *thermosetting* and *thermoplastic* types. A curing process occurs in thermosetting plastics at the molding temperature, and an appropriate period of time must be allowed for this to complete itself. The molding process for thermoplastics, on the other hand, is merely one of consolidation, a process which commences at a temperature somewhat below the molding temperature listed in Table 2.1 and which is complete a short time after this temperature is reached. Having been consolidated, however, a thermoplastic must be cooled below a certain characteristic temperature before it can be stressed without causing severe distortion. Moreover, it must not subsequently be heated above this temperature if severe distortion under stress is to be avoided. This temperature, known as the *heat-distortion temperature*, cannot be defined precisely, because it is dependent on the applied stress and the acceptable strain rate. The figures listed in Table 2.1 are for one standardized set of conditions and strictly are comparative only, although they do give rough indications of the temperatures which these plastics will withstand without severe distortion in metallographic mounting. Note that thermoplastics generally have low heat-distortion temperatures compared with those of thermosetting plastics.

Resistance to Chemical Attack. All the plastics listed in Table 2.1 have adequate resistance to the comparatively mild reagents used for many metallographic etchants. However, they may have unsatisfactory

---

*The items in this list have been selected to represent the plastics of greatest current use in metallography, but the list is by no means comprehensive. Other plastics can be assessed on the same bases.

TABLE 2.1. Typical Properties of Plastics Suitable for Metallographic Mounts

Plastic	Type	Molding conditions			Heat-distortion temperature, °C(a)	Coefficient of thermal expansion, in./in./°C(b)	Transparency	Chemical resistance	Ref
		Temperature, °C	Pressure, lb/in.²	Curing time					
Phenolic molding powder	Thermosetting(c)	170	4000	5 min	140	$3.0–4.5 \times 10^{-5}$	Opaque	Not resistant to strong acids or alkalis	
Acrylic (polymethyl methacrylate) molding powder	Thermoplastic	150	4000	nil	65	$5–9 \times 10^{-5}$	Water white	Not resistant to strong acids	
Epoxy casting resin	Thermosetting(d)	20–40	...	24 h	60(e)	$4–7 \times 10^{-5}$	Clear but light brown in color	Fair resistance to most alkalis and acids. Poor resistance to conc. nitric and glacial acetic acids	
Allyl molding compound	Thermosetting(f)	160	2500	6 min	150	$3–5 \times 10^{-5}$	Opaque	Not resistant to strong acids and alkalis	3
Formvar (polyvinyl formal) molding compound	Thermoplastic	220	4000	nil	75	$6–8 \times 10^{-5}$	Clear but light brown in color	Not resistant to strong acids	4
Polyvinyl chloride molding compound	Thermoplastic(g)	160(h)	3000	nil	60	$5–18 \times 10^{-5}$	Opaque	Highly resistant to most acids and alkalis	5,6

(a) As determined by the method described in ASTM D648-56, at a fiber stress of 264 lb/in.² (b) The coefficient of thermal expansion in most metals is in the range $1-3 \times 10^{-5}$ in./in./°C. (c) Wood-filled grade, preferably with low filler content. (d) A liquid epoxy resin with an aliphatic amine hardener. (e) Depends on the curing schedule; can be as high as 110°C with heat curing. (f) A diallyl phthalate polymer with a mineral filler. (g) A stabilized rigid p.v.c. For example, a mixture of 100 parts of a paste-making grade of p.v.c., 2 parts dibasic lead phosphate and 2 parts tribasic lead sulfate. (h) Must not exceed 200°C.

resistance to stronger reagents sometimes used, for example, for etching refractory metals. Polyvinyl chloride and epoxy plastics suffer least from this disability.

Plastics also exhibit various levels of resistance to the solvents likely to be used for cleaning and drying operations, although, in this instance, a satisfactory solvent can be chosen to match the plastic. For example, all the plastics listed have good resistance to alcohol, but formvar and epoxy are sufficiently attacked by acetone to be liable to stain formation when this solvent is used for drying; acrylic plastics are severely attacked by acetone and by chlorinated hydrocarbons. Allyl plastics have outstanding resistance to solvents.

Fissure Formation at Specimen/Plastic Interface. Only one of the plastics listed — namely, the epoxy casting type — adheres physically to metals; this is a basic requirement for complete absence of fissuring between specimen and mount. Nevertheless, provided that suitable precautions are taken, fissuring can be kept to an almost indiscernible level for some of the remaining plastics.

The first factor of importance with nonepoxy plastics is the relative coefficients of thermal expansion of the plastic and the specimen. The coefficients of expansion of plastics vary from only slightly greater than to considerably greater than the range characteristic of metals (Table 2.1). A large difference generally is desirable because it increases the tendency for the plastic to shrink onto the specimen during cooling from the molding temperature. Likewise, maintenance of the molding pressure during cooling to as low a temperature as possible is desirable, although there is little point in continuing this below the heat-distortion temperature; this precaution consequently will be less effective the higher the heat-distortion temperature of the plastic.

Polyvinyl chloride and acrylic plastics have good characteristics, in ascending order of merit, on these counts. Nevertheless, satisfactory absence of fissuring will still not be obtained with these materials if the shape of the specimen precludes free contraction onto any portion of the surface. For example, satisfactory absence of fissuring may be obtained on the outer surface but not on the inner surface of a transverse section of a tube. If the affected surface is the one of interest, it may then be more effective to use a plastic with a comparatively low coefficient of expansion.

Although epoxy resins adhere to metals and hence have the potential to produce mounts with no interface fissures at all, it does not follow that epoxy mounts are completely immune from fissure formation. This is because stresses induced at the interface during curing, or resulting from differential thermal expansion of the specimen and resin during subsequent thermal cycles, may be sufficient to rupture the interface bond. Specimens whose shapes preclude uniform contraction will, again, be particularly susceptible to this problem. For example, it is commonly encountered with sheet specimens; shrinkage tends to cause the plastic to be drawn tightly against the ends but to pull away from the faces of the sheet (Fig. 2.2a).

FIG. 2.2. Sections of an oxidized sheet of nickel mounted in epoxy. (Hussey, Beaubien and Caplan, Ref 9)

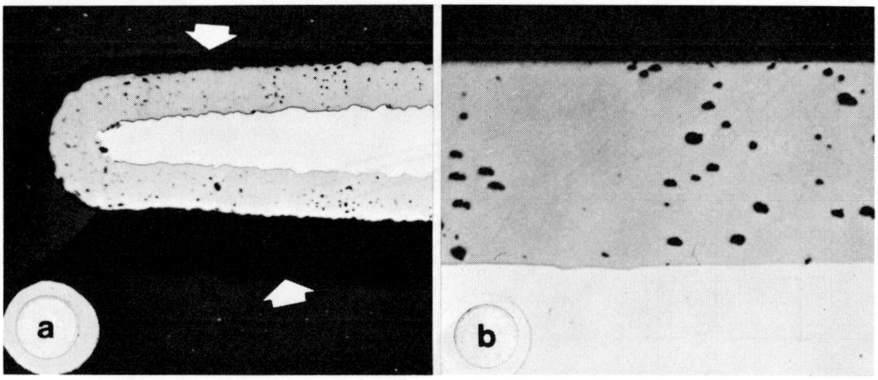

(a) Epoxy cast in the standard way has adhered to the end of the sheet but pulled away from its side faces. Magnification, 75×. **(b)** Good adhesion on the side face of a sheet cast in epoxy by the technique illustrated in Fig. 2.4. Magnification, 400×.

The setting strains result essentially from differential thermal contraction during cooling of the resin from the temperature at which crosslinking occurs.[10] Consequently, the problem is more likely to arise when the thermal coefficient of expansion of the specimen material is low; the differences between the coefficients of thermal expansion of metals and those of epoxy resins can vary by a factor of two or more (Fig. 2.3). The first precaution that can be taken against fissuring, therefore, is to reduce the effective curing temperature to the minimum; this can be done by using minimal amounts of hardener, by refraining from heating the mix, and preferably by cooling the mix (see p. 25). The second is to use a grade of epoxy with a low coefficient of thermal shrinkage; this can be done by choice of supply or by addition of a filler to the epoxy, which has a significant effect in reducing the coefficient of thermal expansion (Fig. 2.3).

Many suitable filler materials are available that reduce the coefficient of thermal expansion of epoxy resins. Probably the most convenient for metallographic purposes are metals and oxides such as silica and alumina (powders of about 300-mesh grade are suitable in all cases), the characteristics of which are summarized in Fig. 2.3. Note that a comparatively large filler addition is necessary to be effective, particularly for specimens of steels, which have low coefficients of thermal expansion. However, addition of filler increases the viscosity of the epoxy resin mix and this sets an upper limit on the permissible addition for the present purposes. The limit is approximately that at which the curves terminate in Fig. 2.3, although the actual limit for a particular application must be determined by trial. Aluminum oxide is very effective on a weight-addition basis but

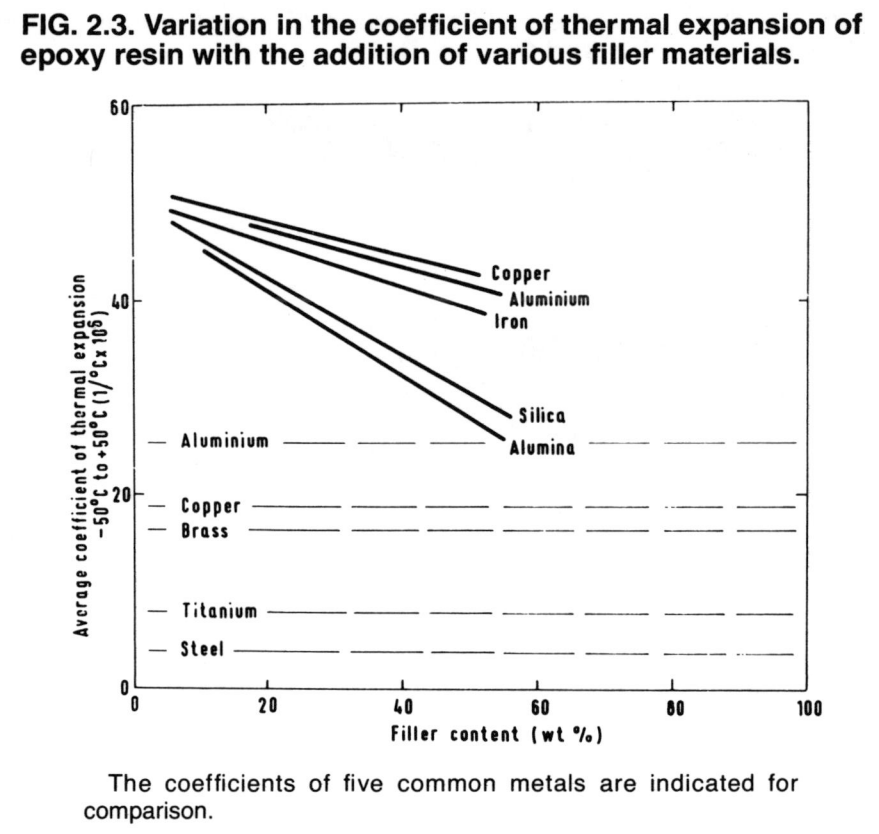

FIG. 2.3. Variation in the coefficient of thermal expansion of epoxy resin with the addition of various filler materials. The coefficients of five common metals are indicated for comparison.

only a little more effective than aluminum on a volume basis. The addition of metals rather than oxides is to be preferred because adding oxides, but not metals, reduces drastically the polishing and abrasion rates of the plastic. The reduction in polishing rate is a disadvantage when rapid removal of a damaged layer is desirable (see p. 237), but the reduction in abrasion rate might be an advantage when good edge retention is desirable (see p. 289). Metal fillers can be matched to the specimen material in terms of etching characteristics as well as of polishing and abrasion rates.

Modified molding techniques can also be adopted to reduce the possibility of fissuring — techniques which are based on, first, reducing to the minimum the volume of epoxy that is polymerized with the specimen and, secondly, transferring the shrinkage fissure to a less-adherent dummy specimen.[9] For example, a sheet specimen can be cast into a slot machined in an epoxy preform, together with a number of stainless steel strips, as indicated in Fig. 2.4. Adhesion between the epoxy and the specimen is then maintained along the full length of the specimen (cf. Fig. 2.2a and 2.2b). Any fissures that develop form along the stainless steel strips.

Finally, if all else fails, it is usually possible to repair an interface fissure

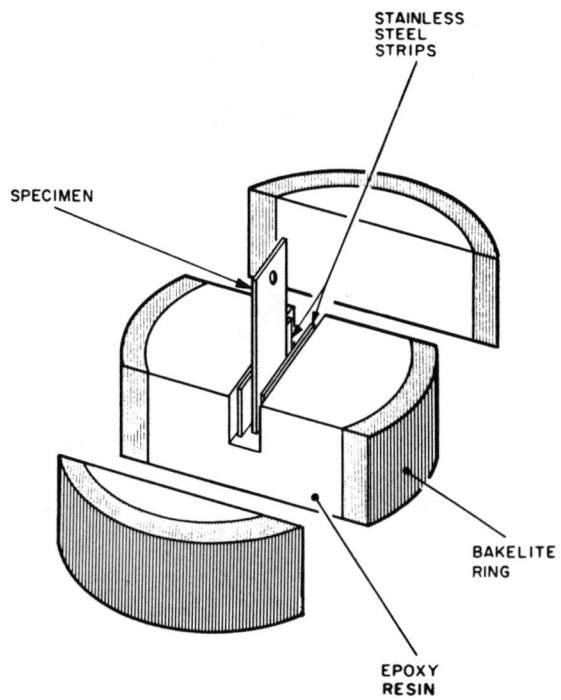

FIG. 2.4. Technique for reducing the tendency for fissures to form along the side faces of sheet specimens cast in epoxy. *(Hussey, Beaubien and Caplan, Ref 9)*

The specimen is cast in a slot machined in an epoxy preform, together with several dummy specimens.

after it has been exposed by a preliminary abrasion operation. A small volume of epoxy casting liquid is placed over the section surface, and the mount is subjected to the vacuum impregnation process discussed later.

Ability to Fill Pores and Crevices. Only liquid casting plastics show any significant tendency to fill pores and crevices in the specimen. Even then, the tendency is only slight, and vacuum impregnation techniques (p. 25) usually have to be employed when this factor is of importance.

Abrasion and Polishing Rates. All plastics have high abrasion rates compared with those of metals (cf. Tables 3.3 and 9.6), but the rates for different plastics differ considerably (Table 9.6, p. 290). On the other hand, the polishing rates of plastics are low, being lower than those of many metals (cf. Tables 5.2 and 9.6). This can mean that the polishing rate of a mounted specimen is restrained by that of the plastic. It then follows from the considerations in Chapter 8 that use of large areas of plastics with low polishing rates is undesirable.

These factors have important effects on retention of the edges of specimens, but discussion of this subject will be left to Chapter 9. It is shown

FIG. 2.5. Methods of making electrical contact with the back surface of a specimen mounted in a nonconducting plastic.

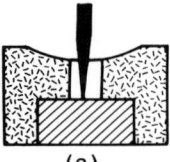
(a)

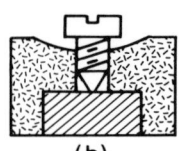
(b)

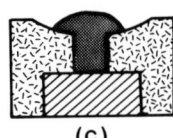
(c)

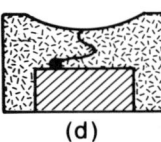
(d)

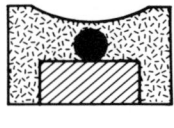
(e)

(a), (b) and (c) A hole is drilled in the back of the mount to expose the back surface of the specimen. Contact is then made using **(a)** a probe, **(b)** a screw, or **(c)** a cast plug of a low-melting-point alloy. **(d)** A wire is connected to the specimen prior to mounting and exposed at the back surface of the mount. **(e)** Rods or balls are set on the back surface of the specimen and exposed by machining away the back of the mount *(Hoffman, Ref 11)*.

there that formvar and polyvinyl chloride are the most satisfactory basic plastics from this point of view; epoxy and phenolic plastics are the least satisfactory. It is also shown that addition of fillers to the plastic can have an important beneficial influence in reducing both abrasion and polishing rates. On the other hand, they can have the severely detrimental effect of wearing out abrasive papers rapidly.

Electrical Conductivity. Plastics characteristically are good electrical insulators, and thus obtaining electrical contact with the specimen is difficult. Contact can be made with the prepared surface by means of a probe, which results in uneven current distribution. Contact can be made to the back surface by various mechanical arrangements, such as those illustrated in Fig. 2.5. All are cumbersome and may damage the specimen.

The easiest solution to the problem is to make the mounting plastic conducting, in which event it is desirable that the resistance from the prepared surface to the back surface of the mount reliably be on the order of 100 ohms. It is frequently stated that this can be accomplished by mixing about 10 vol % of a metallic powder, such as aluminum or copper flake, with the plastic prior to molding. A low-resistance mold will be obtained under these circumstances, however, only if connecting chains of metallic particles form in the plastic by chance. In practice, the resis-

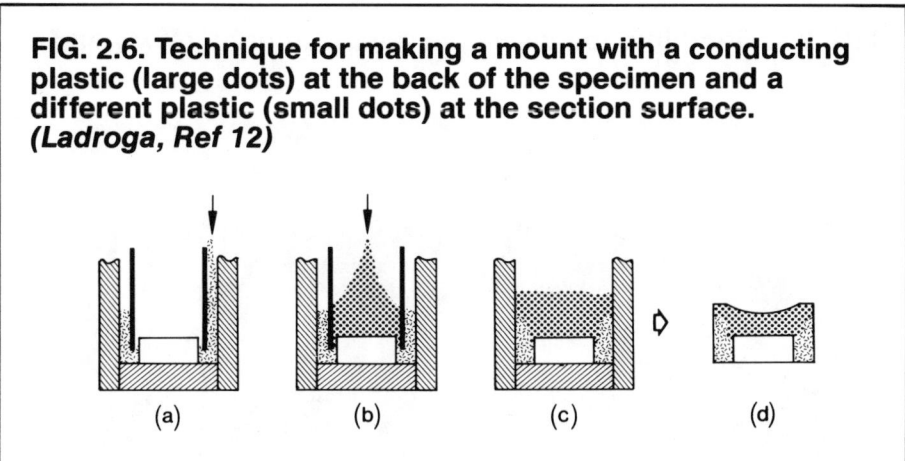

FIG. 2.6. Technique for making a mount with a conducting plastic (large dots) at the back of the specimen and a different plastic (small dots) at the section surface. (Ladroga, Ref 12)

tance of mounts prepared by this technique varies unpredictably from a few ohms to almost infinity.

Molds which are consistently of low resistance can be made by coating the individual plastic particles with a film of conducting material prior to mounting. Details of a method of preparing a polyvinyl chloride powder coated with carbon black and suitable for this purpose are given in Appendix 2-A. Standard mounts made with this powder consistently have a resistance in the range from 100 to 200 ohms. A disk of an appropriate metal may be molded in the back surface of such a mount, and electrical contact can be made to this disk. If, for any reason, the conducting plastic has a detrimental effect on the characteristics of the working face of the mount, a mount with a more appropriate distribution of plastics can be achieved by the method illustrated in Fig. 2.6. The specimen is placed in a normal molding set and a tube is inserted temporarily (Fig. 2.6a). Powder of the plastic required for the section surface of the mount is poured into the annulus between the tube and the mold until the base of the mold is covered and the annulus filled, the back surface of the specimen remaining uncovered (Fig. 2.6a). The tube is then filled with the conducting plastic (Fig. 2.6b), and withdrawn gently (Fig. 2.6c). The mold, when finally processed, has the distribution of plastic illustrated in Fig. 2.6(d).

Transparency. Acrylics are the only plastics listed which are highly transparent. The epoxy and formvar types, although translucent, are sufficiently transparent for many purposes.

Reflectivity. Plastics all have poor reflectivity in vertical bright-field illumination. They thus appear in strong contrast against the edge of a clean metal specimen, but not when a nonmetallic layer (such as an oxide scale or a corrosion product) is present on the surface. It may in the latter case become difficult, particularly in a photographic print, to distinguish clearly the interface between the surface layer and the plastic mount. Deposition of a metallic layer on the nonmetallic layer before sectioning (see p. 300) will obviate this problem. Otherwise, all that can be done

FIG. 2.7. Beryllium foil viewed in polarized light. *(Miley and Calabra, Ref 13)*

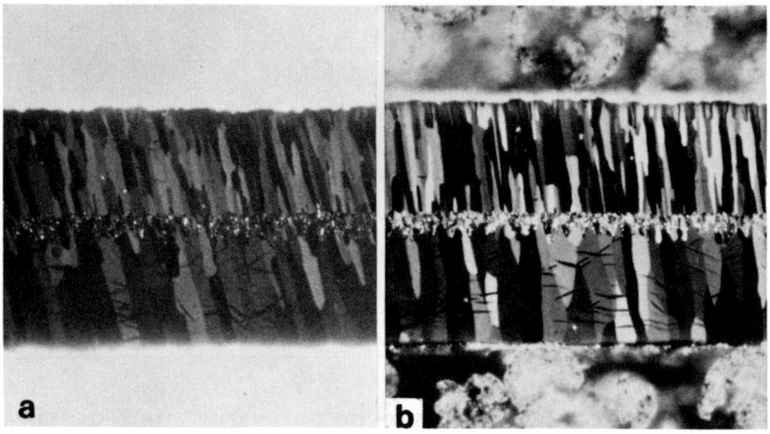

(a) Mounted in epoxy with white aluminum oxide filler. **(b)** Mounted in black epoxy with black aluminum oxide filler. The reduced contrast in **(a)** is due to flare from the mount. Magnification (both), 250×.

is to adjust the photographic technique to reduce the contrast in the final image. In these respects, plastic-mounted specimens are not very different from unmounted specimens although the mismatch in reflectivity with respect to nonmetallic surface layers does tend to be smaller.

The reverse problem arises with translucent plastics, such as epoxy resins, during examination under polarized light. The mount may reflect strongly and the specimen poorly, in which event the flare from the plastic reduces significantly the contrast observed in the specimen (Fig. 2.7). The use of a black epoxy to which has been added a black pelletized aluminum oxide alleviates this problem (cf. Fig. 2.7a and 2.7b).[13]

General Fields of Usefulness of Various Plastics

Although, as we have already noted, the best plastic for a specific application may have to be selected after a detailed consideration of the relative importance of several factors, it is still possible to draw some general conclusions about the fields of application of the various plastics which we have considered, viz.:

Phenolics have little to commend them except low cost and ready availability. They are, nevertheless, quite adequate when the mount is used merely as a holding device.

Acrylics exhibit poor chemical resistance, particularly to solvents, which virtually restricts their use to applications where extreme clarity is required in the mount.

Epoxies find special application as casting resins. Other advantages are that true adhesion to the specimen is obtained, the mount is reasonably transparent, only simple molding equipment is required, and heating of the specimen can be kept to a low level.

Allyls give improved apparent adhesion and specimen-edge preservation compared to phenolics, but are less satisfactory in both of these respects compared to formvar and polyvinyl chloride types. They are expensive and not readily available. Other disadvantages are low polishing rate and poor resistance to strong acids.

Formvar is excellent with respect to both apparent adhesion and edge-retention characteristics. Main deficiencies are unsatisfactory resistance to some strong acids, restricted availability, and a marked tendency to stick in the mold.

Polyvinyl chloride is comparable to formvar with respect to both apparent adhesion and edge-retention characteristics, yet is less expensive and more readily available. It is comparable to phenolics with respect to cost, availability and ease of handling, and consequently should be considered for general-purpose mounting as well.

MOLDING METHODS FOR THERMOPLASTICS AND THERMOSETTING PLASTICS

Many satisfactory presses and molding die sets are manufactured specifically for metallographic purposes. They vary from simple hand presses to automatically controlled hydraulic and pneumatic presses, and the following points are worth considering when selecting from this range:

1. A hydraulic or pneumatic press should be of robust design, because maintenance of the desired pressure for a lengthy period is a severe requirement.
2. A heater of high capacity (about 500 watts) is desirable to ensure the minimum cycle time, particularly when starting from cold. Preheating of preformed blanks reduces the cycle time.
3. An automatic temperature cutoff control is a desirable feature. Mount, specimen and mold all can be severely damaged by accidental overheating.
4. It is desirable that the mold-ejection arrangement load the press symmetrically.

Molding techniques are also straightforward if the equipment manufacturer's recommendations are followed, although the following points should be noted:

1. The specimen should be clean and should be at least 1 cm smaller than the diameter of the mounting cylinder in any lateral dimension. Sharp corners should be eliminated from the specimen, if possible.
2. Sufficient plastic must be placed in the mold to ensure that the upper ram of the molding die set does not contact the specimen.

3. The die set must be cold enough when loaded with plastic powder to ensure that partial setting of the powder does not occur before loading of the mold has been completed.
4. The pressure applied is not critical, provided that it exceeds a certain minimum. Excessive pressures are undesirable, however, because of the increased risk of damaging the specimen. Pressure must be applied immediately upon commencement of mold heating in the case of thermosetting plastics, but it may be delayed with thermoplastic materials; this is even desirable when the specimen is fragile.
5. Control of temperature is more critical than control of pressure. A certain minimum temperature must be exceeded in all cases, although this is to some extent dependent on the curing time allowed in the case of thermosetting plastics. Excessively high temperatures result in charring of the plastic or, in the case of thermoplastics, in the plastic becoming so fluid that it penetrates into clearances in the molding die set. In general, a temperature of 200°C should not be exceeded.
6. Thermosetting plastics may be ejected while hot after they have become fully cured. Slow cooling under pressure is desirable, however, with thermoplastics; this will, for example, reduce the tendency for crack networks to develop in acrylic mounts. Cooling under pressure to well below the heat-distortion temperature of the plastic is always desirable when development of a fissure between specimen and plastic is undesirable.
7. Difficulties in ejection usually are experienced only when the working surfaces of the die set are damaged. It is good practice, however, to treat these surfaces with a silicone mold-release agent, for which purpose pressure-pack sprays are available; this is virtually obligatory with formvar plastics.

The following are common mount defects and their causes:

1. *Radial cracks* usually result from attempts to mold a specimen whose dimensions are too large for the particular size of mount, especially when the specimen contains sharp corners. Such cracking can be alleviated by reducing the molding temperature and by allowing the mold to cool to a lower temperature before removing the applied pressure.
2. *Transverse cracking* usually results from evolution of gases from either the plastic or the specimen; baking out the plastic or the specimen prior to mounting may help alleviate this problem. Transverse cracking may also result from use of a mold which initially was too hot.
3. *Porous friable areas* may result from low molding pressure, short curing time, or charging into an excessively hot mold, either singly or in combination. The cause is almost certainly insufficient time at temperature when the porous area is in the center of the mount; this will be most obvious with transparent plastics.

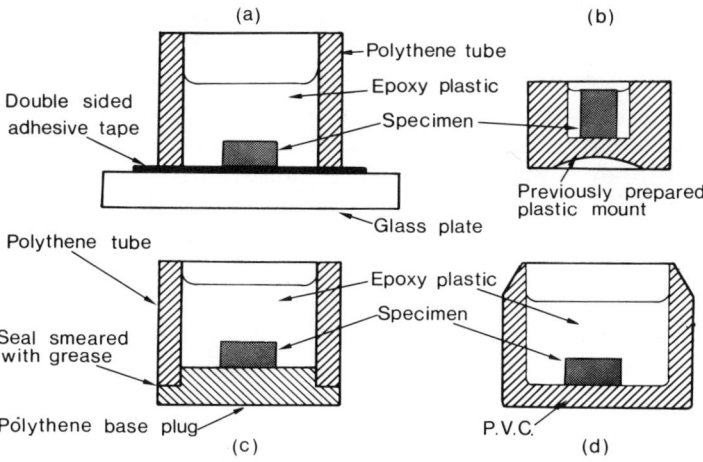

FIG. 2.8. Molds suitable for casting plastics of the epoxy type.

The base of the mold tube in **(a)** must be dressed regularly against an abrasive paper to ensure a leak-free joint. The method of manufacturing the mold in **(d)** is set out in Appendix 2-B; a mold of similar form can be manufactured from silicone rubber by the method set out in Appendix 2-C.

4. *Bulging of front or back surface* is usually caused by insufficient curing time or insufficient pressure while the material is above the heat-distortion temperature.

MOLDING METHODS FOR CASTING PLASTICS

The only equipment required for casting of plastics is a simple receptacle or mold of appropriate shape, although a complication arises from the fact that epoxy resins adhere strongly to many materials. This can sometimes be made a virtue by the use of mounting techniques such as that illustrated in Fig. 2.8(b). Moreover, it can be circumvented by using, in an arrangement such as that sketched in Fig. 2.8(a), a mold consisting of a thin tube of a material such as cardboard to which epoxy adheres strongly; the mold tube then becomes a permanent part of the mount.

Simple molds machined from a metal such as aluminum can be used, provided that the mold surfaces are maintained in a highly polished condition and are treated regularly with an appropriate release agent. However, epoxy plastics contract very little during curing, and the mounts consequently are difficult to eject from a rigid mold of this type unless further mechanical complications are introduced into the mold system. Similar molds can be machined from acrylic plastics, to which epoxy plastics do not adhere, but acrylics also are rigid and thus also cause ejection difficulties. More flexible plastics ease mount ejection and, among these, polyvinyl chloride has acceptable, and polyethylene has excellent, parting characteristics. Coating the molds with a silicone release agent is desirable with these materials but not essential. Satisfactory mold designs employing available solid forms of these materials are sketched in Fig. 2.8(a) and 2.8(c). An even simpler form of mold (Fig. 2.8d) can be formed from a dipping-grade polyvinyl chloride by the method set out in Appendix 2-B (p. 29); this type of mold has a comparatively limited life but is inexpensive and is easy to make in any size.

Silicone rubber, however, is by far the most satisfactory mold material presently available. Parting characteristics are excellent and even thick-wall molds are flexible enough to permit easy mount ejection; silicone rubber molds are simple to make (see Appendix 2-C, p. 29).

Care must be taken in mixing the two constituents of epoxy plastics. The optimum proportions of plastic and hardener for the small batches required for metallographic mounts may be somewhat different from that normally advised by the manufacturer, and must be determined by trial. An inadequate amount of hardener will result in soft mounts; an excessive amount will cause large temperature rises during hardening and perhaps even cracking of the mount. Once established, the proportions must be measured accurately and the ingredients must be very thoroughly mixed; otherwise, locally uncured regions may develop in the mount. This vigorous mixing inevitably entraps small air bubbles, elimination of which requires that the mixture be allowed to stand before casting for sufficient time to permit most of these bubbles to escape. If this is not sufficiently effective, the mold, after casting, can be treated in a simple evacuation chamber such as that illustrated in Fig. 2.9; this treatment also removes most air bubbles trapped near the specimen during casting. The vacuum obtainable from a water-ejection pump is suitable. In fact, the mixed

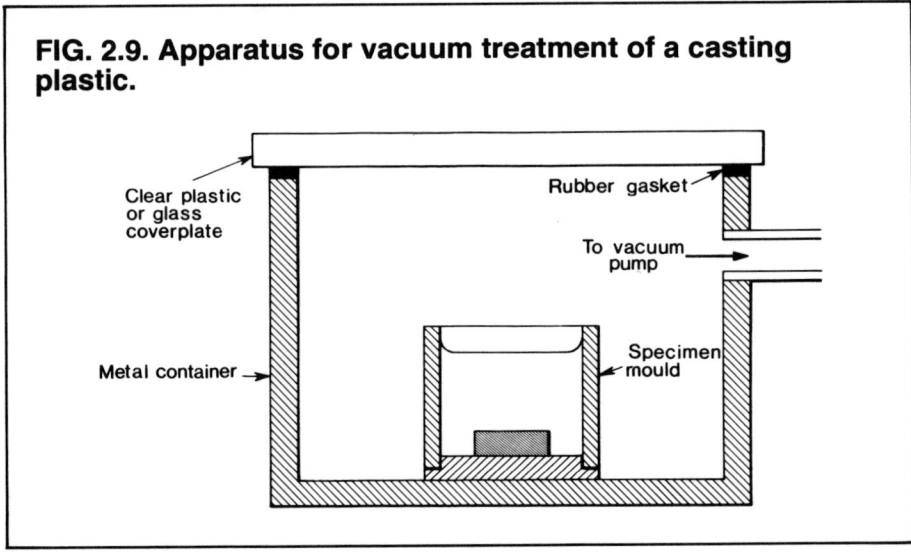

FIG. 2.9. Apparatus for vacuum treatment of a casting plastic.

plastic, while in the liquid state, should not be subjected to a vacuum exceeding about 200 mm Hg, because severe frothing otherwise occurs. Some grades of epoxy can be stored for several months after mixing, if refrigerated; the supplier should be consulted for advice on this point.

The curing time required for epoxy resins can be reduced considerably by heating to a slightly elevated temperature (50 to 75°C) but, because the curing process is exothermic, control of temperature becomes difficult with the small volumes cast in metallographic mounts. Development of excessive temperatures in the curing plastic causes frothing, cracking, and development of fissures between specimen and plastic. However, the plastic may safely be heated to these temperatures if it is first allowed to cure at room temperature for about 1 h after mixing; curing is then completed after a further 15 to 30 min at 70°C. Heat curing also has the advantage of increasing the heat-distortion temperature of the plastic. On the other hand, it may be difficult by any of these means to prevent excessive heating in large mounts 7.5 to 15 cm in diameter; cooling in the air blast from a room air conditioner is then recommended.[14]

The immediately preceding remarks are concerned with temperature increases in the plastic itself. It may also be necessary to keep to a minimum the temperature increase in the specimen being mounted when a greater increase might induce artifact structures. It is difficult to quantify the temperature rise likely to occur under a specific set of circumstances, because many parameters are involved. Suffice it to say that temperatures of 40 to 50°C might easily be attained, and temperatures as high as 80 to 100°C are possible. Factors conducive to small temperature increases are resin mixes that give long setting times, small volume ratios of plastic to specimen, and molds with good thermal conductivity. Forced cooling, such as in the air blast of an air conditioner, also helps discourage specimen heating.[14] It should be possible with good practice to keep the temperature increase in the specimen below 10°C.

VACUUM IMPREGNATION WITH CASTING PLASTICS

Specimens containing areas which are so porous as to cause difficulties during polishing or etching may profitably be impregnated with an epoxy resin. An impregnation treatment is also desirable when the specimen contains surface defects which are to be studied, because bubbles may otherwise form in the mount adjacent to these defects.

Two types of impregnation techniques may be used. In the first, the specimen is placed in an apparatus, made up from standard laboratory glassware, of one of the types illustrated in Fig. 2.10. In the case of epoxy resins, the mixed liquid should be previously degassed by pumping down the apparatus before the specimen is immersed in the liquid plastic. The vacuum in the main apparatus must not be reduced to a level which would cause frothing of the plastic. This pressure is almost invariably adequate to ensure satisfactory impregnation, but if a higher vacuum is required the mixed liquid must be given a preliminary treatment at the pressure concerned until all frothing ceases. The specimen is removed before the

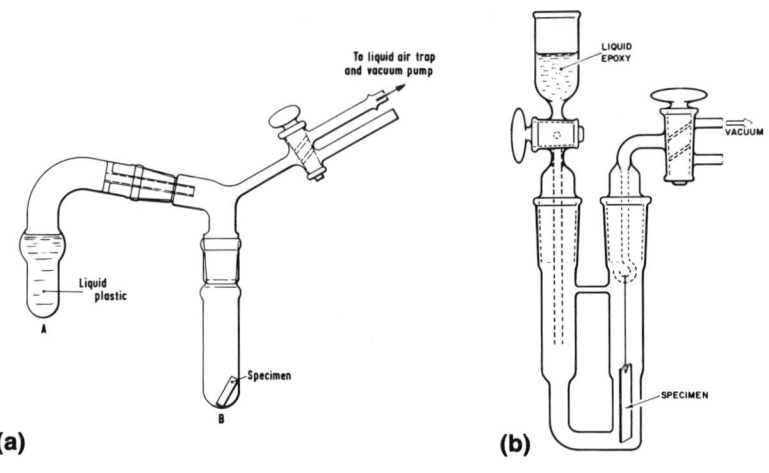

FIG. 2.10. Apparatus for vacuum impregnation of a porous specimen, using standard chemical glassware.

In **(a)**, the apparatus is evacuated in the position shown, the cock is shut, and the apparatus is then tilted to flood the specimen with liquid plastic *(Caplan and Beaubien, Ref 15)*. In **(b)**, the epoxy liquid is admitted after the specimen is outgassed without being outgassed itself *(Hussey et al, Ref 9)*.

plastic cures, the plastic is allowed to cure, and the impregnated specimen is then mounted in the normal way.

In the second technique, the specimen mold is set up in an apparatus of the type illustrated in Fig. 2.9, to which is added a small ladle that can be rotated by a remote-control handle. The ladle is filled with liquid plastic, the apparatus is evacuated, and the mold is then filled by tilting the ladle after the liquid plastic has degassed. Apparatus has been designed to facilitate mounting of large numbers of specimens by this means.[16]

MOUNT DIMENSIONS

Molds are commonly available for producing mounts with diameters of 1, 1¼ and 1½ in. Smaller-diameter mounts tend to rock excessively during hand abrasion and polishing operations; larger-diameter molds give reduced abrasion and polishing rates for a given applied load. A diameter of about 1 in. is optimum unless specimen dimensions dictate otherwise. The thickness of the mount ideally should be about half its diameter; thinner mounts are more difficult to handle and thicker ones tend to rock during manipulation.

In spite of the foregoing, very large mounts 6 in. or more in diameter can be handled, but only with fully mechanized equipment.

MOUNTING OF THIN SPECIMENS

One practical difficulty encountered in mounting is that of keeping thin specimens upright and undistorted so that they can be sectioned at exactly the required orientation. Following are several methods of overcoming this problem, five of which are illustrated in Fig. 2.11:

1. Bend a balancing tag at one or both ends.
2. Support the specimen between two suitably bent sections of shim stock[17] or some similar springlike device.
3. Assemble several sheet specimens together by clipping them between folds in a strip of soft lead [18] or between the coils of a closely wound spring.
4. Assemble a number of sheet specimens together by cementing them to spacers, using an epoxy cement.[19]
5. Mount the specimen on the flat, section and remount.[20]
6. (Not illustrated). Machine slots or holes of appropriate dimensions in a preformed blank and cement the specimens into the slots or holes using an epoxy cement.[21]

FIG. 2.11. Sketches illustrating methods of mounting thin specimens so that truly perpendicular sections can be obtained.

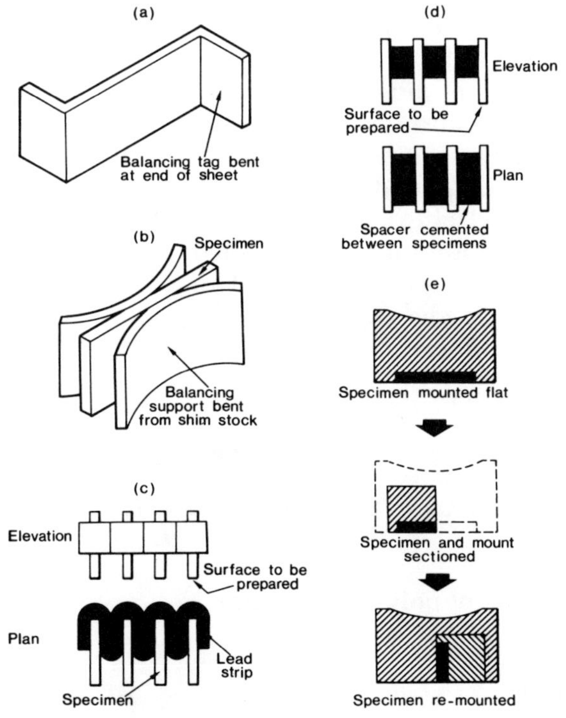

7. (Not illustrated). In the case of metals of appropriate melting point, fuse specimens into glass before sectioning and mounting; for example, thick-wall tubing can be collapsed onto thin wire specimens.[22]
8. (Not illustrated). Mount specimens in a mineral-filled grade of epoxy resin which has a puttylike consistency when mixed.[23] A blob of the mix is pressed against a flat plate which has suitable release characteristics, and the specimen is then pushed through the blob to the plate at the desired orientation. The consistency of this epoxy material is such that the specimen retains its original orientation while the epoxy sets. After setting, the blob can be remounted by a standard technique.

MARKING FOR IDENTIFICATION

Plastics can be readily engraved by power-driven burrs or vibratory tools, or even by hand scribers. Many can be written on with a nib pen using a waterproof ink to which has been added 1 ml of ethyl alcohol per 15 ml.[24] In the case of transparent plastics, identifying information can be written on a slip of paper which can then be molded in the mount so that it is located just beneath an external surface.[25]

REMOVAL OF SPECIMENS FROM PLASTIC MOUNTS

In cases where the plastic does not adhere to the specimen — and this is so for all but the epoxy plastics — the mount plastic can usually be fractured mechanically without damaging the specimen, particularly with the assistance of judiciously placed saw cuts. Epoxy plastics must be removed by chemical means, the bulk first being cut away by machining. Proprietary compounds are available for this purpose. If a suitable proprietary compound is not available, the mount may be immersed in boiling dimethylsulfoxide (BP > 189°C) for 1 to 2 min and then immersed quickly in either liquid nitrogen or an acetone – dry ice mixture, the treatment being repeated several times.[26] *Vapors of dimethylsulfoxide are very toxic and capable of diffusing through the human skin; great care must be taken in handling this material.*

APPENDIX 2-A
Method of Preparing a Conducting Plastic

Ingredients:	Parts by Weight
Plastersol grade of polyvinyl chloride	100
Tribasic lead sulfate	2.5
Dibasic lead phosphite	2.0
Carbon black	15

Ball mill ingredients together for 24 h.

Molding Conditions:
   Pressure: 2500 to 3000 psi
   Temperature: 160°C (must not exceed 200°C)
   Cooling: Cool to below 40°C before ejecting mount from mold.

Note:
   Some separation of the ingredients may occur during storage after several years. In this event, the resin can be reconstituted by further ball milling.

APPENDIX 2-B
Method of Manufacturing a Mold for Epoxy Resins From a Polyvinyl Chloride Dipping Compound

Make a smoothly finished metal mandrel about 20 cm long whose diameter is equal to that intended for the mount. Heat the mandrel to about 120°C (250°F) and dip it to a depth of about 2.5 cm in a previously prepared paste of polyvinyl chloride dipping compound for about 30 s. Remove, and allow the plastic cup which has formed to reach the temperature of the mandrel, by which time the plastic will have changed from a milky white to a translucent light brown color. Alternatively, the mandrel may be replaced in an oven held at 120°C until this color change occurs. Finally, strip the cured cup of polyvinyl chloride when the mandrel has cooled, preferably by directing a jet of compressed air along the mandrel/plastic interface.

APPENDIX 2-C
Method of Manufacturing a Mold for Epoxy Resins From a Cold-Cure Silicone Rubber

A reasonably fluid grade of elastomer should be chosen and the catalyst addition adjusted to give a pot life of at least 20 min. Mixing can be carried out in a paper cup or a metal, glass or plastic container and by either simple hand or mechanical mixing, care being taken to keep entrapment of air to a minimum. Entrapped air should then be removed by vacuum treatment in a device such as that illustrated in Fig. 2.9. The mix is then poured into a mold of the type illustrated in Fig. 2.12. The dimensions of the pattern in this mold are those required of the final mount, and the pattern must be clean because all imperfections will be reproduced; a gap of at least 5 mm should be allowed around the pattern. A light oil or grease, such as petroleum jelly, should be applied to all mold surfaces to ensure good release; standard silicone release compounds may not be effective in this respect, but polyethylene types are. The catalyzed rubber mix is poured slowly into the outer periphery of the mold and allowed to flow around the pattern to reduce the probability of air being trapped. The top of the pattern should be covered by at least 5 mm of rubber. The assembly should then be allowed to stand for at least 24 h at room temperature to allow the rubber to cure, followed (if possible) by heating in an air oven at 70°C for 16 h.

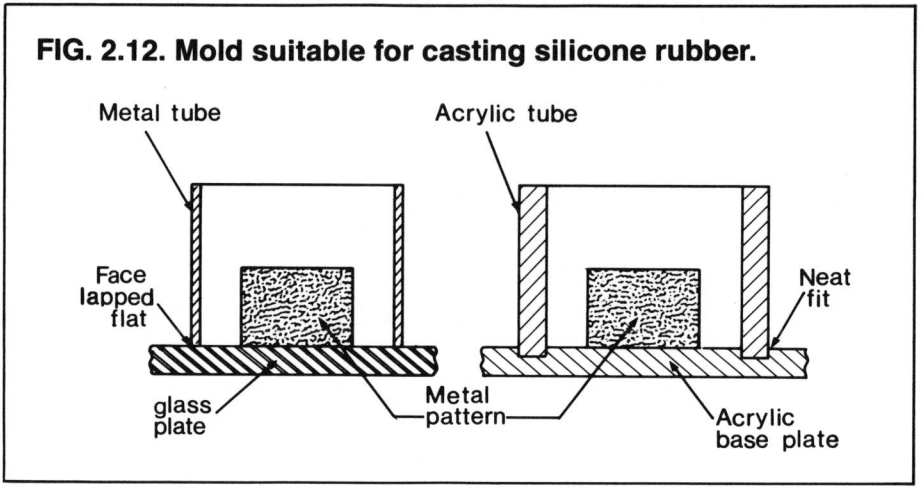

FIG. 2.12. Mold suitable for casting silicone rubber.

REFERENCES

1. J. R. Vilella, "Metallographic Technique for Steel", American Society for Metals, Cleveland, 1938.
2. H. S. Link, "Symposium on Methods of Metallographic Specimen Preparation", ASTM Special Technical Publication No. 285, 1960, p. 8
3. C. A. Godden, *Metal Progress*, 1961, 79 (1), 122
4. J. V. Hardy and A. D. Hopkins, *Metallurgia*, 1955, *51*, 209
5. G. L. Gibbon and J. V. Furth, *Metal Progress*, 1963, *84* (3), 118
6. C. W. Weaver, *Metallurgia*, 1964, *69*, 195
7. L. E. Samuels, *J. Inst. Metals*, 1955, *83*, 359
8. F. M. Cain, Jr., "Symposium on Methods of Metallographic Specimen Preparation", ASTM Special Technical Publication No. 285, 1960, p. 37
9. R. J. Hussey, P. E. Beaubien and D. Caplan, *Metallography*, 1973, *6*, 27
10. H. Lee and K. Neville, "Handbook of Epoxy Resins", McGraw-Hill, New York, 1967.
11. C. G. Hoffman, *Metal Progress*, 1963, *84* (3), 118
12. W. L. Ladroga, *Metal Progress*, 1963, *83* (2), 108
13. D. V. Miley and A. E. Calabra in "Metallographic Specimen Preparation", edited by J. L. McCall and W. M. Mueller, Plenum, New York, 1973, p. 1.
14. N. J. Gendron in "Metallographic Specimen Preparation", edited by J. L. McCall and W. M. Mueller, Plenum, New York, 1973, p. 121.
15. D. Caplan and P. E. Beaubien, *Canadian J. Technol.*, 1952, *30* 211
16. P. B. Petretsky, *Microstructural Science*, 1977, *5*, 273
17. D. J. Rahn, *Metal Progress*, 1959, *76* (2), 109
18. G. Sproule, *ibid.*, 1944, *46*, 484
19. S. J. Broderick, *ibid.*, 1944, *46*, 1276
20. R. L. Duffner and E. S. Norris, *ibid.*, 1946, *50*, 658
21. C. G. zur Horst, *ibid.*, 1945, *47*, 509
22. U. E. Wolff and L. B. Fradette, *ibid.*, 1959, *76* (2),111
23. E. Wade, *Metallurgia*, 1968, *78*, 81
24. H. F. Bartell, *Metal Progress*, 1945, *47*, 940
25. W. Koppa, *ibid.*, 1945, *47*, 274
26. A. Baczewski, *Metallography*, 1970, *3*, 481

CHAPTER 3

Abrasive Machining: Principles of Material Removal

MOST OF THE CUTTING and preliminary flattening stages used in the preparation of metallographic specimens employ devices in which a number of abrasive particles are held firmly together. They may be sintered together in a vitreous bond, as in a cutoff or grinding wheel; they may be cemented to a backing, as in a coated abrasive paper; or they may be buried in a layer of deposited metal, as in some forms of diamond lap. The individual particles of abrasive (commonly called *grits*) usually are formed by crushing larger grains; consequently they are single crystals of irregular shape. Only one point of an abrasive particle contacts the surface of the specimen and this point can be regarded as a machining tool. It is, however, a machining tool that does not have the carefully controlled geometry characteristic of the machining tools used, for example, in lathe turning or milling. Three types of abrasive need to be considered, namely:

1. *Silicon carbide* (SiC). An artificial product made by the reaction between silica and carbon at high temperatures. Crystal structure, hexagonal-rhombohedral. Hardness, approximately 2500 HV.
2. *Aluminum oxide* (Al_2O_3). Usually produced by fusing refined bauxite. Crystal structures: α form, hexagonal; γ form, cubic. Hardness, approximately 2000 HV. The α form is also available in an impure natural product known as *corundum,* and in an even more impure natural product known as *emery.* The α form is referred to unless otherwise stated.
3. *Diamond.* The crystalline form of carbon available as an artificial product or as a natural product. Crystal structure, cubic. Hardness, approximately 8000 HV.

The characteristics of these abrasives have been described comprehensively by Coes.[1]

All three of these abrasives are supplied commercially in a range of grades in which the diameter of the abrasive particles is controlled to within specific limits. The grading system used for silicon carbide and aluminum oxide abrasives is outlined in Appendix 3-A; note that changes have recently been made in the grading system following metrication of the size ranges. Note also that the sizing system changes between the 220 and P240 metricated grit numbers. Some of the data used throughout this book were obtained using the older premetrication grades of abrasive, and some were obtained with the metricated grades. For grade numbers of 240 or larger, the prefix P indicates a grade in the metricated microgrit range; the absence of the prefix with a grade number of 240 or larger indicates a premetrication grade; no distinction is made for grades of 220 and smaller.

At first sight, the preliminary abrasive machining (*abrasion*) processes appear to be simple and straightforward, involving only the manipulative complications involved in positioning the section plane at the desired place and producing a flat surface with as fine a finish as possible. But in fact these stages can have disastrous consequences in the final result because they plastically deform, and occasionally heat, a surface layer. The microstructure of this layer may be recognizably different from the true structure of the specimen. This leads to one of the most critical problems in the design of an integrated preparation sequence, a topic that will be developed in Chapter 8. It will emerge there that two separate features of an abrasive machining process must be considered, namely: (*a*) the rate at which material is removed from the specimen (the abrasion rate); and (*b*) the nature and depth of the plastically deformed layer that is produced in the surface being formed. Consideration of these two parameters must be based on an understanding of the physical processes that occur when the point of an abrasive particle contacts and moves across the specimen surface.

A Model of Abrasive Machining

THE BASIS OF THE MODEL

The morphology of a real abrasive particle is extremely complex and irregular (Fig. 3.1), but only a small portion of one of its points actually makes contact with the specimen surface (Fig. 3.2). It is only the shape of this portion that really matters. Obviously, then, the first step in the development of a model of the action of an abrasive particle is to choose a simple shape to represent this active portion. The simplified shape must be both adequately realistic and simple enough to make it possible to analyze the process.

A sphere is perhaps the simplest first choice but is not sufficiently realistic to be acceptable. Many independent experimenters have established that points of spherical shape do not remove material from the surface by producing ribbonlike chips, and all abrasion processes remove chips in profusion (see, for example, Fig. 3.18). The next simplest possibility is to characterize the abrasive point as a pyramid moving in a direction perpendicular to one of its faces. The discussion below is based on this model.[2-4]

Possible modes of interaction between a point modeled in this way and a workpiece surface are illustrated in Fig. 3.3, where the abrasive particle is represented by a cube of clear plastic and the specimen by a slab of modeling clay. Three modes of interaction can be recognized, namely:

FIG. 3.1. Typical particle of 220-grade silicon carbide abrasive.

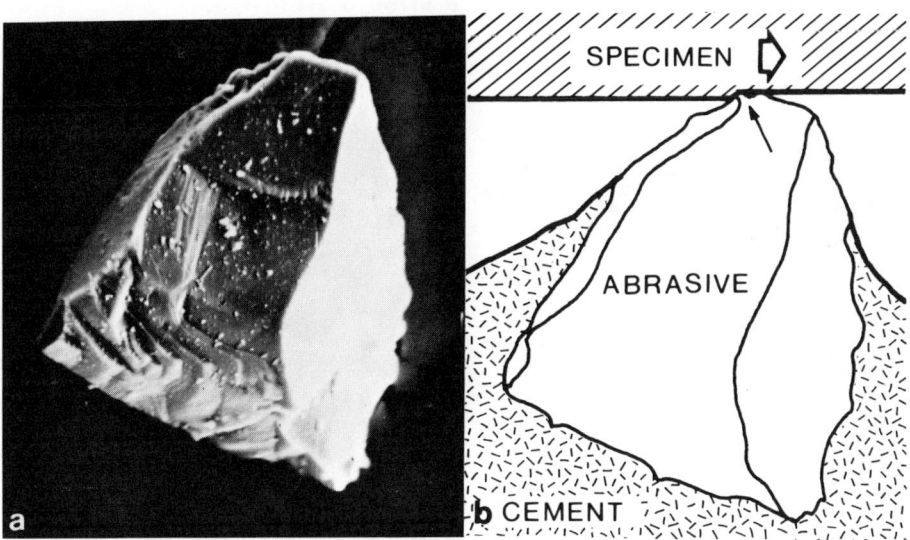

(a) Scanning electron micrograph. Magnification, 1000×. (b) Sketch illustrating the manner in which this particle would be mounted in a practical abrasive device; the small arrow indicates the portion of the particle that would interact with the surface of a specimen.

FIG. 3.2. Detail of the contacting point of an abrasive particle in a 60-grade silicon carbide abrasive paper.

The paper was worn against a specimen of hard steel and then used to abrade a specimen of 70:30 brass. The arrow indicates a brass chip that adheres to the cutting edge, illustrating that only a small section of the edge of this particle was truly active during abrasion. Scanning electron micrograph. Magnification, 600×.

FIG. 3.3. Model illustrating possible modes of interaction between an abrasive particle and the surface of a specimen.

(A) A loose, tumbling particle. (B) A fixed particle plowing a groove. (C) A fixed particle cutting a chip.

Mode A. The abrasive particle is loose and rolls between the workpiece and another parallel surface. A corner of the particle digs into the workpiece, the particle tumbles onto an edge, and then another corner contacts the workpiece, and so on. A track of angular indentations is produced in the surface. No material removal occurs as a primary mechanism, although some material could be removed by a secondary mechanism such as removal of the ridges at the sides of the indentations. Nevertheless, the efficiency of material removal approaches zero.

Mode B. The abrasive particle is fixed in one orientation with a relatively acute included angle between the advancing face and the specimen surface. A groove is produced in the surface. A standing-wave bulge forms in front of the advancing particle, and material is displaced into a ridge at each side of the groove (see Fig. 3.4, top). No material is removed by a primary mechanism, although secondary mechanisms are again possible. The efficiency of material removal approaches zero.

Mode C. The abrasive particle is fixed but with a less acute included angle between the advancing face and the specimen surface. A groove is produced in the surface and a ribbonlike chip of material is separated from the surface (see Fig. 3.4, bottom). In the simplest case, all of the material that was in the groove is removed from the surface. The efficiency of material removal then approaches 100%.

Thus, machining with a vee-point tool is different from orthogonal machining. A chip* is almost required to separate in orthogonal machin-

---

*This should strictly be called a *machining chip*, to distinguish it from a *fracture chip* (see Chapter 7).

FIG. 3.4. Comparison of the plowing and cutting modes of a vee-point tool.

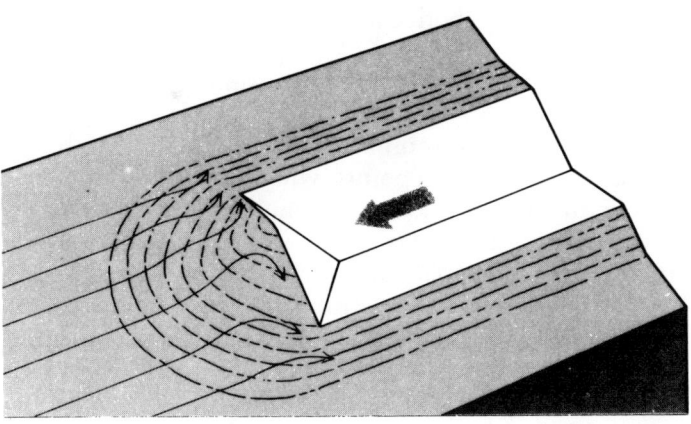

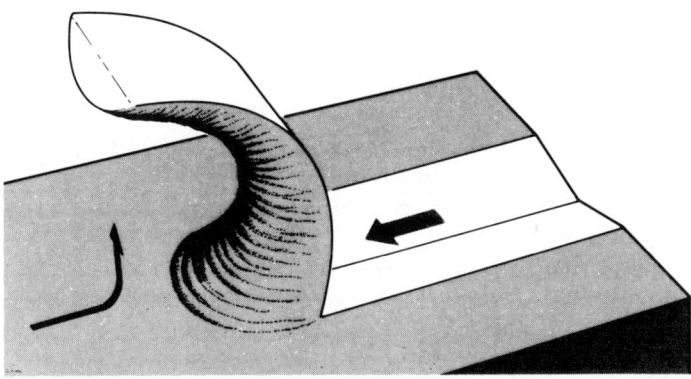

In plowing (top), material in the surface layers of the specimen first moves upward ahead of the rake face and then moves around it into side ridges. In cutting (bottom), a ribbon of material is separated from the specimen and moves upward past the rake face of the tool until it breaks off. (In normal machining with orthogonal tools, the workpiece material does not have the option of flowing around the tool; a chip virtually has to separate.)

ing, but in machining with a vee-shape tool the surface layers have the opportunity of flowing around the point rather than separating as a chip (Fig. 3.4).

An abrasion process in which the specimen is rubbed against a glass plate covered with a loose abrasive powder is represented by mode A: this is known as *three-body* abrasion. A matte surface comprising a series of angular indentations is produced, and little or no material is removed.* This type of process is not suitable for preparation of metallographic specimens. If, however, the glass plate is replaced by a softer material (such as lead, cast iron or wax), some of the abrasive particles may embed in the plate and so become fixed, at least temporarily. The result is a mixture of three-body and *two-body* (see below) abrasion. Such processes have limited use in metallography, but only if two-body abrasion can be made to predominate. The abrasion device is then commonly known as a *lap*.

Therefore, the first requirement of a metallographic abrasive device generally is that those abrasive points which contact the specimen surface should operate in either mode B or mode C. In most cases, this is achieved by cementing the abrasive particles to each other and to a backing, as illustrated diagrammatically in Fig. 3.1(b). The working surface of a practical abrasion device consequently consists in principle of a two-dimensional array of points which are fixed in space. A second requirement is that as many of the abrasive points as possible operate in mode C, thus achieving the highest possible abrasion rate. Consequently, we now need to consider what determines which of the two modes will operate, and this turns out to be the angle of inclination between the advancing or working face of the abrasive point and the specimen surface.

Experiments can be carried out with model points in which a hard tool with a pyramidal point is made to traverse a softer metal surface under a constant applied load. The angle of incidence of the working face of the tool can then be varied. The inclination of the working face is best defined by the terminology used in machining practice — namely, as a *rake angle*. This is the angle between the face and the normal to the workpiece surface; the angle is regarded as positive if the working face slopes away from the normal with respect to the direction of motion, and negative if it slopes toward the direction of motion. The terminations of the grooves produced in the specimen surface are inspected after each experiment to determine whether or not a chip has been cut.

Results typically obtained in experiments of this nature are illustrated in the left-hand column of photographs in Fig. 3.5. A chip is cut wherever the point has a sufficiently positive rake angle (Fig. 3.5a), but a standing-wave prow is formed when the point has a more negative rake angle (Fig. 3.5b, c and d), the prow becoming flatter the more negative the rake angle. There often is, as in the experiment illustrated, a fairly sharp transition from cutting to plowing, so that a *critical rake angle* (α_c) for the transition can be defined. In the case illustrated, $\alpha_c \simeq 0°$. The transition from plowing to cutting is not always as sharp as that illustrated in Fig. 3.5 (see, for example, Fig. 3.14), but to a first approximation the critical-angle concept can be taken to be a general one. The value of α_c differs considerably for different metals and alloys (Table 3.1).

The series of photographs in the right-hand column of Fig. 3.5 confirms that the concept of a critical rake angle for cutting has validity. These photographs are of scratch terminations obtained in an experiment of a

---

*These remarks do not apply to brittle specimen materials (see Chapter 7).

FIG. 3.5. Terminations of scratch grooves in high-purity aluminum.

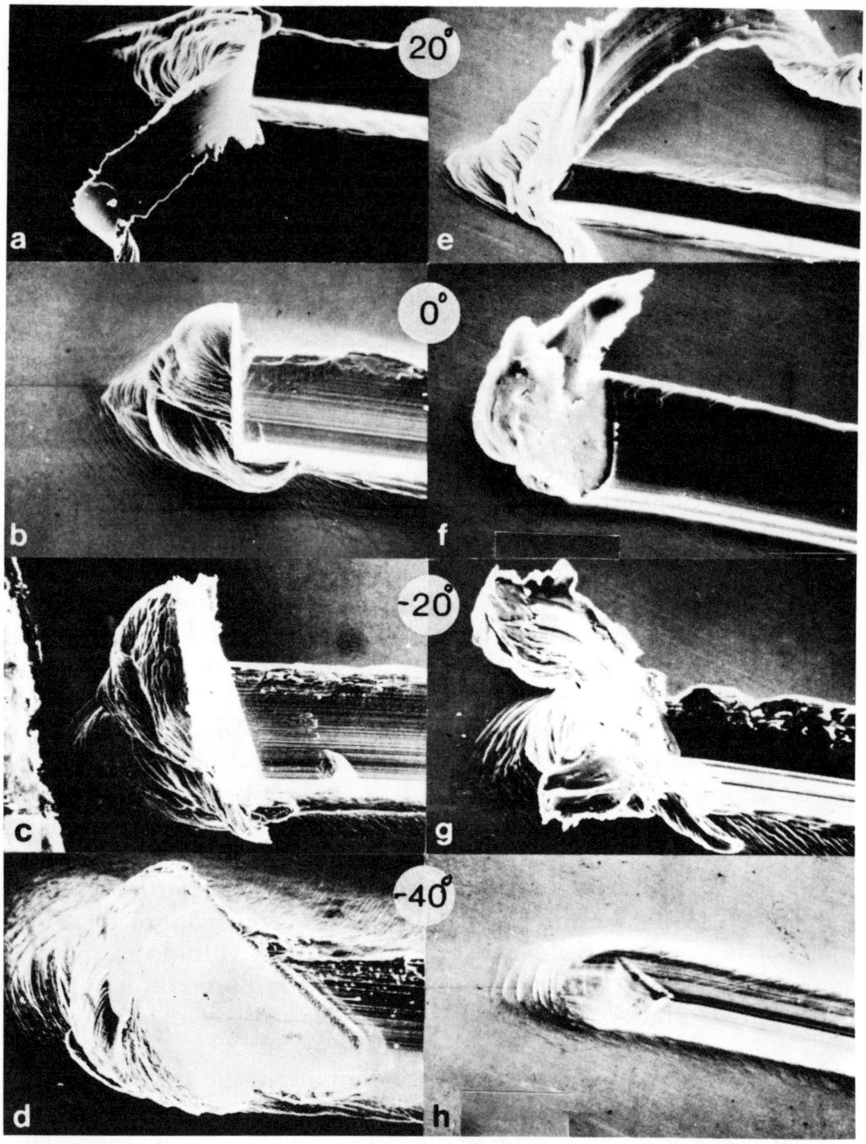

(a) to (d) Grooves produced by vee-point tools. Magnification, 125×. (e) to (h) Grooves produced by the abrasive particles in 220-grade silicon carbide abrasive paper. Magnification, 500×. In all cases, the approximate rake angles of the points are indicated. Hardness of aluminum, 25 HV.

type that will be referred to several times later. The surface of a specimen is polished and then the specimen is placed on an abrasive paper; a normal load is applied and the specimen is traversed across the abrasive

TABLE 3.1. Critical Rake Angle for Cutting, and Fraction of Contacting Points That Cut, for Unused 220-Grade Silicon Carbide Abrasive Paper

Metal or alloy	Hardness, HV	Critical rake angle	Fraction of cutting points
Lead(a)	4	−35°	0.7
Aluminum(a)	25	−5°	0.3
Copper(a)	50	−45°	0.75
α Brass(a)	70	−35°	0.7
Nickel(a)	130(a)	−25°	0.5
Steel	150(b)	−10°	0.4
	255(a)(c)	0°	0.25
	855(d)	−80°	>0.95

(a) See Ref 4. (b) Mild steel. (c) Cold drawn 0.35%C; see Ref 2. (d) Quench hardened; 1.4%C.

paper in the usual way but only for a few millimetres. The terminations of the scratch grooves so produced are then examined to determine whether or not a chip has been cut. Often, the imprint of the working face of the abrasive point can be seen at the end of the abraded groove; an approximate estimate can then be made of the rake angle of the working face of the abrasive point that has produced the groove. Points with rake angles similar to those of the model points used for Fig. 3.5(a) to (d) have been selected for illustration in Fig. 3.5(e) to (h). It can be seen that the model points and the real points behave in very similar ways.

Lest it be thought that these concepts apply only to the softer materials, examples are given in Fig. 3.6 of scratch terminations produced by an abrasive paper in steels with a range of hardnesses. Chips of normal morphology have been produced in all cases, the scratch grooves and the chips simply being smaller in cross section the harder the specimen material. Expectations based on bulk behavior cannot be applied to chip cutting on this scale because only a small volume of material is being deformed by each point; moreover, this volume of material is subjected to hydrostatic stresses as well as shear stresses during separation of the chip.

There is a direct implication in photographs of the type shown in Fig. 3.5(e) to (h) that real abrasive points have a range of effective rake angles. It is consequently to be expected that the over-all efficiency of material removal would be determined by the fraction (f) of these points that have rake angles suitable for cutting a chip. This in turn would be determined by the distribution of point rake angles in the particular abrasive device and by the value of the critical rake angle for the material being abraded. The volume of material removed by one of the cutting points would then be determined by the length of traverse, which is known, and the cross-sectional area of the groove produced. Consequently, the parameters that control the cross-sectional area of the groove now need to be considered.

Consider first the static indentation made by a contacting point, regarding it as a hardness indenter. Assuming that all points carry the same

FIG. 3.6. Terminations of scratch grooves produced in steel by 220-grade silicon carbide abrasive paper.

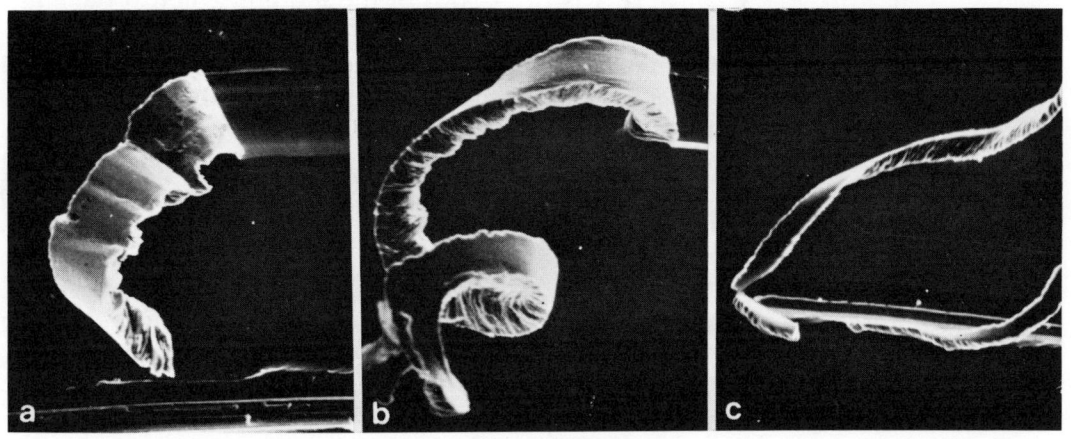

(a) Hardness of steel, 140 HV. Magnification, 600×. (b) 450 HV; 500×. (c) 880 HV; 600×.

FIG. 3.7. Variation in cross-sectional area of grooves produced by points of different geometric shapes.

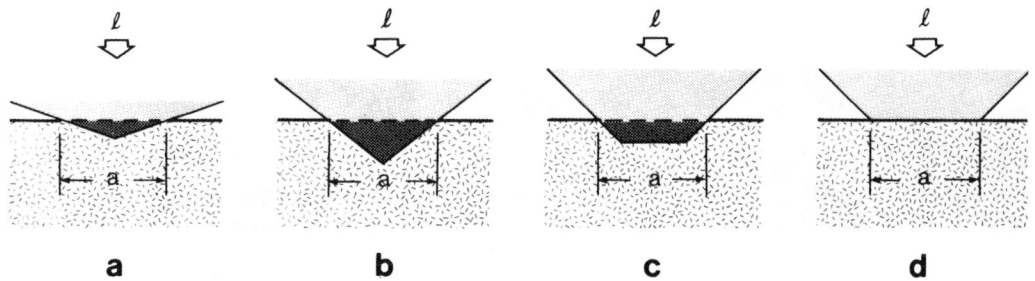

With specimen hardness and applied load held constant, all points indent to the same projected area, but there is considerable variation in cross-sectional area (dark shading) among the grooves they produce.

load, each will indent plastically into the workpiece until the projected area of the indentation attains a characteristic value — a value that is related to the indentation hardness value of the surface layers of the specimen. Now assume that this depth of indentation is maintained when the point moves. The cross-sectional area of the groove will then be smaller the more obtuse the angle of the point (cf. Fig. 3.7a and b). Minor perturbations in the geometry of the point will not matter much, but the volume of the groove will be much reduced if there is a flat on the point (cf. Fig. 3.7b and c). It will eventually reach zero when the flat becomes large enough for the point to be supported elastically (Fig. 3.7d). However, it is necessary in a simple first-approximation model to assume that the points are sharp and triangular in cross section, and to define their form in terms of the included angle of the point. We can define a form factor $\phi = x/y$, where x is the depth of indentation when producing a groove of width y.

The above static analog is not exact. Once a point begins to move it is supported only on its working face by either the standing-wave prow or the chip that it forms. The depth of indentation then changes, but experiments indicate that this complication can also be ignored in a first-approximation model.[2,3] Moreover, in practice the abrasive points produce grooves in a surface that already contains grooves of the same general dimensions and geometry as those of the new grooves being produced; the geometry of interaction between the points and the specimen surface is consequently much more complex than that implied in Fig. 3.7. The corrections that thereby need to be made to the model being developed here are those of magnitude rather than of principle, so that they too can be ignored in our first-approximation model.

Characteristics of Coated Abrasive Papers

We have by now identified two important parameters by which an abrasive point can be characterized — namely, its rake angle and its form factor. We have also seen that the points in real abrasive devices have a range of rake angles and that only a fraction of these will have a rake angle suitable for removing material from a particular specimen material by cutting a chip. We now need to see how these concepts can be related to practical abrasive devices, and to do this we need to study their structure in more detail.

The abrasive device most commonly used in metallographic practice is a sheet of paper that has been covered uniformly with a layer of graded abrasive, the abrasive particles having been cemented to each other and to the backing. The process by which the papers are manufactured usually ensures that a long axis of each abrasive particle is aligned approximately normal to the backing. Generally, the papers are known as *coated abrasive products*. Commonly they are also known as *abrasive papers* when the backing is a paper, and as *waterproof abrasive papers* when the cement is resistant to water and the paper is impregnated with a water-resistant resin. It will be convenient to consider this type of device first, not only because it is the device most commonly used in metallography but also because it constitutes the simplest case of abrasive machining. The geometry of motion is simple, and abrasion is carried out under an approximately constant applied load.

A range of grades of silicon carbide abrasive papers are shown in silhouette in Fig. 3.8, and one of them is shown in section in Fig. 3.9. The abrasive particles exposed at the working surface of an unused paper are covered with a layer of resin (Fig. 3.9), but this layer is soon removed from contacting points during use (Fig. 3.2 and 3.11b). This, incidentally, provides a positive method of identifying points that have been in contact with the specimen. The first point to note is that, even though an abrasive

FIG. 3.8. Silicon carbide abrasive papers viewed in silhouette.

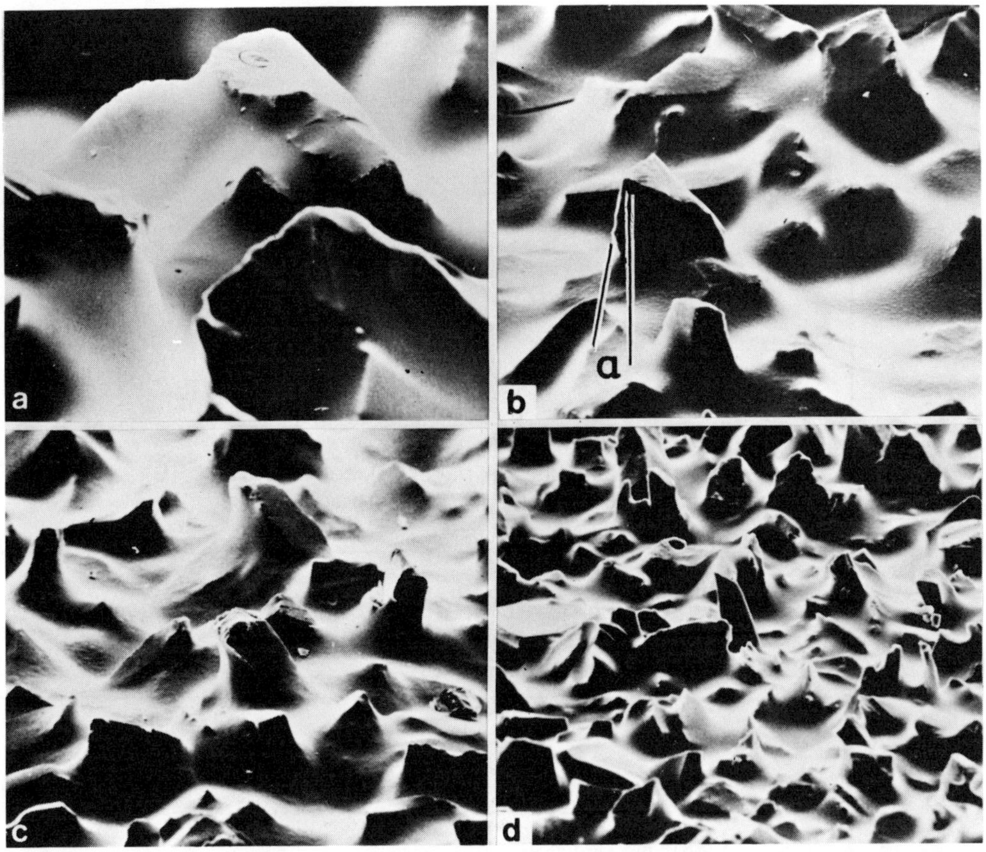

(a) 80-grade paper. (b) 150-grade paper; the manner in which the rake angle of a point might be defined for a specimen moving from left to right is indicated. (c) 220-grade paper. (d) P600-grade paper. Scanning electron micrographs. Magnification (all), 175×.

paper in principle presents a two-dimensional array of points to the specimen surface, the points are not quite coplanar. Consequently, only a fraction of them contact the surface of a specimen placed against the working surface of the paper: up to four or five in a hundred is a representative figure (Table 3.2). The remainder play no part in the abrasion process, although some may do so later in the life of the paper if any of the first generation of contacting points are removed in use. The areal density of contacting points has to be determined by experiment. A simple method of doing this is to place a prepolished specimen against the surface under an appropriate load, translate the specimen a few millimetres, remove it, and determine the number of grooves produced per unit area of surface. Values so obtained for typical silicon carbide papers are given in

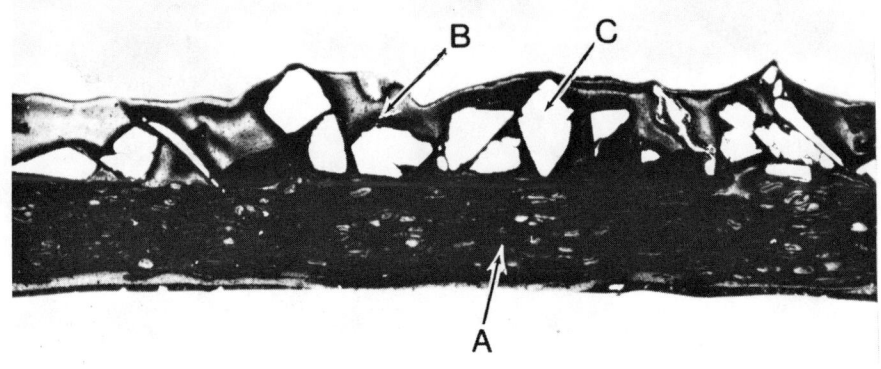

FIG. 3.9. Section of a 220-grade silicon carbide abrasive paper. *(Mulhearn and Samuels, Ref 2)* Arrows indicate backing paper **(A)**, cementing resin **(B)**, and one of the abrasive particles **(C)**. Magnification, 100×.

TABLE 3.2. Some Characteristics of Silicon Carbide Coated Abrasive Papers Used Against Steel With a Hardness of 255 HV *(Mulhearn and Samuels, Ref 2)*

Characteristic	Grade of paper				
	150	220	280	400	600
Number of particles visible at working surface/cm²	...	100×10^2	150×10^2	550×10^2	2000×10^2
Number of contacting points/cm²:					
Unused	125	130	195	580	1000
After 400 traverses	150	200	225	550	...
Fraction of contacting points that cut (f):					
Unused	0.20	0.23	0.18	0.13	0.12
After 400 traverses	0.12	0.11	0.09	...	...
Number of cutting points/cm²:					
Unused	25	30	35	75	120
After 400 traverses	18	22	20	...	...
Mean apex angle of contacting points:					
After 50 traverses	135°	155°	155°	165°	165°
After 400 traverses	135°	155°	155°	165°	...
Mean form factor of contacting points (ϕ_m):					
After 50 traverses	3.3	4.5	4.5	8.0	8.0
After 400 traverses	3.3	4.5	4.5	8.0	...

Table 3.2. Note that the number of contacting points is larger the finer the grade of paper. The number also increases slightly with use of the papers. This is the first basic parameter by which a practical abrasive paper has to be characterized.

The second important parameter is the distribution of rake angles of these contacting points. Quantitative statistical estimates of rake angles can be made from photographs, such as those in Fig. 3.8 and 3.9. However, such estimates can only be indicative, because there are always doubts regarding the specific points, and the exact portion of the points, that will contact the specimen surface. Moreover, only a two-dimensional estimate of rake angles can be made. Nevertheless, such an assessment gives a guide to the performance that can be expected of an abrasive paper, a typical distribution of rake angles being given in Fig. 3.10. The fraction (f) of contacting points that will cut a chip can now be estimated, by the method illustrated in Fig. 3.10, if the critical angle for cutting is known (Table 3.1). Values of f determined in this way for a steel abraded on a range of grades of silicon carbide paper are given in Table 3.2, and values for a range of materials abraded on a particular grade of silicon carbide paper are given in Table 3.1. These figures can be confirmed by experiments of the type used to determine the number of contacting points. The terminations of the scratch grooves are examined and an estimate is made of the proportion of these grooves that have chips attached to them. The agreement is good.[2] Note, incidentally, that many points

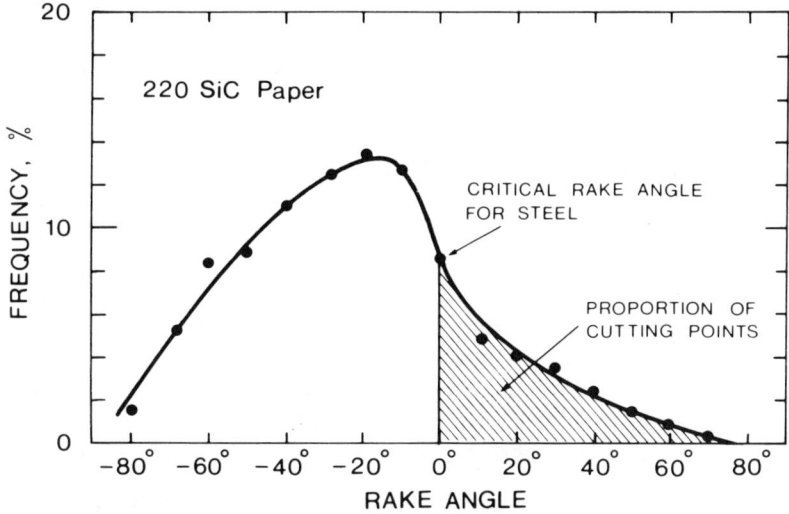

FIG. 3.10. Distribution of rake angles of contacting abrasive points in unused 220-grade silicon carbide abrasive paper. *(Mulhearn and Samuels, Ref 2)*

Measurements were made on sections such as that shown in Fig. 3.9. The proportion of points that would cut chips in a material with a critical cutting angle (α_c) of 0° (e.g., steel) is indicated by the fraction of the area below the curve that is shaded.

do have positive rake angles, but this is true only because some particles are tilted off normal (Fig. 3.8 and 3.11f). For example, the tilted particle shown in section in Fig. 3.11(f) presents a highly positive rake angle to a specimen moving from left to right, although a highly negative rake angle to one moving in the opposite direction. It is only these tilted particles that can provide cutting points for materials for which α_c is either positive or negative but close to 0°.

The figures in Table 3.2 show dramatically that, of the many abrasive particles visible on the surface of a coated abrasive paper, only one or two in a hundred remove material by cutting a chip. About the same proportion produce scratches in the surface by plowing. The remainder do nothing.

The next parameter needed to characterize a real coated abrasive paper is one that describes the shape of the contacting points projected in a plane normal to the direction of motion — i.e., a factor that describes the cross-sectional area of the grooves that the points will produce in the surface of the specimen. The simplest way of doing this is to assume that the points have a sharp "vee" shape. A form factor (ϕ) can then be ascribed to each point in terms of the included angle of the vee (see discussion of Fig. 3.7, pp. 39-40). Measurements of the statistical distribution at this factor can be made on photographs such as those in Fig. 3.8 and 3.9, but experience indicates that an estimate of the mean value (ϕ_m) often suffices. Typical values of ϕ_m are listed in Table 3.2. The value of ϕ_m differs for different grades of paper, a fact which can be recognized qualitatively in Fig. 3.8.

Earlier discussion indicated that the volume of the groove will be overestimated by the above analysis if the contacting point does not have a sharp point. Later discussion will indicate that it is reasonable to assume that the points in fresh papers are sharp but not always so for worn papers.

We have so far considered only unused abrasive papers. Experimentally determined values of f and ϕ_m are given in Table 3.2 for papers on which a steel of moderate hardness has been abraded several times over one track. Marked changes in the characteristics of the papers have occurred, particularly in fractions of cutting points. This must have been due to changes in the geometry of the cutting points. It is indeed known that changes in the geometry of the points do occur in use by one of a number of mechanisms, namely:

1. *Fracture:* A major portion of the particle may break off (Fig. 3.11a), in which event the particle no longer provides a contacting point. On the other hand, small fragments may break away from the contacting region of the point (Fig. 3.11b), in which event the point may remain in contact but its geometry as a cutting tool may change.

2. *Point flattening:* A flat is formed on the point (Fig. 3.11c) either by attritious wear or by microfracturing. Care is necessary to distinguish worn flats of this nature (a) from gross fractures which happen to produce flats and (b) from caps (see below). Point flattening may change the rake angle of a point. It certainly will reduce the

FIG. 3.11. Various modes in which the contacting points of abrasive particles deteriorate in use.

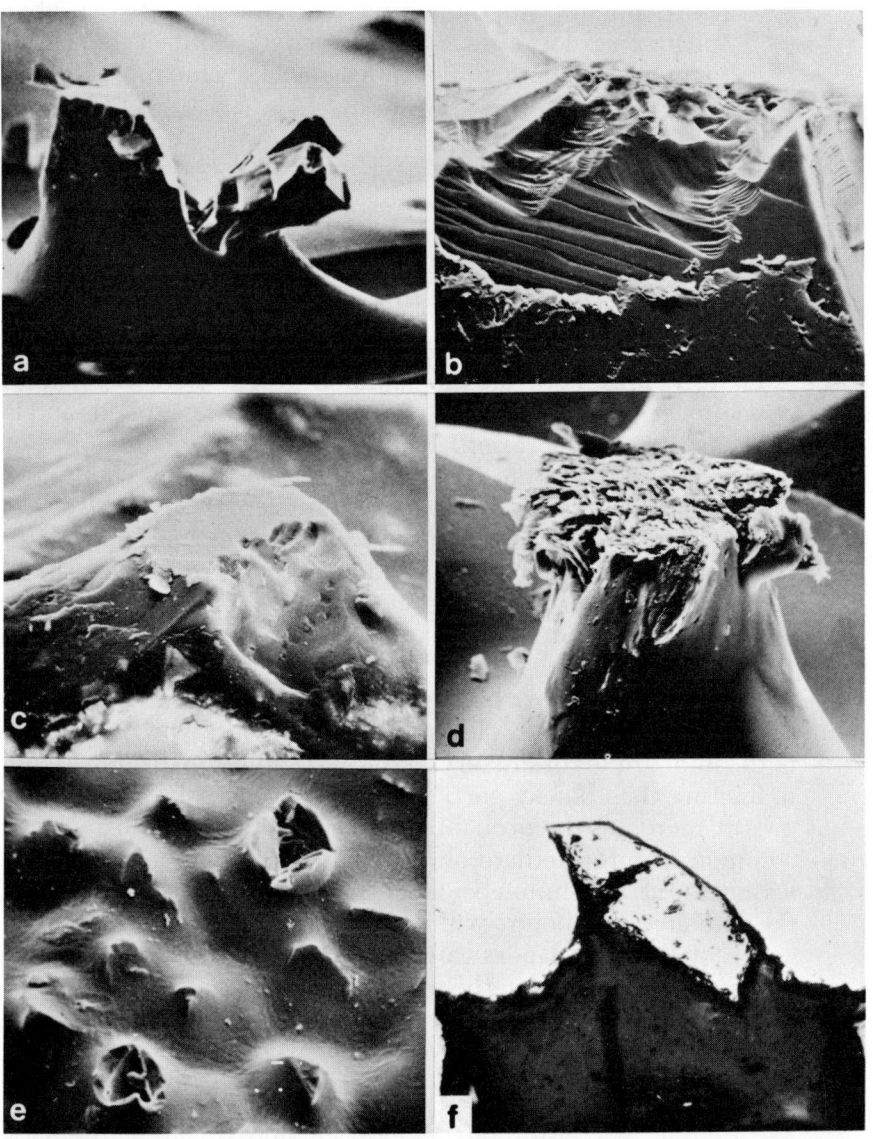

(a) Large fragment removed by fracture. Silicon carbide paper (220 grade) abraded by soft aluminum. Scanning electron micrograph. Magnification, 860×. (b) Small fragments removed by fracture. Silicon carbide paper (80 grade) abraded by a hard steel. SEM; 620×. (c) Flat formed by wear. Aluminum oxide paper (180 grade) abraded by a hard steel. SEM; 1000×.

(d) Point capped with a layer of specimen material. Silicon carbide paper (80 grade) abraded by brass. SEM; 320×. (e) Entire abrasive particle removed due to loss of bonding to cement. Silicon carbide paper (220 grade) soaked in water for five days and then abraded by brass. SEM; 100×. (f) Cement eroded away. Silicon carbide paper (220 grade) abraded by steel. Section; 500×.

volume of the scratch groove that is formed in the specimen (see Fig. 3.7, p. 39).
3. *Capping:* A cap of specimen material accumulates on the contacting point, possibly to the extent that contact between abrasive and specimen is lost (Fig. 3.11d). In effect, contact then is from metal to metal, and both surface finish and abrasion rate are deteriorated.
4. *Shelling:* An entire abrasive particle is removed from the coated layer due to loss of adhesion between the particle and the cement holding it in place (Fig. 3.11e).
5. *Erosion:* The cementing resin erodes away (Fig. 3.11f), and the abrasive particles become undermined to the extent that they eventually become detached. A new set of abrasive particles is thereby exposed. The process is thus regenerative until the coated layer is removed completely.

Point fracture is by far the commonest and most important degenerative process occurring with the comparatively low speeds and loads used in metallographic abrasion. Point flattening is uncommon, but capping does occur under some circumstances (see p. 62). Shelling is likely if papers have been soaked in water for any length of time (see p. 59). Erosion is likely when papers have been used for long periods. However, the relative importance of these deterioration processes is different under different abrasion conditions. For example, point flattening seems to be a dominant process in many industrial abrasion operations.[5-7]

In general, silicon carbide is the most easily fractured of the three abrasives being considered. Thin splintery particles, which are more common in the finer grades of silicon carbide and aluminum oxide, are more prone to fracture than blocky particles. The presence of cracks in the particles greatly increases the probability of large chunks fracturing off; cracks are common in intermediate and finer grades of silicon carbide and, to a lesser extent, aluminum oxide. It is not yet possible, however, to predict the rate at which points will deteriorate in use. Consequently, deterioration of abrasive papers can be quantified only after the event by determining the change in the number of contacting points and the fraction of cutting points (Table 3.2).

Mathematical Development of the Model

We are now in a position to develop an approximate mathematical model for both the surface finish and the abrasion rate achieved by a coated abrasive paper. Characterizing surface finish by the mean depth (y_m) of the scratch grooves, it can be shown that:[2]

$$y_m = 1/\phi_m \cdot \sqrt{L/(PCA)} \qquad \text{(Eq 1)}$$

where L is the load applied to the specimen; P is the indentation hardness value of the layers of specimen material involved, expressed as a mean pressure; C is the number of contacting points per unit area; A is the area of the specimen; and ϕ_m is the mean form factor of the contacting points, as defined previously.

This analysis implies that the depth of the scratches produced does not depend directly on the diameter of the abrasive particles with which the paper is coated. This can be understood by reconsidering Fig. 3.1 and 3.2, which indicate that the interaction between the abrasive and the workpiece is confined to, and solely dependent on the shape of, a small region of the contacting point. Two particles having points with the same shape would, when carrying the same load, produce a groove of the same cross section, irrespective of the shape and dimensions of the rest of the particle, including its diameter. The role of the rest of the particle thus is only that of holding the active point in place — that is, it has the same role as the shank of a lathe tool.

However, the diameter of the abrasive particles can have indirect effects on surface finish. It becomes possible to pack a larger number of particles into a unit area of the surface of an abrasion device (hence, obtaining a large value of C) when the diameter of the abrasive particles is small. The value of y_m will be small when C is large. Whether this potentially larger value of C is actually achieved in practice depends on the nature of the abrasive device and how it is made — that is, on how well the abrasive particles are packed together. In practice, C is indeed always largest for the finest grades of commercial abrasive papers (Table 3.2). Moreover, the crushing and grading processes used to produce the different grades of commercial abrasives are such that the particles in different grades have points with different shapes. Consequently, they have different values of ϕ_m (Table 3.2). The shape of a contacting point may also change as the point is worn in service. The general trend to be expected from these differences in point shape is for the finer grades of paper to produce deeper scratches, which is not the trend found in practice. It follows that the dominating factor is the number of contacting points.

Turning now to abrasion rate, the mass of material removed during a particular traverse (m_n) in a simple rectilinear abrasion operation can be shown to be:[2]

$$m_n = f_n \rho DL/(P \phi_m) \qquad \text{(Eq 2)}$$

where ρ is the density of the specimen; D is the distance traversed; f_n is the fraction of cutting points during the nth traverse; and ϕ_m, L and P are as for Equation 1. This formula may need to be modified to describe behavior during repeated traverses over the same track, which occurs in metallographic practice, because the characteristics of an abrasive device may change. The deterioration can in the general case be described in terms of an exponential decrease in the fraction of cutting points, i.e.:

$$f_n = f_0 \cdot e^{-\beta n} \qquad \text{(Eq 3)}$$

where f_0 is the fraction of cutting points in an unused paper and β is the deterioration constant of the particular device. Then the total loss in mass up to the nth traverse is:

$$M_n = \rho DL/(P \phi_m \beta) \cdot f_0(1 - e^{-\beta n}) \qquad \text{(Eq 4)}$$

With some metals, however, little deterioration of the abrasive papers occurs, at least after an initial breaking-in period (see p. 56). In this event, a simple modification of Equation 2 can reasonably be used:

$$M = f_0 \rho DL/(P \phi_m) \quad (Eq\ 5)$$

The analysis implies that the only abrasive parameters of importance in abrasive rate are the form of the contacting points (ϕ_m), the fraction of cutting points in the unused paper (f_0), and the rate at which the fraction of cutting points changes with use.

INFLUENCE OF SPECIMEN MATERIAL ON ABRASION RATE

The material of the specimen can be expected to have an influence on abrasion rate in at least three ways — namely, through its hardness, through its effect on critical rake angle, and through its effect on the actual efficiencies of the plowing and cutting processes. Practical abrasion-rate tests also indicate that the specimen material has a major influence on the rate at which abrasive papers deteriorate with use, but this will be discussed later.

Hardness

It is commonly asserted either implicitly or explicitly that the one specimen property that determines abrasion rate is hardness. Equations 1 to 4 above indicate that hardness is certainly involved, but they suggest that it is only one of several factors. A unique relationship with bulk specimen hardness is not even to be expected because, as we shall see in Chapter 4 (p. 100), the surface layers of a specimen become severely work hardened soon after abrasion has commenced. Further interaction between abrasive and specimen occurs within this work-hardened layer. Consequently, the hardness correlation to be expected is the correlation with the maximum hardness that can be achieved in the specimen material by work hardening. This is the value that should be used for P in Equations 1 to 4 above.

The extensive experiments of Krushchov and Babichev[8] are usually quoted to support the proposition that there is a direct correlation between abrasion rate and specimen hardness. However, the hardness values that they reported for a number of the metals are so unusual as to raise doubts about either the purity of the metals or the hardness determinations that were made on them. A similar set of experiments carried out later by Richardson[9] do not suffer from these doubts, and it is these results that are summarized in Fig. 3.12. Even so, the correlation observed between abrasion rate and hardness of unalloyed metals is not a very convincing one.

There is a second anomalous feature of the results of Krushchov and Babichev[8] and of Richardson[9]. They both found that increasing the hardness of a given metal by cold working, alloying or heat treatment, singly or in combination, did not reduce the abrasion rate nearly as much as would have been expected from the presumed primary correlation (Fig. 3.12). The small influence of work hardening can be explained on the

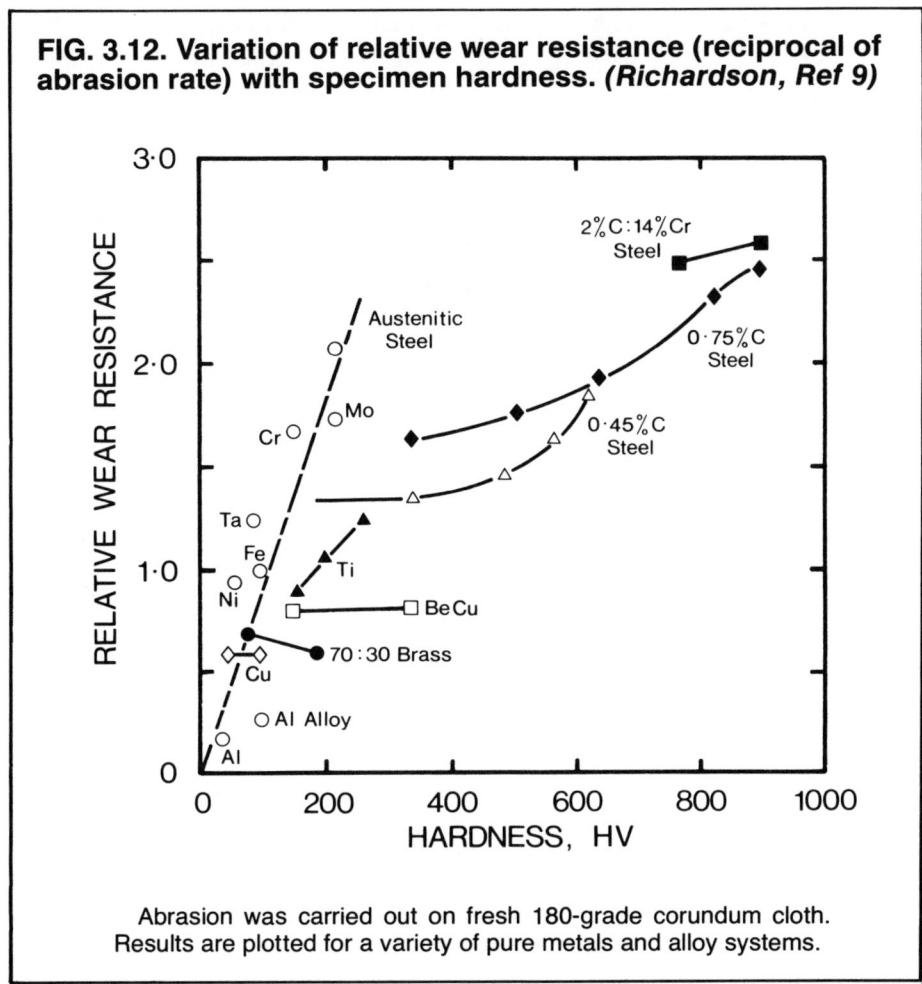

FIG. 3.12. Variation of relative wear resistance (reciprocal of abrasion rate) with specimen hardness. *(Richardson, Ref 9)*

Abrasion was carried out on fresh 180-grade corundum cloth. Results are plotted for a variety of pure metals and alloy systems.

basis outlined above — namely, that it is the maximum hardness that can be achieved in the specimen material by cold working that should be considered. Both Krushchov and Babichev[8] and Richardson[9] recognized this. The implication then is that the abrasion rate of a metal is determined by its exponent of work hardening as well as by its bulk hardness. Larsen-Basse[10] has extended this concept to explain the specific correlations between abrasion rate and hardness of quench-hardened and tempered steels. He also proposed that the volume fraction of cementite in the structure has an influence. In general, the small influence of alloying and heat treatment remain unexplained, but it is already apparent that much more than bulk hardness has to be taken into account when predicting abrasion rate.

Nevertheless, the greatest difficulty with these and most similar investigations of the parameters affecting abrasion rate is that they were carried out entirely on fresh tracks of abrasive paper. This is not representative of metallographic practice, where specimens invariably are traversed repeatedly over one track. Abrasive papers then deteriorate in a manner that varies with specimen material. Significant deterioration often occurs very early in the life of a paper, and this deterioration can be the dominant factor determining the performance of a paper. Consequently, the performance of unused paper is, at best, of academic interest in the present context.

As a general conclusion at this stage, therefore, it can be said that there is a tendency for abrasion rates to be lower with harder materials, but that more specific predictions cannot be made.

Critical Rake Angle

The model of abrasion described earlier implies that the critical rake angle for cutting is a parameter that has a major influence on abrasion rate. Moreover, it is known that the critical rake angle (α_c) differs considerably for different materials (Table 3.1), and that it varies in a manner that is not simply related to specimen hardness or any other known bulk material property. Theoretical two-dimensional analyses[4,11] suggest that the friction at the rake face of the tool and the shear flow stress of the specimen material are the factors that determine α_c. However, a three-dimensional analysis really is required to describe the behavior of a vee-point tool, because three-dimensional flow is involved in the plowing option that such a tool can adopt.

Values of α_c have been determined by experiments with model tools (as illustrated in Fig. 3.5) for a few materials (Table 3.1), most of them being comparatively soft and in the annealed condition. The latter is in itself a limitation. It is α_c for the severely deformed fragmented layer on an abraded surface (see p. 100) that needs to be known, because succeeding abrasive points will interact almost entirely within this layer. Moreover, it has not been possible to determine α_c for many hard materials because of experimental difficulties: the model points fracture too easily, even when made in diamond. An exception is found with very hard steels for which α_c has a highly negative value, much more negative than for softer steels (Table 3.1). This perhaps explains the small dependence of abrasion rate on hardness for steels, as found by Krushchov and Babichev[8] and by Richardson[9]. The decrease in abrasion rate that would be expected to result from the reduction in the depth and volume of the scratch grooves produced in harder steels may be partly counterbalanced by an increase in the fraction of cutting points.[12] The advantage of a highly negative value of α_c would persist, however, for only as long as a contacting abrasive point remained intact. It might be expected that the points would fracture more readily with harder specimens. This is indeed found to be so for aluminum oxide and silicon carbide abrasives (see p. 65). Diamond abrasive points, on the other hand, do not fracture so readily with even the hardest steels (see p. 73), and so retain the rake-angle advantage.

Thus there are many uncertainties about the value that should be ascribed to α_c in a particular set of circumstances. But this does not negate its significance in determining the efficiency of an abrasion process.

Efficiency of the Cutting and Plowing Processes

The basic model of abrasion described earlier assumes that no material at all is removed by a plowing point (efficiency $\simeq 0\%$) and that all of the groove swept out by a cutting point is removed (efficiency $\simeq 100\%$). The correctness of these assumptions can be tested approximately in experiments in which the contour across the grooves produced by model vee-point tools is determined.

The simplest method of determining surface contours involves recording the magnified vertical movement of a stylus as it is made to traverse the surface. The generalized contour of a groove to be expected in the type of experiment under discussion is sketched as an insert at top left in Fig. 3.13. The ratio of the combined volumes of the two side ridges to that of the groove can be estimated from the contour, and the ratio should be zero when a chip is cut with 100% efficiency; it should be one when a groove is plowed with 0% efficiency.

The specimen material which was the subject of the experiment illustrated in Fig. 3.13 was a comparatively soft steel, and the tool was lubricated with carbon tetrachloride. Grooves made with tools which had

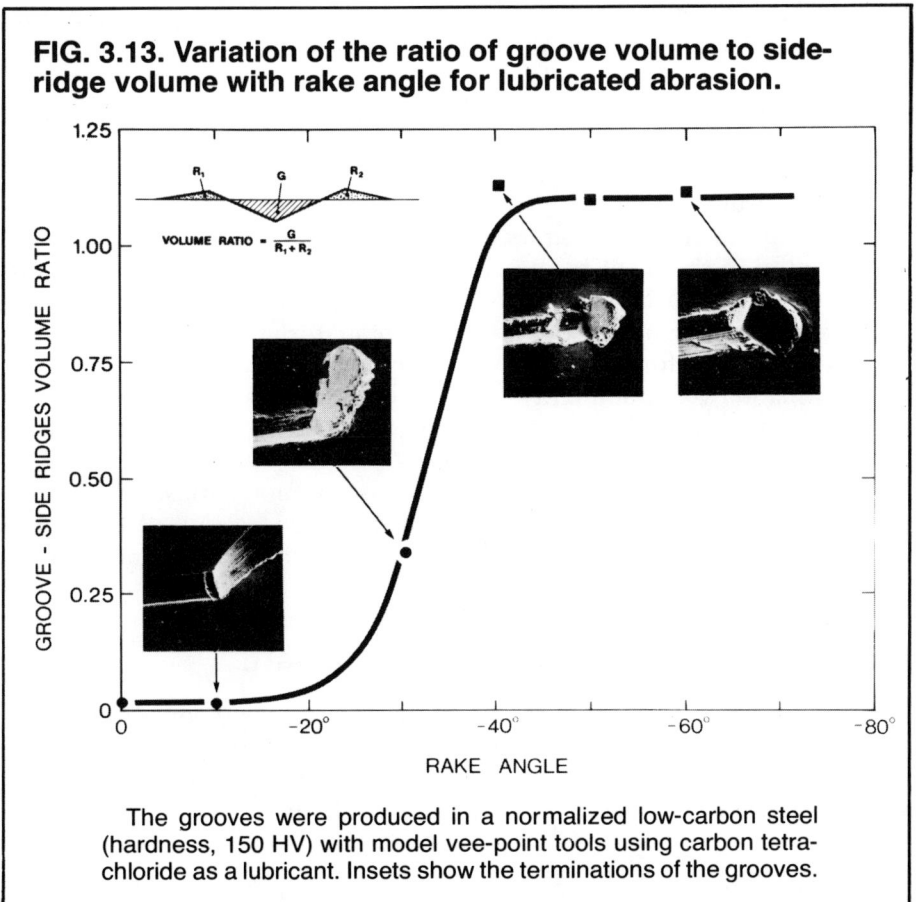

FIG. 3.13. Variation of the ratio of groove volume to side-ridge volume with rake angle for lubricated abrasion.

The grooves were produced in a normalized low-carbon steel (hardness, 150 HV) with model vee-point tools using carbon tetrachloride as a lubricant. Insets show the terminations of the grooves.

rake angles less negative than $-10°$ had chips at their terminations. The ridge-to-groove ratio was then close to zero: that is, all of the material of the groove had been removed in the chip. Tools with α more negative than $-20°$ produced grooves with prows at their terminations. The ridge-to-groove ratio was then close to one:* that is, all of the material in the groove had been displaced into the ridges. The transition between the two limiting modes was reasonably sharp, being spread over only ~10°. Grooves produced by tools with rake angles within the transition range terminated in an irregular feature intermediate between a chip and a prow. The ridge-to-groove ratio was then between 0 and 1: that is, the material of the groove had been distributed between the chip and the ridges. This intermediate type of chip would eventually be removed from the surface and so would contribute to the efficiency of material removed. In principle, this transition is another factor that should be incorporated in an abrasion model, but it would be difficult to do so.

The behavior of the same steel without an active lubricant was different in several respects (Fig. 3.14). First, the ridge-to-groove ratio for points which definitely cut chips was about 0.1: that is, most of the material in the groove was removed in the chip but some was displaced into the ridges. The efficiency of cutting consequently was only ~90%, a reduction in efficiency which is not of great consequence in a first-approximation model. More importantly, the ridge-to-groove ratio for points which produced prows was much less than one. This means that a significant amount of the groove material, approaching 25%, was removed by points that operated essentially in a plowing mode. The reason for this is that the prow, which is quite voluminous in steels, was removed periodically by shear fracture at its base. A succession of prows so removed is illustrated in the inset in the center of Fig. 3.14. The prow was most voluminous and fractured off most frequently when the rake angle was close to α_c. Consequently, the efficiency of material removal was as high as for chip cutting for a range of rake angles close to α_c. A slip-line field analysis made by Challen and Oxley[11] predicts that shear fracture at the base of a prow is a possibility under particular combinations of rake angle, interface friction and specimen hardness.

Significant material removal by plowing is common with soft materials. It can be of sufficient importance to warrant modification of Equations 4 and 5. The difficulty is that any correction factors would again have to be determined empirically.

INFLUENCE OF ABRASIVE ON ABRASION RATE

We have so far assumed that the abrasive grits are rigid compared to the workpiece material: that is, that the abrasive deforms only elastically while the workpiece deforms plastically. A simple theoretical analysis suggests that this assumption is valid only if the hardness of the abrasive

---

*The ratios measured in this experiment were actually slightly greater than one due to instrumental deficiencies.

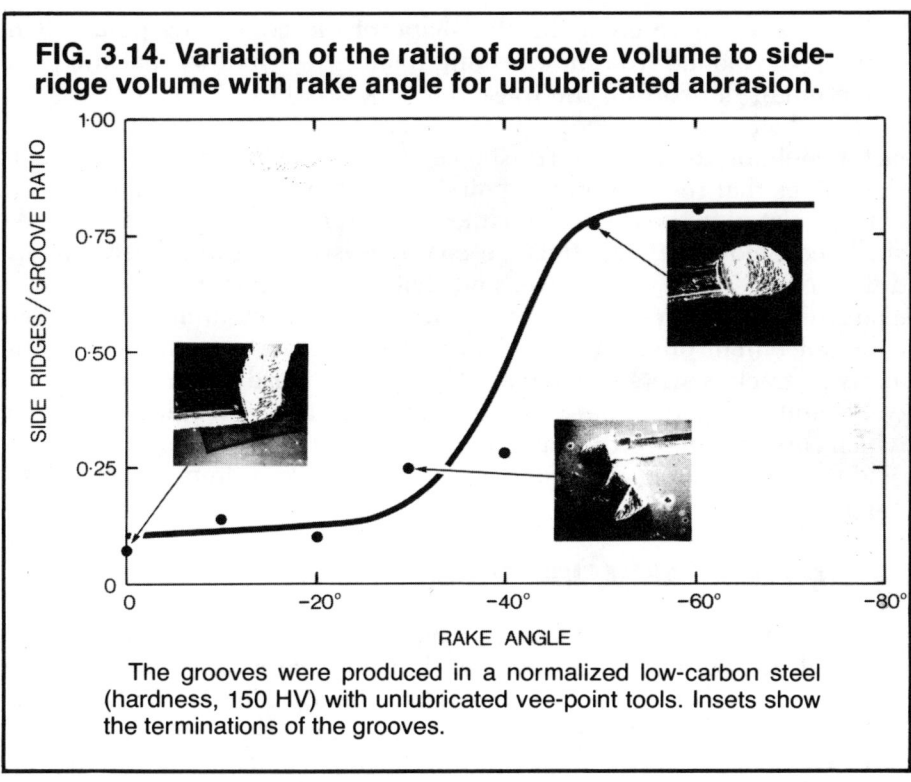

FIG. 3.14. Variation of the ratio of groove volume to side-ridge volume with rake angle for unlubricated abrasion.

The grooves were produced in a normalized low-carbon steel (hardness, 150 HV) with unlubricated vee-point tools. Insets show the terminations of the grooves.

is more than about two and a half times that of the workpiece. This is a criterion which is met for workpieces with hardnesses up to 1000 HV for silicon carbide, up to 800 HV for aluminum oxide, and up to 3000 HV for diamond. For silica and glass, incidentally, the corresponding values would be only 400HV and 250 HV, respectively. Some reduction in groove volume is to be expected when the workpiece material is harder than these limiting values. On the other hand, the groove volume should be independent of abrasive hardness when the hardness of the workpiece is less than the limiting value. Again, the value ascribed to the workpiece in all of these considerations should be that of the material in the fully work-hardened condition (see p. 100).

These predictions have been only partly supported by experiments done by Richardson,[9] experiments in which materials with a range of hardnesses were abraded on a continuously renewed track of papers that were coated with a number of different abrasives. Richardson found that some material removal occurred as soon as the hardness of the abrasive exceeded that of the workpiece, that the material removal rate then increased rapidly until the hardness ratio was about 1.2 to 1, and that it then increased slowly but significantly with further increase in the ratio (these results are also reviewed by Moore[13]). However, this search for a simple correlation with hardness is again probably overly optimistic in that it overlooks several other potentially important characteristics of the abrasive.

The first of these could be the shape of the contacting points. This shape is developed from intersecting fractures. It follows that the point shape might depend on the fracture characteristics of the abrasive material, which differs among various commercial abrasives. The second factor could be the fracture resistance of the contacting points in use and the shapes that re-form on the points when they do fracture. This again will be related to the fracture characteristics of the abrasive material. We shall see below that the fracture characteristics of the abrasive points differ considerably in practice, and that this probably is a determining characteristic. Moreover, point fracture commences immediately upon commencement of relative movement, so that the concept of a "fresh" abrasive track is strictly a hypothetical one. A third factor could be the coefficient of friction between the abrasive and workpiece material, which could influence the value of the critical rake angle for cutting. No evidence is available on which to assess the likely importance of this parameter.

INFLUENCE OF ABRASION FLUIDS

Abrasive machining devices of most types are flooded with a fluid during use, and this fluid could behave as a lubricant, as a coolant, or as a medium to flush abrasion debris away from the abrasion track. Under the conditions used in metallographic preparation, the cooling role and, more particularly, the flushing role are the important ones. There is probably no lubricant role. Hence the general term *abrasion fluid* will be used instead of the more commonly used *abrasion lubricant*.

Liquids such as water, kerosine, alcohol and soluble cutting oils do not influence the critical rake angle for cutting. The one simple liquid known to do so is carbon tetrachloride, which is known to be a most effective genuine lubricant in machining at low cutting speeds. For example, it reduces α_c for aluminum from about $+10°$ to $-40°$ and has a substantial influence on steels (cf. Fig. 3.13 and 3.14). However, carbon tetrachloride should never be used in practical abrasion because it is a serious hazard to health. Chlorinated extreme-pressure lubricants might also be similarly effective, but are equally impractical in metallography.

USEFULNESS OF THE MODEL

It is apparent by now that the model of abrasion described is a highly simplified first-approximation model and has many limitations. It does, however, provide a basis for understanding the mechanisms involved in abrasion. It is also a guide to the parameters that are likely to influence abrasion rate and surface finish. In fact, it exposes the fact that so many parameters influence practical abrasion that it is questionable whether the development of a significantly more precise model is worthwhile, or perhaps even possible.

Abrasion rate is determined by a complex of interacting parameters. The possibility of predicting, except in the most general way, the relative abrasion rates of different specimen materials from known bulk physical or mechanical properties consequently seems to be remote. Certainly, hardness cannot be expected to be a sole reliable indicator, and this is a correlation that metallographers should erase from their minds. Thus, the abrasion rates to be expected in practical metallographic specimen preparation processes need to be established experimentally. This will be our concern in the following sections of this chapter.

Abrasion Rates Achieved in Practice

MEASUREMENT OF ABRASION RATES

The thickness of the layer removed during a known distance of traverse while a known pressure is applied to the specimen is basically the information that a metallographer needs to know. The change in this abrasion rate with increase in the number of traverses over a given track of the abrasive paper also needs to be known, but the number of traverses and the abrasion distance are not independent variables. The number of traverses will be considered to be the primary variable here. One traverse always represents a track length of 50 cm in the figures quoted. Finally, a practical operator needs to have some idea of the time required to remove a given thickness of material from the specimen, but this again is not an independent variable. The results quoted below were obtained on a wheel rotating at 200 rpm and a surface speed of 100 m/min; this is representative of metallographic practice. The removal rate in μm/min thus is 100 times that expressed in μm/m. The results quoted were obtained under controlled conditions of abrasion using the apparatus and the loss-of-weight method described in Appendix 3-B. The specimens were cylinders, 1.2 cm in diameter, to which a load of 500 g was applied. This corresponds to a pressure of 395 g/cm^2, which is about that used in practice. Unless otherwise stated, the abrasion track was flushed with a stream of water.

CHARACTERISTICS OF WATERPROOF SILICON CARBIDE AND ALUMINUM OXIDE PAPERS

The abrasives in these papers are cemented to each other and to the backing with a waterproof resin, the backing also being impregnated with a waterproofing resin (cf. papers discussed on p. 69). The papers consequently can be used in conjunction with water, as well as with many other liquids. Silicon carbide papers are readily available in grades from about 80 to P1200, and aluminum oxide papers in grades from 240 to P1200.

The abrasion rates obtained with papers of these types possibly always decrease in an exponential manner given a long enough period of use. Even if this is so, the deterioration constants vary so widely with different abrasion conditions and different specimen materials that it is convenient in practice to divide materials into three groups. These groups can be

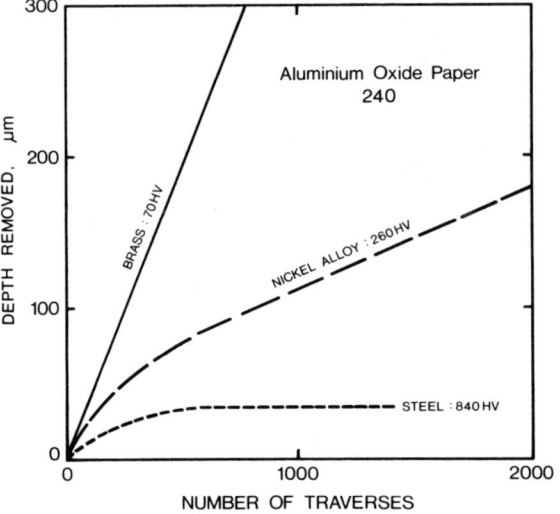

FIG. 3.15. Variation of the depth removed with number of traverses for alloys representing three groups of materials.

For brass, the abrasion rate does not change significantly with increasing use of the paper. For nickel alloys, the abrasion rate decreases initially but then stabilizes to a steady value. For steel, the abrasion rate decreases rapidly to zero.

characterized by the abrasion-rate-deterioration curves illustrated in Fig. 3.15. This division is, however, arbitrary to some extent because a full spectrum of deterioration curves between the two limiting cases shown in Fig. 3.15 is found in practice. The characteristics of the three groups selected are:

Group 1: No significant deterioration in abrasion rate occurs over many thousands of traverses, over which period the abrasion rate is approximately constant. The material can be characterized by this abrasion rate. Materials in group 1 include copper and copper alloys; aluminum and aluminum alloys; silver; and titanium.

Group 2: Significant deterioration in abrasion rate occurs over the first few hundred traverses but little occurs thereafter. After the settling-in period, the abrasion rate remains approximately constant for many thousands of further traverses. For practical purposes, the materials can best be characterized by this constant abrasion rate. Materials in group 2 include nickel-chromium austenitic steels; nickel and nickel alloys; titanium alloys; chromium; and gold.

Group 3: Marked deterioration in abrasion rate occurs in an exponential manner until, after about 1000 traverses, effectively no further material removal occurs. For the present purposes, the simplest way of characterizing the material then is by the maximum thickness that can be removed on a track of paper ($T\infty$). Materials in group 3 include ferritic steels, platinum and tungsten.

Materials That Cause Little Deterioration in the Abrasion Rate (Groups 1 and 2)

The following parameters significantly influence the abrasion rates that are obtained with materials having these characteristics. Unless otherwise stated, the working track of the paper is assumed to be flushed with a stream of water during abrasion.

1. *Applied pressure:* Abrasion rate increases linearly with applied pressure, as predicted by Equation 5, p. 48.
2. *Type of abrasive:* The abrasion rates obtained with silicon carbide papers are greater than those obtained with aluminum oxide papers, sometimes substantially so (Table 3.3). The only exceptions

TABLE 3.3. Abrasion Rates Obtained With Various Metals That Cause Little Deterioration of Abrasive Papers

Metal or alloy		Abrasion rate(a), μm/m		
Description	Hardness, HV	Silicon carbide, P240	Aluminum oxide, P240	Diamond, 220
Aluminum:				
High purity, annealed	24	2.61	1.93	1.76
Alloy, heat treated	150	1.29	0.85	0.65
Cadmium, commercial purity, annealed	21	2.4	2.0	...
Chromium, high purity, annealed	200	0.25	0.20	0.16
Copper:				
High purity, annealed	50	0.61	0.28	0.19
Brass 70:30, annealed	70	0.81	0.72	0.40
Brass 60:40, leaded	155	2.06	1.48	0.77
Aluminum bronze	200	0.78	0.66	0.18
Gold, high purity, annealed	22	0.26	0.16	0.08
Lead, commercial purity	4.2	1.3	1.3	...
Nickel:				
Commercial purity, annealed	130	0.08	0.17	0.14
Alloy (Nimonic 100), heat treated	260	0.10	0.21	0.06
Steel, austenitic, type 304	155	0.09	0.36	...
Silver, high purity, annealed	35	1.17	0.41	0.21
Tin, high purity	9	3.5	3.45	...
Titanium:				
Commercial purity, annealed	200	0.25	0.15	0.11
Alloy (6Al/4V), heat treated	295	0.15	0.07	0.07
Zinc, commercial purity, annealed	35	1.24	1.22	...

(a) Abrasion rates obtained after about 500 traverses on a track of paper. Load applied to specimen, 395 g/cm². Abrasion rates in μm/min for a specimen abraded on a track 16 cm in diameter on a wheel rotating at 200 rpm can be obtained by multiplying these figures by 100.

among the alloys listed in Table 3.3 are nickel and nickel-base alloys.

3. *Grade of abrasive:* The maximum abrasion rate is achieved with abrasive grades in the P240 to 150 range (Fig. 3.16); both finer and coarser grades give reduced abrasion rates. This general type of relationship has also been found in a number of investigations of fresh papers.[13-15] With fresh papers, however, it is usually found that the abrasion rate either is constant or increases slightly with the coarser grades. Even so, it is still apparent that grades coarser than 150 grade should not be used in metallography, because a coarser finish and a deeper damaged layer are obtained (see p. 112) without any compensating benefit. Further attention will therefore be confined mostly to the P240 grade, which achieves about the optimum combination of maximum abrasion rate and minimum depth of damage.

4. *Specimen material:* The abrasion rates achieved with a representative range of materials are listed in Table 3.3. Softer materials tend to have higher abrasion rates, but not necessarily so (e.g., gold). Some comparatively hard alloys have unexpectedly high abrasion rates (e.g., leaded 60:40 brass). The use of hardness values for the specimen material in the fully work-hardened condition

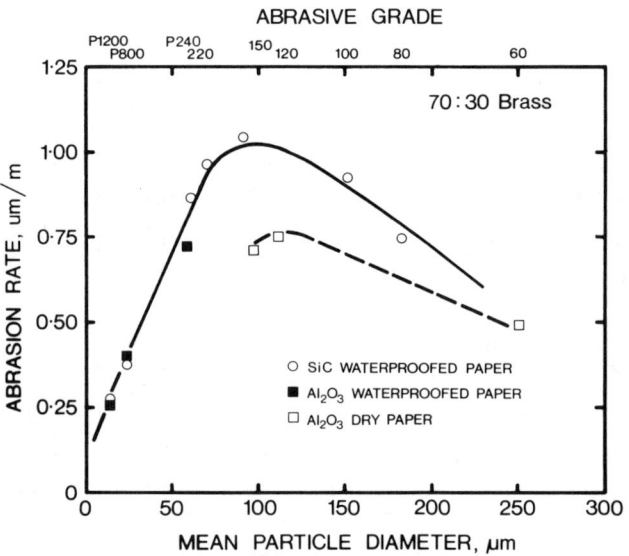

FIG. 3.16. Variation of abrasion rate with grade of abrasive paper for 70:30 brass.

The waterproof grades of silicon carbide and aluminum oxide papers (solid curve) are discussed on p. 57. The aluminum oxide papers that are used dry (dashed curve) are discussed on p. 69.

TABLE 3.4. Effect of Lubricant on the Relative Abrasion Rates Obtained When 70:30 Brass Is Abraded on P240-Grade Silicon Carbide Paper

Lubricant	Relative abrasion rate
Flowing water	1.0
Silicone fluid	1.0
Extreme-pressure lubricating oil	0.7
No lubricant (dry)	0.5
Waxed(a) but no liquid	0.4
Stagnant water	0.4

(a) Interstices between abrasive grits filled by rubbing paraffin wax against the paper.

would not significantly improve the correlation between abrasion rate and hardness.

5. *Abrasion fluid:* Maximum abrasion rate generally is achieved only when the operating track of the abrasive paper is flushed with a stream of fluid; the stream must, moreover, have sufficient velocity to wash away the abrasion debris as it forms. Water does this effectively, and generally is the only fluid that can be used in the required way in practice. The abrasion rate is reduced when water, or almost any other fluid, is present only as a stagnant film of liquid (Table 3.4). An exception is a silicone fluid of the type marketed as a mold-release compound; the abrasion rate then is the same as when the paper is flushed with water (Table 3.4). This type of silicone is available in an aerosol pack. The abrasion rate is also reduced when papers are used either dry in their natural state or dry and loaded with a wax (Table 3.4). The exception here is lead. The highest abrasion rate with lead is obtained when the paper is loaded with wax, because this prevents severe embedding of abrasive fragments in the specimen surface (see p. 131). The influence of the abrasion fluid becomes progressively less marked with increasing coarseness of abrasive grade, and is not significant for most materials with abrasive grades of 100 or coarser.

6. *Soaking the papers in water:* The abrasion rate decreases progressively when papers are soaked in water for any period of time.[16] It may be halved after soaking for 4 or 5 days, which is a possibility in small laboratories where papers may be stored in water between uses. Soaking in water appears to weaken the bond between the abrasive and the polymer cement to the extent that whole abrasive grains are torn out when the paper is used (Fig. 3.11e).

7. *Batch and brand of paper:* The abrasion rate may vary by about ±10% among papers from different batches from one manufacturer, and among papers from different manufacturers. For group 2 ma-

terials abraded on aluminum oxide papers, however, some brands may survive many more specimen traverses than others before the stable-abrasion-rate state is reached. All the values of abrasion rate reported in this chapter are for the one brand of paper.*

The abrasion model discussed earlier goes some way in explaining these effects. It would explain the smaller abrasion rates obtained with aluminum oxide papers if the contacting points of these papers were of more obtuse shape than those of silicon carbide papers. They would then have a less favorable distribution of rake angles (i.e., a distribution of rake angles displaced to more negative values) and a less favorable form factor. The variation of abrasion rate with abrasive grade could be explained on the basis of the interaction of two opposing factors. The coarser grades of abrasives, as they happen to be made, tend to have more obtuse points (this is qualitatively apparent in Fig. 3.8 but can also be established quantitatively — see Table 3.2). Hence they would have a smaller fraction of cutting points and a less favorable form factor. The trend that would result from this factor would be a continuous increase in abrasion rate with increasing fineness of abrasive grade. However, the spiky points of the finer 400 (P600) and 600 (P1200) grades can be expected to fracture more readily early in use. In the event of such fractures, blunter points would be produced and these blunt points would become the operating points. At a certain stage, the second factor would predominate and the abrasion rate would fall.

The differences in abrasion rate obtained with different materials are more difficult to explain, although it is to be expected from the earlier discussion of the abrasion model that the complex interaction of several factors would be involved. We shall now see that even more complexities than those already discussed are in fact involved. These concern the manner in which abrasive papers deteriorate in use.

The main mechanism by which papers deteriorate under the conditions being used is that of point fracture. Generally, a large fragment breaks away from the point, as illustrated in Fig. 3.11(a), and a point can fracture in this way during its first interaction with the specimen. This accounts in a general way for the decrease in abrasion rate that often occurs early in the life of the paper. The abrasion rate achieved thereafter will depend on the rate at which the points continue to fracture and on the shapes of the points that become the new contacting points.

With some of the softer specimen materials (e.g., high-purity aluminum, gold, and 70:30 brass), deterioration of abrasive papers results from formation of rough flats on the surface of the paper. This appears to occur by what is, in effect, a machining process (Fig. 3.17a). Large portions of the abrasive particles are fractured off level with the flats (Fig. 3.17b), with the result that a new generation of points is brought into contact with the specimen. The process consequently is partly regenerative as long as new grits remain in the coated layer. The most likely explanation

---

*Norton "Tufback Durite".

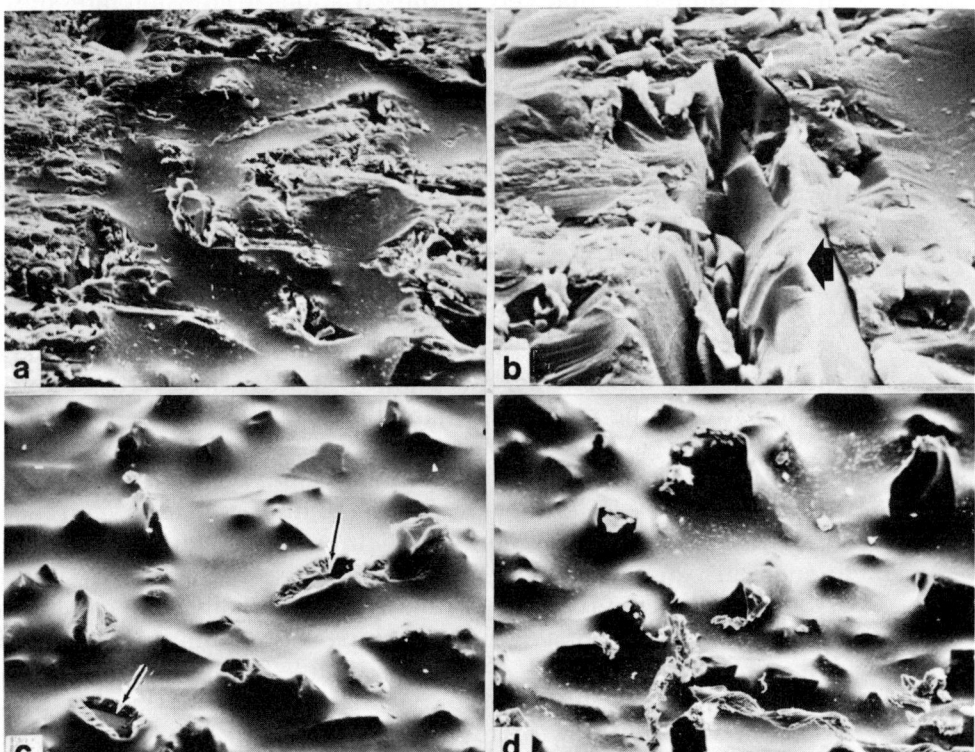

FIG. 3.17. Abrasive papers used for 2000 traverses, viewed in silhouette.

(a) P240-grade aluminum oxide paper used in abrading high-purity aluminum; rough flats have been machined on the surface of the paper. Scanning electron micrograph. Magnification, 110×. (b) Detail of one of the flats shown in (a); the abrasive particle (arrow) has fractured level with the flat. SEM; 1300×. (c) P240-grade silicon carbide paper used in abrading silver; typical fractured grits are indicated by arrows. SEM; 110×. (d) P240-grade silicon carbide paper used in abrading a leaded 60:40 brass; the grits are undamaged. SEM; 140×.

of this effect is that fragments fractured from the abrasive on first contact become embedded in the surface of the specimen. These embedded particles would then cause even more extensive fractures in the grits that are left in the surface of the paper. They would also machine away the cementing resin. This effect occurs to varying degrees with the materials mentioned, but in a way that is not apparently related to abrasion rate.

With the majority of specimen materials, however, deterioration occurs only by simple point fracture; some typical fractured points are indicated by arrows in Fig. 3.17(c). Generally, with these materials, fracture occurs well towards the base of an abrasive particle. The particle is thereby removed from the abrasion system, so permitting a new generation of points to become contacting points. Reasonable abrasion rates are thus maintained for as long as the abrasive-containing layer remains. Measurements of the changes in the contacting points would be necessary to account quantitatively for the variations in abrasion rates observed in practice.

There are a few notable exceptions to the above general behavior. Some materials cause little or no point fracture (Fig. 3.17d). These are the materials that have unexpectedly high abrasion rates for their hardness levels, such as leaded 60:40 brass. It is possible that the forces on the abrasive points are lower for these materials due to reduced frictional forces.

The papers illustrated in Fig. 3.17 were all flushed with a stream of water during use, and little or no abrasion debris remains on the paper. Much abrasion debris accumulates on the paper, however, when the paper is used either dry or with a stagnant fluid. This includes stagnant water but excludes silicone mold-release fluids. In severe cases, debris accumulates to the extent that metallic patches become visible on the track (Fig. 3.18a); these patches are packed agglomerates of abrasion machining chips which have accumulated between groups of abrasive grits (Fig. 3.18b). In less severe cases, the interstices between the abrasive grits are only loosely filled with chips of abrasion debris, but often to such an extent that the contacting points are barely exposed (Fig. 3.18c); the clogging then may or may not be visually discernible. In less severe cases, again, small groups of chips may accumulate on individual contacting points (Fig. 3.18d). It is reasonable to suppose, however, that, in all cases, contact between the specimen surface and the abrasive point would be interfered with to some extent. The abrasion rate would thereby be reduced to varying degrees. The papers do not clog with debris when a silicone fluid is present. Presumably, this fluid prevents the particles of debris from being welded together and to the paper.

The importance of the effects of clogging was first recognized by Wilman and his coworkers,[17,18] although they studied emery papers only. It is difficult to avoid clogging with emery papers under any circumstances. For this reason, and for others that will emerge later (see pp. 112 and 134), emery papers are not recommended for preparation of metallographic specimens.

Materials That Cause Severe Deterioration in the Abrasion Rate (Group 3)

Ferritic steels, in either the annealed or the hardened-and-tempered condition, are the most important examples of group 3 materials. They cause severe deterioration of abrasive papers in grades of 100 or finer. The characteristics of the steel illustrated in Fig. 3.19 (hardness, 840 HV) are typical. The rate of deterioration varies with hardness, increasing as hardness increases; the abrasion rate falls to a value close to zero within several hundred traverses for a hard steel and within several thousand traverses for a soft steel. Abrasion systems which behave in this way are best characterized for our present purposes by the maximum depth that can be removed ($T\infty$) on one track. This is affected significantly by the following parameters:

FIG. 3.18. Clogging of P240-grade silicon carbide papers on which 70:30 brass has been abraded under various conditions of lubrication.

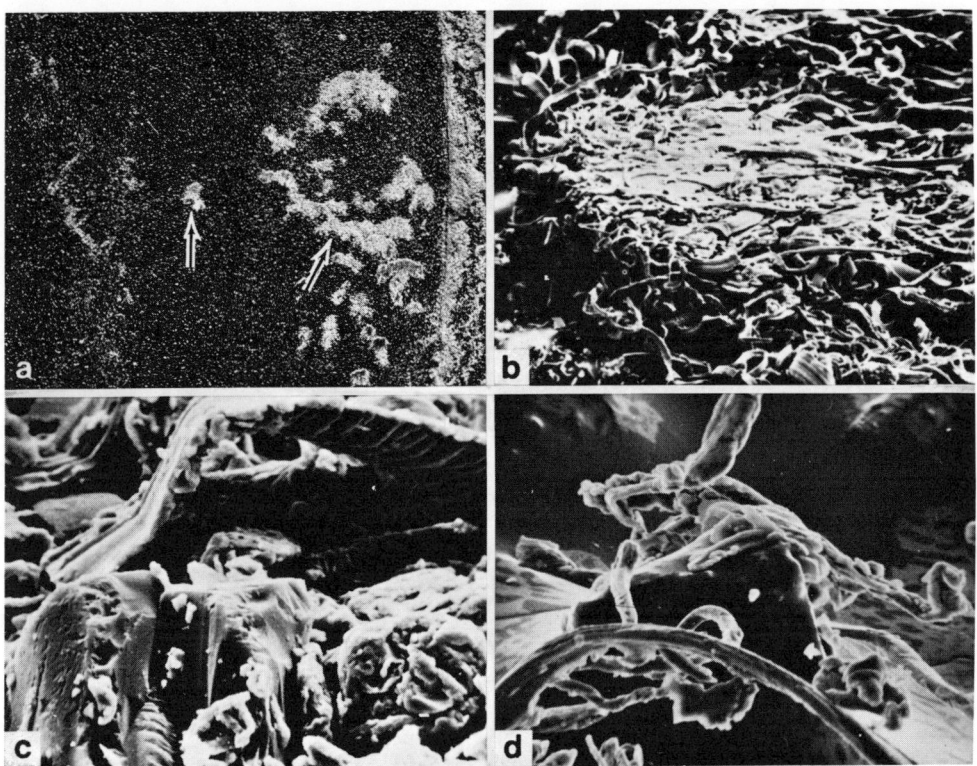

(a) Clogged patches (arrows) on a track on which carbon tetrachloride was used as the lubricant. Magnification, 4×. (b) Scanning electron micrograph of one of the clogged patches shown in (a). Magnification, 130×. (c) An accumulation of abrasion debris on a track lubricated with stagnant water. Magnification, 1440×. (d) A group of abrasion chips attached to a contacting point in a track that was filled with wax and used dry. Magnification, 1040×.

1. *Applied pressure:* $T\infty$ increases approximately linearly with applied pressure for both types of abrasive and for steels of all hardnesses.
2. *Type of abrasive:* The performance of aluminum oxide papers is always markedly superior to that of silicon carbide papers. The characteristics illustrated in Fig. 3.19 are typical.
3. *Grade of abrasive:* $T\infty$ increases with increasing coarseness of paper grade in the range P1200 to P240 for both types of abrasive. Aluminum oxide papers coarser than P240 are not considered here (but see p. 70). The abrasion rates achieved with coarser grades of silicon carbide paper vary with the hardness of the steel. $T\infty$ increases, but not very rapidly, with softer steels and decreases slightly with harder steels (Fig. 3.20). The 220 grade of silicon carbide paper gives anomalously low values of $T\infty$ with steels of all

hardnesses. However, systematic variation of $T\infty$ with mean diameter of abrasive particle is not necessarily to be expected, because it is the shape of the contacting points and the manner in

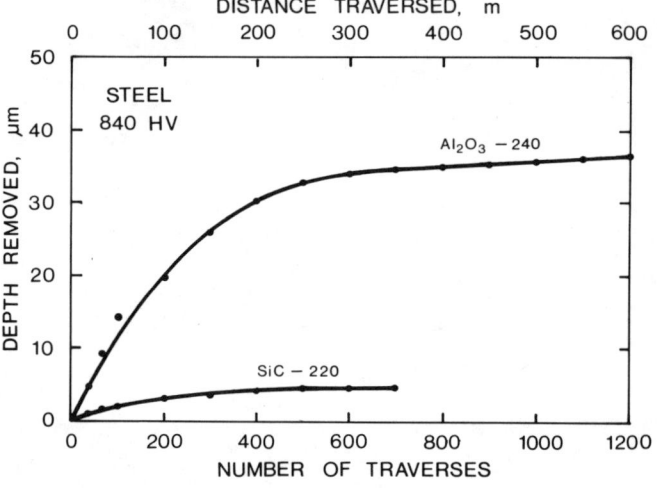

FIG. 3.19. Variation of depth removed with number of traverses for a hard steel (1.4% C; 840 HV) abraded on equivalent grades of waterproof silicon carbide and aluminum oxide papers.

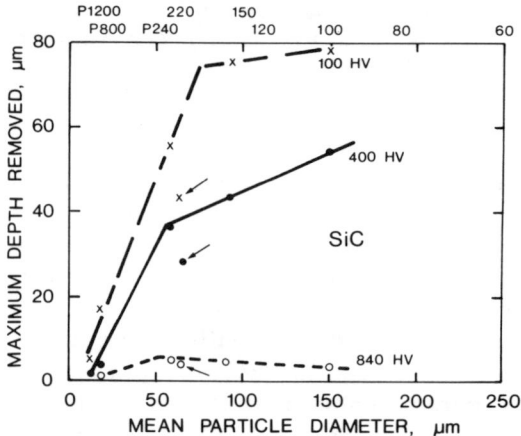

FIG. 3.20. Variation with grade of silicon carbide abrasive paper of the maximum depth of material that can be removed from steels of various hardnesses.

The points indicated by arrows draw attention to the poor performance of the 220 grade compared with that of the P240 grade.

which the points deteriorate in service that determine the abrasion rate, and not the diameter of the abrasive grits. These factors are not necessarily related.

4. *Specimen material:* $T\infty$ decreases generally with increases in hardness, although the exact nature of this relationship depends on several factors, including the nature of the steel and the type of abrasive. The general type of correlation obtained between $T\infty$ and hardness for a range of carbon steels is indicated in Fig. 3.21. Even these trends, however, cannot necessarily be extrapolated to other types of steel. For example, one result is shown in Fig. 3.21 for a high-alloy high speed tool steel; the value of $T\infty$ is much lower than for carbon steels of comparable hardness. These differences must be attributed to the effects of the microstructural constituents on the deterioration of the abrasive.

5. *Abrasion fluid:* The maximum value of $T\infty$ is achieved only when the abrasion debris is continuously flushed away from the abrasion track by a stream of fluid (usually water). The proportional decrease in abrasion rate under other circumstances is about the same as for group 1 materials (Table 3.2). Date and Malkin[19] have shown that severe clogging and capping occur when steel is abraded on fine grades of aluminum oxide papers when they are used dry. This occurs in the same way as was illustrated earlier for group 1 and group 2 materials (Fig. 3.17). The decrease in $T\infty$ can be attributed to the clogging. The exception is when the abrasion track is sprayed regularly with a silicone mold-release compound from an aerosol pack; as for group 1 materials, the full abrasion rate is then obtained.

6. *Soaking the papers in water:* The consequences are the same as for materials in groups 1 and 2 — namely, the abrasion rate is progressively and significantly reduced as soaking is continued over several days.[16]

7. *Batch and brand of paper:* $T\infty$ may vary by about $\pm 10\%$ among different batches of silicon carbide papers from one manufacturer, and by about the same amount among papers from different manufacturers. Quite substantial differences may be found, however, between aluminum oxide papers from different manufacturers; $T\infty$ may differ by a factor of two or three.

The manner in which both silicon carbide and aluminum oxide abrasives deteriorate when used to abrade steels is different from that for group 1 and group 2 materials. For example, Fig. 3.22(a) and (b) compare an unused aluminum oxide paper and one that had been used to abrade brass — a group 1 material. Many of the grits have fractured close to their bases in a manner similar to that previously discussed. On the other hand, very few grits fractured close to their bases when the same paper was used to abrade a hard steel (Fig. 3.22c). This was so even when abrasion had been carried to the stage where the abrasion rate had been reduced virtually to zero. Thus many undamaged points remained in the paper at this stage. However, some grits had fractured so as to produce approximately flat surfaces on their points; one such point is arrowed in Fig. 3.22(c) and one is shown in detail in Fig. 3.22(d) (a small metal cap also adheres to this point). The surface of the specimen must have settled onto, and been supported entirely by, these flats. Moreover, the flat points could not have been cutting chips. Experiments of the type described earlier in connection with Fig. 3.5 confirmed that they were not. Date and Malkin[19] found that coarser grades of aluminum oxide papers reached a similar state when used dry to abrade steels. No explanation is available for the differences that occurred in the fracture behavior of the grits when different types of material were abraded.

There are two possible reasons why fractured, flattened points do not cut chips. First, the rake angles of the new points might be much more negative than those of the original points. The fraction of contacting points does indeed decrease with use (Table 3.2), which can account quantitatively for the decrease in cutting rate during the early stages of deterioration.[2] Secondly, the flat points would produce grooves of reduced volume if they did cut (Fig. 3.7). Ultimately, a flat point would remove no material at all when the flat became large enough to be supported elastically by the specimen surface. The harder the steel, the smaller the flat that is required for this situation to arise. Both factors probably contribute in practice to the reduction in abrasion rate, the second being dominant in well-worn paper.

FIG. 3.21. Variation with hardness of the maximum depth of material that can be removed from various carbon steels by abrasion on P240-grade aluminum oxide paper.

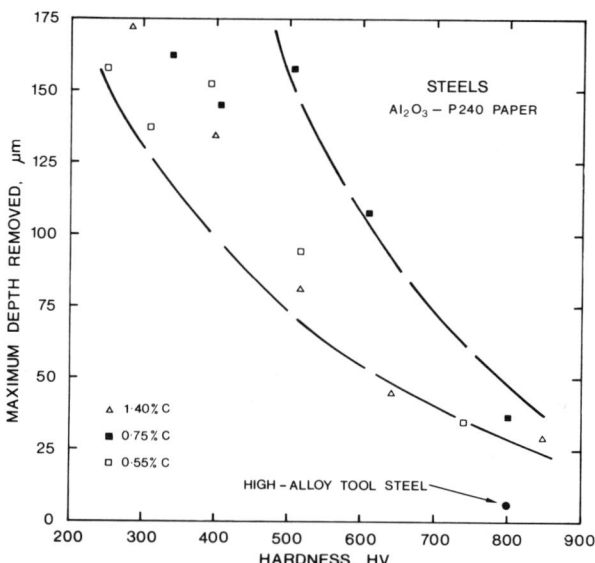

One result for a high-alloy high speed tool steel is plotted at lower right.

FIG. 3.22. P240-grade aluminum oxide papers viewed in silhouette before and after use.

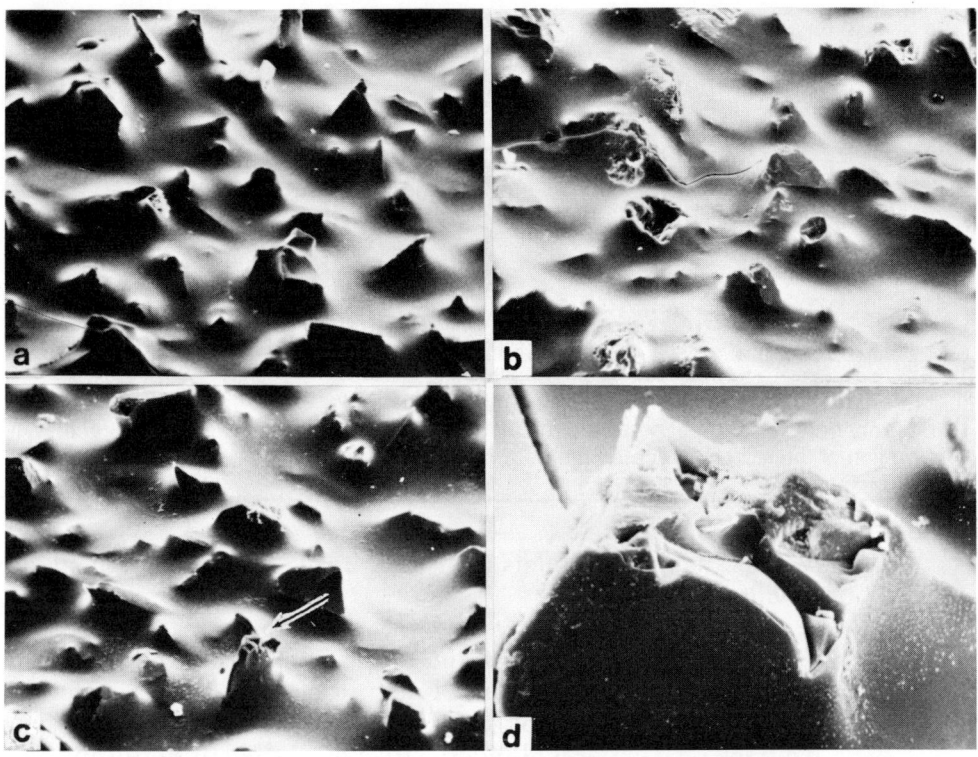

(a) Unused. Scanning electron micrograph. Magnification, 130×. (b) After 2000 traverses against 70:30 brass; many grits have fractured close to their bases. SEM; 125×. (c) After 2000 traverses against steel (hardness, 840 HV); few grits have fractured close to their bases, but flats have formed on some of the points by fracture (one flat is indicated by an arrow). SEM; 125×. (d) Same as (c), showing a fracture flat in more detail. SEM; 650×.

The stress distribution in abrasive particles loaded close to their points has been analyzed.[20] This analysis indicates that fracture can be expected to initiate slightly away from the point, then to propagate for some distance along a plane perpendicular to the working face, and eventually to run off in an irregular manner. This type of fracture path is followed in the model points illustrated in Fig. 3.23, and in a real point illustrated in Fig. 3.11(a). Thus a point which initially had a highly negative rake angle would regenerate a point with an adequate clearance angle. It would in effect remain unchanged as a cutting tool, as illustrated by the model point in Fig. 3.23(a). However, fracture of a point with a positive (or nearly positive) rake angle would produce a new point without clearance, as illustrated in Fig. 3.23(b) and (c). The points that originally were able to cut a chip and remove material consequently are likely to be the

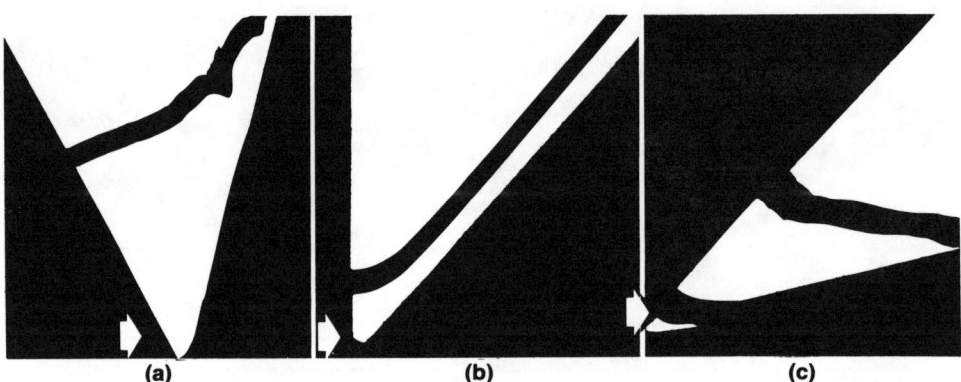

FIG. 3.23. Fractures in model glass tools used for cutting lead at various rake angles. *(Patterson and Mulhearn, Ref 20)*

(a) Rake angle, −20°. **(b)** Rake angle, 0°. **(c)** Rake angle, +45°. In all cases, fracture first propagates in a direction perpendicular to the rake face of the tool. The direction of motion of the lead is indicated by the arrows.

very ones that develop flats after fracture, and so become ineffective in removing material.

It is not clear, however, why points sometimes fracture close to their bases, as occurs with group 1 and group 2 materials, and sometimes do not, as with group 3 materials. Fracture close to the base clearly is an advantage because it enables a new set of points to become active.

If a track on a paper is abraded with a hard steel until the abrasion rate has been reduced to zero, a nearly normal abrasion rate is obtained when a softer group 1 or group 2 metal is abraded on the same track. Moreover, a softer steel behaves as though the track had not been used. Points that have become so flat that they cannot cause indentations in a hard steel can still do so in a softer material. In addition, the regenerative processes that normally occur with group 1 and group 2 materials can commence. A new set of abrasion points can then be exposed, and so the abrasion process can continue in the normal way for the particular specimen material.

The superior performance of aluminum oxide papers over that of silicon carbide papers can, in the above model, be attributed to the greater resistance of the aluminum oxide to the development of flats on the contacting points in use. The original point shape and the number of contacting points must, if anything, be less favorable for the aluminum oxide particles than for the silicon carbide particles, because slightly lower abrasion rates are achieved with group 1 materials. The points must last longer when abrading group 3 materials. The superior performance of some manufacturers' supplies of aluminum oxide over those of others could be explained in similar terms. However, these hypotheses are unconfirmed.

It remains now to discuss the 80-grade waterproof silicon carbide papers, which deteriorate only very slowly and which can be represented by an average abrasion rate over, say, the first 2000 traverses (Table 3.5). The depth removed is two to three times greater than that achieved with 100 and 150 grades in the same number of traverses. The 80-grade abrasive particles are found not to fracture in a gross way in use. Instead, small irregular fragments are chipped from the contacting points (Fig. 3.11b), and so the cutting points remain effective (Fig. 3.2). The 80-grade grits are obviously more resistant to fracture than the finer grades, perhaps because they have a blockier shape (cf. Fig. 3.8a and 3.8b), and perhaps also because they contain fewer fracture-initiating defects, such as cracks.

TABLE 3.5. Characteristics of Steels Abraded on 80-Grade Waterproof Silicon Carbide Papers

Steel	Hardness, HV	Abrasion rate(a), $\mu m/m$
0.08%C, annealed	100	0.26
0.4%C, annealed	230	0.22
0.4%C, quench hardened and tempered	400	0.11
1.4%C:		
Quench hardened and tempered	600	0.07
Quench hardened	840	0.01

(a) Load applied to specimen, 395 g/cm^2. Abrasion rates in μm/min for a specimen abraded on a track 16 cm in diameter on a wheel rotating at 200 rpm can be obtained by multiplying these figures by 100.

Summary

Determinations of the abrasion rates achieved under practical conditions confirm that a large number of parameters affect abrasion rate. The number is even larger than that predicted by the abrasion model discussed earlier. In particular, the manner in which the abrasive paper deteriorates in use, and the influence that the specimen material has on the mode and rate of this deterioration, are dominant.

Quantitative information that takes into account the deterioration of the paper is available for only a limited range of metals and alloys (Table 3.3). These materials are representative enough to provide a guide, but clearly it is dangerous to generalize too far. Actual measurements of abrasion rates, even·if only semiquantitative, need to be made in any unusual situation when the relative merits of various abrasion systems have to be compared. Nevertheless, the following broad generalizations can be made about the conditions necessary to achieve maximum abrasion rate.

For steels and like alloys which cause severe deterioration of abrasive papers:

1. Use aluminum oxide papers.
2. Use the P240 grade for fastest cutting.
3. Flush the abrasion track with a stream of water.
4. Apply as heavy a load as practicable.
5. Discard the papers after about 1000 traverses on the one track.
6. Do not soak the papers in water for more than a day or so.

For other alloys:

1. Use silicon carbide papers.*
2. Use the P240 grade, or perhaps the 150 grade, for fastest cutting. Do not use coarser grades under any circumstances.
3. Flush the abrasion track with a stream of water.
4. Apply as heavy a load as practicable.
5. Do not soak the papers in water for more than a day or so.

CHARACTERISTICS OF SOME NONWATERPROOF TYPES OF ALUMINUM OXIDE PAPERS

A range of coated papers is available in which aluminum oxide abrasive is cemented to a heavy paper backing by means of a waterproof resin, but in which the paper itself is untreated. The papers consequently must be used dry. They are available in grades 320 to 36.

Comparatively low abrasion rates are obtained when group 1 and group 2 metals are abraded on these papers, one set of results being given in Fig. 3.16. Consequently they are not generally recommended for these materials, particularly in the coarser grades.

The characteristics of these papers used with one particular steel (a group 3 material) have been thoroughly investigated by Date and Malkin.[19] This investigation showed that finer grades have no advantage with this type of material either; the abrasion rate approaches zero within 1000 traverses, with $T\infty$ being about the same as for the waterproof abrasive papers that have already been discussed. This is because, since they have to be used dry, the papers clog and cap in the same way as discussed earlier for fully waterproof papers. However, significant clogging and its associated capping does not occur with the 180 or coarser grades. Improved performance is then obtained. Behavior that is characteristic of group 1 materials is obtained for softer steels (hardness up to about 400 HV), and behavior characteristic of group 2 materials for hard steels (Fig. 3.24).

The coarser grades of these papers can therefore reasonably be considered for an early roughing stage for group 3 materials, but not for group 1 or group 2 materials.

---

*Except when embedding of abrasive is a serious problem, when it may be advantageous to use aluminum oxide papers (see p. 133).

Abrasive Machining: Principles of Material Removal / 71

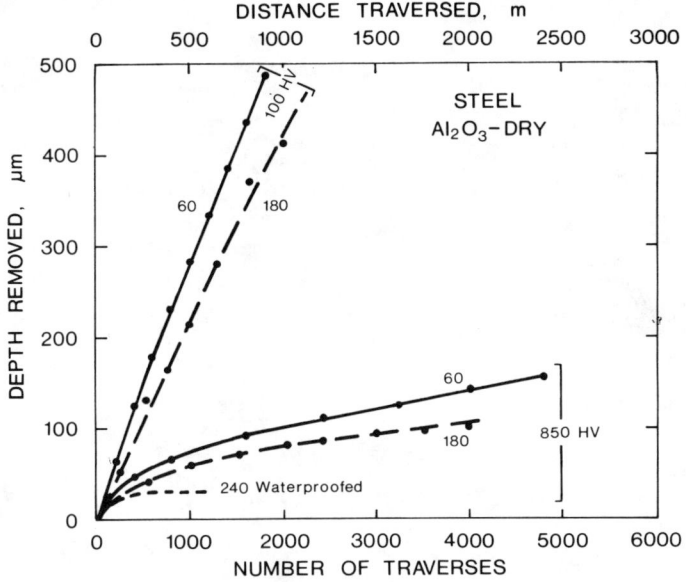

FIG. 3.24. Variation of depth removed with number of traverses for abrasion of steels on nonwaterproof aluminum oxide papers.

The characteristics of two grades of paper (60 and 180) used with steels of different hardnesses (100 and 850 HV) are illustrated. A comparable curve for a finer, waterproof grade of paper is included for the 850-HV steel.

CHARACTERISTICS OF DIAMOND LAPS

Disks coated with diamond grits are available commercially, most taking the form of a metal disk covered with a layer of graded diamond grits buried in a layer of deposited metal (usually nickel). The diamond grits are held mechanically in the deposited layer. They are uniformly distributed, and many have points suitably oriented for contacting the specimen surface (Fig. 3.25a). The grits tend, however, to be somewhat more sparsely distributed than those of silicon carbide and aluminum oxide papers (cf. Fig. 3.8a and 3.25a), and thus a somewhat lower density of contacting points can be expected. A density of contacting points of about 10/cm^2 is typical for a commercial 220-grade lap; this is about one-tenth the density of coated abrasive papers of the same grade (cf. Table 3.2). For this reason (see p. 46), a rather coarser finish is obtained with diamond laps than with corresponding grades of conventional abrasive papers. The points themselves are comparatively blocky but are composed of facets (Fig. 3.25b) which can constitute effective cutting points.

For comparatively soft specimen materials that exhibit group 1 and group 2 behavior when abraded with conventional abrasives, abrasion

FIG. 3.25. Diamond lap in which 220-grade grits are held on a metal disk by a deposit of nickel.

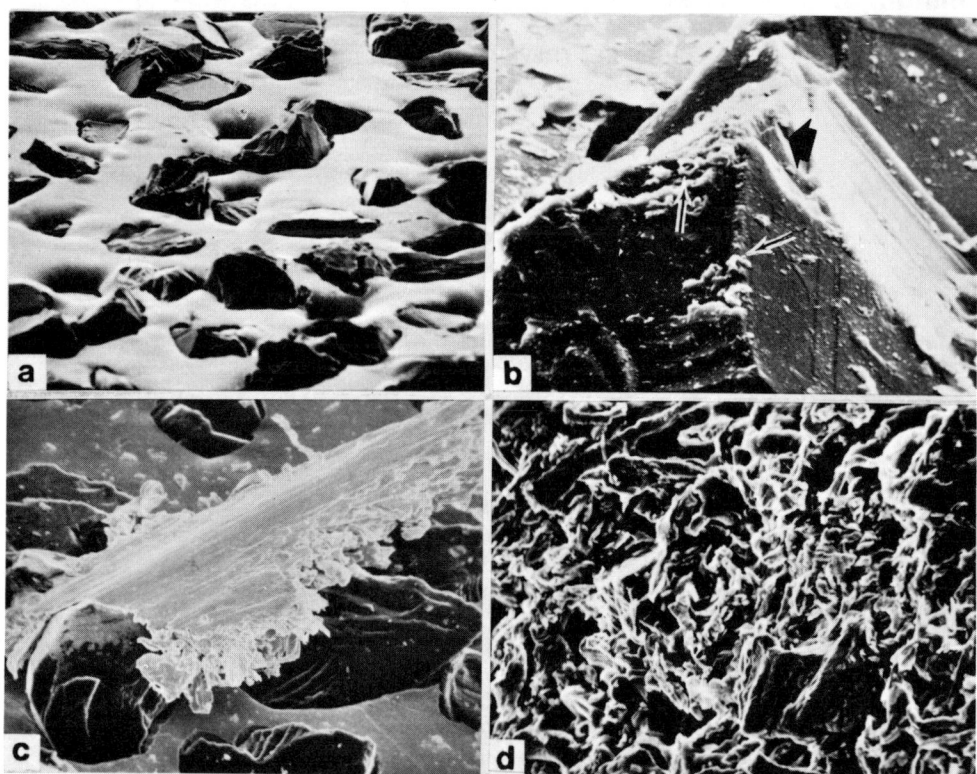

(a) General view showing the distribution and alignment of the diamond particles. Magnification, 100×. (b) Detail of the point of one diamond particle following abrasion of a hard steel. Cleavage facets (large arrow) produce sharp corners in the point comparable in size to the chips that have been cut (small arrows). Magnification, 110×. (c) Clogging of a lap on which a specimen of annealed brass has been abraded. The abrasion debris has packed into solid masses between groups of diamond particles. Magnification, 250×. (d) Surface of a lap that has been filled with wax and then used to abrade annealed brass as in (c). Abrasion debris has accumulated but has not packed into solid masses. Magnification, 450×.

rates obtained with diamond laps are usually somewhat lower than those obtained with the conventional abrasives. They are, however, still of a useful order of magnitude (Table 3.3).* This implies that the points of the diamonds have a larger distribution of negative rake angles than the points of either the aluminum oxide or the silicon carbide used in coated papers.

Moreover, abrasion rates of the order listed in Table 3.3 are obtained with most soft materials only if precautions are taken to prevent the lap

---

*The information in this section applies specifically to "Dialox" laps manufactured by Dunnington Co. (formerly The Glennel Corporation).

surface from becoming clogged with abrasion debris. If the laps are used in the usual way, either flooded with a stagnant liquid or flushed with a stream of water, abrasion chips accumulate and pack into solid masses between groups of diamond particles (Fig. 3.25c). Contact between specimen and diamond point clearly is greatly impaired. The abrasion rate may then be reduced by a factor of two or three. Clogging of this nature can be prevented by filling the surface of the lap with wax or soap, or by spraying the abrasion track regularly (once per minute, for example) with a silicone mold-release compound from an aerosol pack. Abrasion chips then accumulate on the abrasion track (Fig. 3.25d), but not in packed masses covering the contacting points. These accumulations of chips can readily be removed from the abrasion track and do not interfere with material removal.

A precise indication of the specimen materials that are likely to cause severe clogging of diamond laps cannot be given. The occurrence of clogging will be made apparent to the naked eye by the development of metallic-color patches on the abrasion track, similar to those illustrated in Fig. 3.18(a). When clogging has occurred, the lap can be restored easily by vigorous scrubbing or ultrasonic cleaning.

The main advantage to be expected of diamond laps, however, is in the abrasion of hard materials from which little or no material can be removed with conventional abrasives. The behavior of a new diamond lap in abrading a hard steel is illustrated in Fig. 3.26. This shows that the rate of material removal decreases somewhat over the first few thousand traverses, but thereafter remains constant at a high value for very long periods.* Once this stabilized state has been reached, a constant abrasion rate is obtained with other materials, including much harder materials. All of the abrasion rates quoted elsewhere in this chapter were obtained on laps which had reached this stabilized condition. It seems that a few of the less firmly held grits are removed bodily from the lap surface when the lap is first used. Thereafter, the morphology of the contacting points appears not to change. The points may not fracture at all. If they do, they must fracture on a small scale and regenerate points of similar morphology. Certainly, massive fracturing of the type that occurs with conventional abrasives (cf. Fig. 3.25b with Fig. 3.11 and 3.17) does not occur.

The abrasion rates obtained in abrading a range of steels[†] of various hardnesses are illustrated in Fig. 3.27. There is a good correlation between abrasion rate and bulk specimen hardness, but the correlation is different for different steels. The highest abrasion rates in these tests were obtained with the steels of highest carbon content, which contain the largest volume fraction of free cementite. This is difficult to explain. For all of these steels, the abrasion rates obtained with diamond laps are

---

*The behavior of papers coated with conventional abrasives in abrading this same hard steel is illustrated in Fig. 3.19.

†Diamond abrasives do not perform well in grinding steels at high surface speeds. This is usually attributed either to degradation of the diamond to graphite, perhaps catalyzed by the iron, or to chemical interaction between diamond and iron at the high temperatures attained during grinding. Whatever the reason, this problem does not arise at the low surface speeds used in metallographic abrasion.

FIG. 3.26. Variation of depth removed with number of traverses for a hard steel (1.4% C) abraded on a diamond lap.

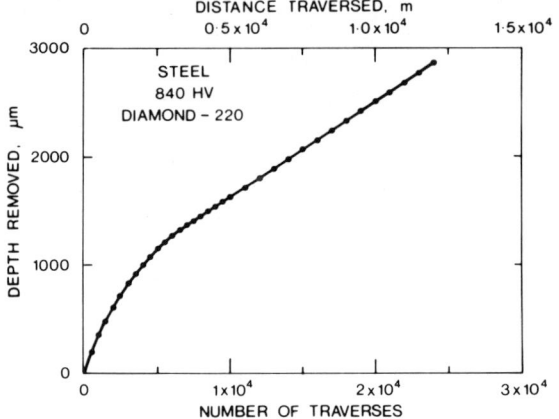

Note the considerable differences between the scales here and those in Fig. 3.15, 3.19, 3.24 and 3.28.

FIG. 3.27. Variation of abrasion rate with hardness for a range of carbon steels abraded on a 220-grade diamond lap.

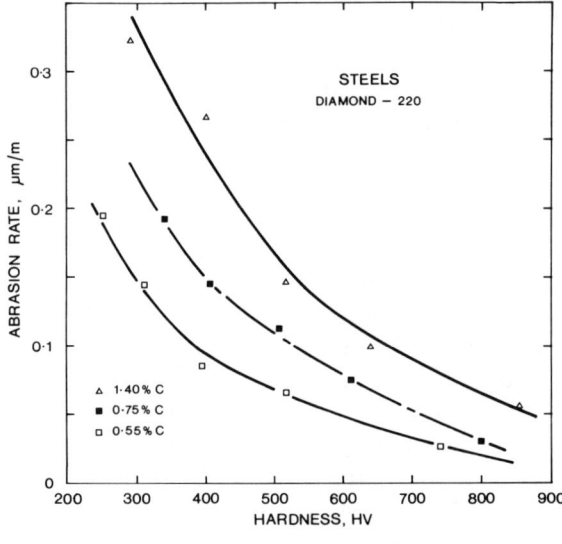

lower than those obtained with fresh aluminum oxide and silicon carbide papers of equivalent grades. This again indicates that the contacting points of diamond laps have slightly more negative rake angles. However, these abrasion rates greatly exceed those of conventional papers that have been used for a few hundred traverses (Fig. 3.28). The total thickness of

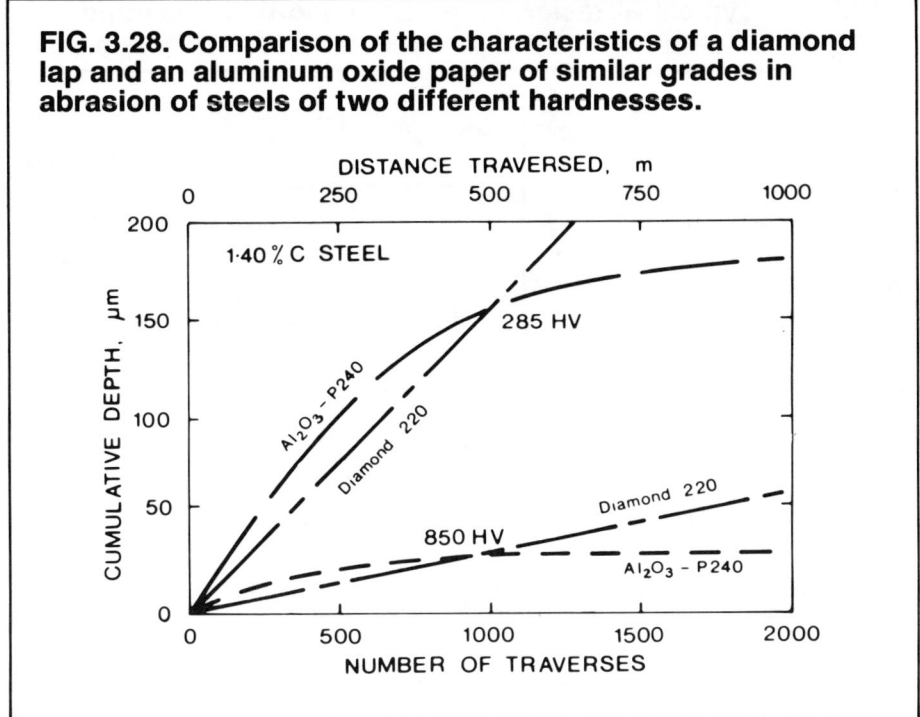

FIG. 3.28. Comparison of the characteristics of a diamond lap and an aluminum oxide paper of similar grades in abrasion of steels of two different hardnesses.

material that can be removed thus is potentially almost infinitely greater, and abrasion can be continued with the assurance that material is continuing to be removed. Metallographers tend to continue to use conventional coated papers long after the papers have stopped cutting because this condition is not obvious. Thus, although initial costs are high, the use of diamond laps can have long-term economic advantages with even moderately hard materials that exhibit group 3 characteristics when abraded with conventional abrasives.

Abrasion rates obtained for two grades of diamond lap in abrasion of several materials from which little material can be removed by conventional abrasives are listed in Table 3.6. Note that there are no great differences in the abrasion rates for the two grades of lap. The exception is tungsten carbide, which has a high abrasion rate for its hardness. This is because tungsten carbide behaves as a brittle material and because material removal occurs by a fracturing rather than a machining mechanism (see Chapter 7). The abrasion rate does decrease slowly when laps are used with these very hard and difficult materials, but significant removal rates are obtained for very long periods of time.

Because these laps deteriorate so slowly in use, and because they cause very little embedding of abrasive in the specimen surface, they are more suitable than conventional coated abrasive products for use in assessing whether or not there is a direct relationship between abrasion rate and specimen hardness. The abrasion rates presented throughout this chapter are plotted against hardness in Fig. 3.29. No correlation is evident, except perhaps to the upper limit of abrasion rate that can be obtained.

TABLE 3.6. Typical Abrasion Rates Obtained With Diamond Laps

Material	Hardness, HV	Abrasion rate(a), μm/m	
		100 grade	220 grade
Chromium – 2% Ta alloy	200	...	0.01
Steel, 0.75%C	400	0.02	0.09
Steel, 0.75%C	800	0.04	0.03
Steel, 1.4%C	850	...	0.06
Steel, high speed tool (M2)	800	...	0.14
Steel, high speed tool (M4)	980	0.02	0.11
Tungsten	330	...	0.05
Tungsten carbide (6%Co)	1550	0.07	0.04

(a) Load, 395 g/cm². Abrasion rates in μm/min for a specimen abraded on a 16-cm-diam track at 200 rpm can be obtained by multiplying these figures by 100.

Below that, the scatter is virtually random. This confirms that specimen hardness is one of the factors that determines abrasion rate but is by no means the only one.

It is also possible to manufacture diamond-plastic laps in the laboratory; details of suitable methods are given in Appendix 3-C.[21,22] Diamond abrasive grades as fine as 6 μm mean diameter can be used effectively in this way with hard, brittle materials such as hardened steels and tungsten carbide (see p. 316).

FINE FIXED-ABRASIVE LAPS

The surfaces produced by the finest abrasive papers available are somewhat undulating, because the contacting abrasive points are not strictly coplanar. The finish is also rather rougher than would be desired, because the density of contacting points that can be achieved is limited. Consequently, there have been many attempts to develop abrasive devices that demonstrate improvements in these respects. Although some success has been achieved, it must be said at the outset that even the most successful of these devices have limitations and are somewhat difficult to use. Their use consequently is justified only in special instances where results of the highest quality are desired.

Examples of the improvements in finish that can be achieved with fine laps are illustrated in Fig. 3.30. Comparison of Fig. 3.30(a) and (b) reveals improvement in the retention of edges; comparison of Fig. 3.30(c) and (d) demonstrates the reduction in the relief developed between constituents of widely differing abrasion characteristics; and comparison of Fig. 3.30(e) to (h) indicates the reduction in damage to nonmetallic inclusions and constituents. Note particularly that scratch grooves are cut in brittle constituents by the fine laps; extensive chips tend to develop in these constituents during normal abrasion. The reason for this is explained in Chapter 7.

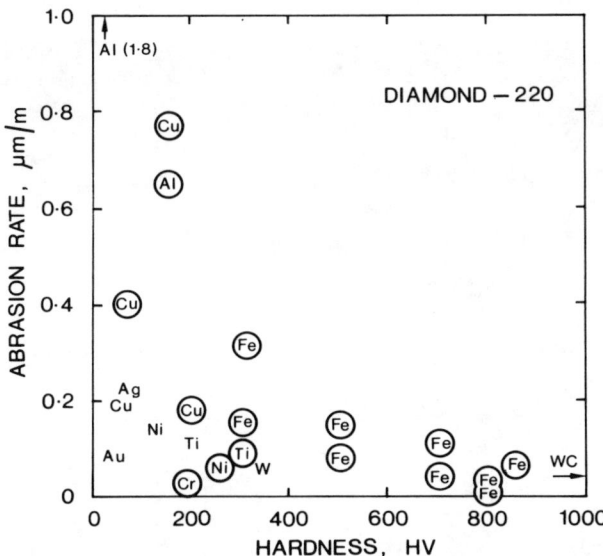

FIG. 3.29. Abrasion rate plotted against bulk hardness for specimens abraded on a 220-grade diamond lap.

Circled symbols are for alloys of the base metal; the remainder are for commercially pure metals in their annealed conditions.

A high density of contacting points is needed to improve the surface finish. This can be achieved by using finer grades of abrasive, by packing the grits closely together, and by ensuring that the contacting points of the grits are closely coplanar. However, other problems are thereby introduced. First, fine laps tend to clog rapidly with abrasion debris irrespective of whether or not they are flooded with liquid (the undesirable consequences of clogging have already been discussed). It is therefore essential that the laps be easy to either clean or re-form. Secondly, loose abrasive particles must be kept to a minimum on the lap surface; loose particles will, for example, rapidly remove nonmetallic inclusions. The abrasive particles must therefore be firmly embedded into a backing, and the backing must be strong enough to hold the particles in a fixed orientation so that they can cut. On the other hand, it must not be hard enough to scratch the specimen surface severely. The backing also needs to be soft so that only a small load can be applied to each abrasive particle. Necessarily, therefore, the formulation of a lap involves compromises.

Two general types of fixed abrasive lap have been developed which may be designated as externally charged and internally charged laps.

Externally Charged Laps

The earliest developments consisted of a block of cast iron, lead, or paraffin wax on which a flat working surface had been prepared by machining. A spiral groove was usually cut in this surface. The abrasive was then

FIG. 3.30. Comparison of finishes obtained with a 600-grade silicon carbide paper (left column) and with a 10-20-μm-grade aluminum oxide – wax lap (right column).

(a) and **(b)** Edge of a corroded aluminum alloy specimen. Magnification, 100×. **(c) and (d)** Aluminum – 13% silicon alloy. Magnification, 250×. **(e) and (f)** Duplex silicate-oxide inclusion in wrought iron. Magnification, 500×. **(g) and (h)** Flake graphite in gray cast iron. Magnification, 500×.

spread over and forced into the working surface by means of a steel roller.[23-26] This type of lap is now generally considered to be too difficult to prepare and maintain for metallographic use.

A more satisfactory arrangement consists of a cloth (usually a fine linen) stretched over a normal polishing head and then impregnated with melted wax.[27] The lap is charged by rubbing abrasive into the surface with the tips of the fingers, the excess being blown off. A lap of this type can be cleaned by swabbing the surface with a solvent for the wax. Simpler developments include the use of strips of lead foil,[28] or sheets of high-quality paper,[29,30] which are charged in the same way but discarded when they become clogged.

The practical problem with all such laps is the difficulty of charging them in such a way that the surface is completely free from loose abrasive. Because these laps usually are effective for preparation of one, or at most two, specimens, the problem is such a major one that use of such laps becomes a marginal proposition. Consequently, this general type is to be recommended only when a lap is required occasionally. The references quoted should then be consulted for operating details.

Internally Charged Laps

An internally charged lap consists of a mass of abrasive particles cemented together with wax.[31] The characteristics of the working face of the lap are therefore established during manufacture and not by the personal skill applied during the charging operation. The laps do clog rapidly, but a new working surface can be prepared by removing the deteriorated surface layers. This is simple to do. All of the subsequent discussion refers to laps of this type.

The methods of manufacturing and using laps of this type are described in Appendix 3-D (p. 83). Laps charged with aluminum oxide abrasives are most suitable for general use.

VITREOUS-BONDED ABRASIVE WHEELS AND LAPS

A wide range of commercial abrasive devices are available which consist of random closely packed arrays of graded abrasive particles sintered together by vitreous bonding. Grits are exposed at the operating surface, and the contacting points of the grits are formed by fracture *in situ*, either during a so-called dressing operation or during use. These devices typically are used in metallographic practice in three ways, namely:

1. As cut-off wheels, where the device is a thin disk and the interaction with the specimen occurs mainly at the periphery of the disk.
2. As grinding wheels, where the device is a thick disk and the interaction with the specimen occurs at either cylindrical or side-face surfaces, as in normal workshop practice.
3. As flat plates or laps

The mechanisms of interaction are the same for these devices as for those already discussed, although the geometrical details may be different. The volume of the groove swept out by a contacting point in the specimen surface obviously depends on the relative geometries and motions of the working surface of the device and those of the specimen. It is a comparatively simple problem in geometry to deal with these parameters. The form of the contacting points may also be expected to be different. This is a more difficult problem to deal with and one about which little is known.

No quantitative information is available on the material-removal rates that are achieved with these devices in metallographic practice. It is to be expected, however, that the contacting abrasive points will deteriorate in use in the same general manner as has already been described. Although a characteristic of these devices is that a new set of abrasive points can readily be exposed by a dressing operation, information is needed on the frequency with which these dressing operations should be carried out.

APPENDIX 3-A
Abrasive Grain Size

Silicon carbide and aluminum oxide abrasives are now almost universally supplied in a metricated series of grades which differ only slightly in size from the premetrication grades. In some instances, however, the grade or grit reference number has been changed. The new numbering system will be used throughout this book, the equivalents to the old system being indicated in Table 3.7. Two grading methods are used. The determination of particle-size distribution of an abrasive that is described as being a *screen grit size* abrasive is made by comparison with a mastergrit in a standard sieving machine, the abrasive being graded according to wt % retained on each sieve. The figures listed in Table 3.7 are the sieve-aperture sizes that pass about half the weight of the abrasive. Standard specifications should be consulted for more detailed information. Finer grades are referred to as *microgrits* and are specified by median particle size as determined in a standardized sedimentation apparatus. This is the figure listed in Table 3.7.

APPENDIX 3-B
Methods of Determining Abrasion Rates

The apparatus illustrated in Fig. 3.31 is suitable for determining changes in the abrasion rate after repeated traverses over one track of abrasive paper. Other types of apparatus have been developed which permit the abrasion rate to be determined on a constantly replenished track of paper.[9,15,16]

The specimen (A) is cemented to a specimen holder (B) which is locked in a chuck (C) in such a way that the holder can be removed and returned to precisely its old location and orientation. The chuck assembly is at-

TABLE 3.7. Approximate Median Sizes of Various Grades of Silicon Carbide and Aluminum Oxide Abrasives

Grit number		Median diameter, μm
Metricated	Premetrication	
Screen grit size range		
60	60	250
80	80	180
100	100	150
120	120	106
150	150	90
180	180	75
220	220	63
Microgrit range		
P 240	240	58.5
P 280		52.2
P 320	280	46.2
P 360	320	40.5
P 400		35
P 500	360	30.2
P 600	400	25.75
P 800		21.8
P 1000	500	18.3
P 1200	600	15.3

tached to a counterbalanced arm (D), to which can be added a weight (E) located vertically over the specimen. The weight can be changed to vary the pressure applied through the specimen surface. The fulcrum post (F) can be moved laterally on a slide (G). A disk of the abrasion device under test (H) is attached to a well-balanced rotating wheel.

The fulcrum is locked at one position on the slide, the specimen holder clamped in its chuck, and the specimen run against the abrasive device under test until contact is achieved over the full specimen area. The fulcrum is then translated across the slide and locked into position so that the specimen now contacts the intended test track on the abrasive disk. The specimen holder is removed from its chuck, weighed, returned to the chuck, the chosen weight loaded on the arm, the wheel run for the time necessary to achieve the desired number of traverses, and the specimen holder removed and reweighed. The procedure is repeated as often as necessary. The depth of the layer removed from the specimen surface during an abrasion sequence can be calculated from the loss in weight, if the surface area and density of the specimen are known; values for density obtained from reference data books are adequate for this purpose.

The loss-of-weight method described above can generally be applied only to unmounted specimens, because most mounts absorb and retain variable amounts of any liquid applied to the abrasion track. An alter-

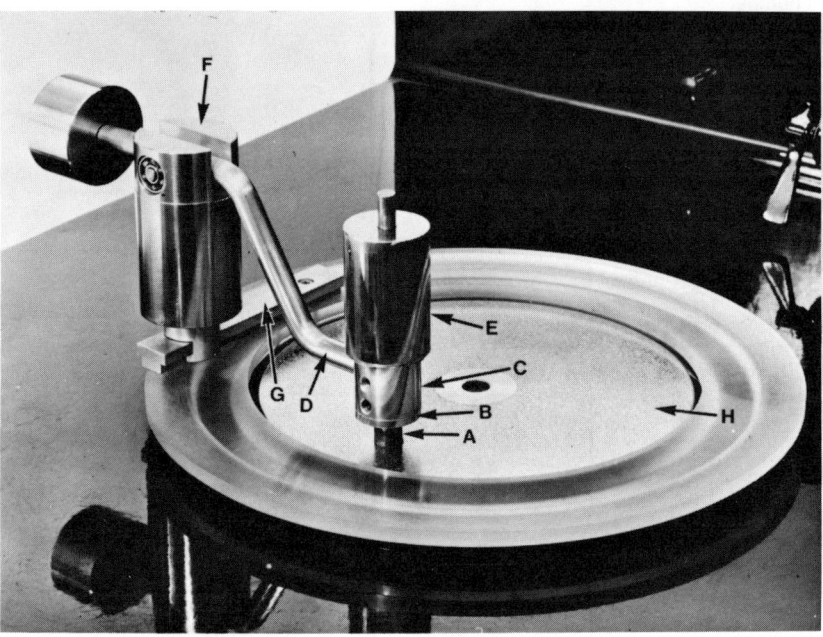

FIG. 3.31. Apparatus for abrasion under controlled conditions with repeated traverses on one track.

native method of measuring the depth removed is to make a Vickers hardness indentation on the test surface after it has been bedded in. Any irregularities around the indentation are then removed by a further short period of abrasion, the diagonals of the indentation are measured, the test abrasion is carried out, and the diagonals of the indentation are remeasured. The depth removed is one-seventh the decrease in diagonal length. A number of indentations distributed over the specimen surface should be followed to check the uniformity of material removal. This indentation method is suitable for measuring the removal of only comparatively thin layers, because the indentations are soon abraded away.

APPENDIX 3-C
Plastic-Diamond Laps

A MOLDED PHENOLIC-DIAMOND LAP (Samuels[21])
Method of Manufacture
A simple mold must be constructed in which a 7.5-cm-diam, or preferably a 15-cm-diam, disk of phenol formaldehyde plastic can be molded. The

mold may be of a type which can be clamped under pressure at room temperature and then subsequently heated to the curing temperature.

A blank disk about 1.5 cm thick is first prepared. A mixture is then prepared in the proportions of 4 g of phenolic powder which has passed through a 300-mesh sieve to 1 g (5 carats) of diamond abrasive; these quantities would produce an impregnated layer about 1 mm thick on a 7.5-cm-diam lap. One surface of the blank is roughened with coarse abrasive paper, the blank is set up in the mold and the plastic-abrasive mixture is spread as evenly as possible over the intended working surface. The mold is reassembled, reheated to the curing temperature and then cooled slowly to room temperature.

Method of Use

The lap surface is flooded with a nonvolatile fluid, such as kerosine or an extreme-pressure lubricant, and the specimen is rubbed against it by hand. A circular motion may be desirable because a reciprocating motion may form marked undulations on the surface — particularly with hard, brittle materials. The laps deteriorate only slowly in service. They may be restored by lapping the working surface against a coarse silicon carbide abrasive paper.

A MOLDED POLYTHENE-DIAMOND LAP (Haddrell et al[22])

Method of Manufacture

The diamond abrasive is mixed with an equal amount of a fine polythene powder. Water containing a wetting agent is added until a thin paste is produced, the paste then being painted evenly over the surface of a preformed disk of polythene about 1.75 cm thick. The painted surface is then covered with a sheet of plastic film, and the disk is placed in a mold similar to that described above and remolded. A molding pressure of about 3000 lb/in.^{2} and a curing temperature of 105°C are required. One carat (0.2 g) of abrasive produces a layer about 0.1 mm thick on a 7.5-cm-diam disk.

This lap is less wear-resistant than the bakelite laps. However, it can be used with more corrosive solutions when polish-attack techniques are employed.

APPENDIX 3-D
An Abrasive-Wax Lap for Fine Abrasion (Samuels[31])

Constituents

Microcrystalline wax
    (a hard grade with softening point of 80 to 90°C) .... 100 g
Aluminum oxide abrasive
    (10-20-μm grade) .... 300 g

Note: The lap will not function properly unless a closely sized grade of abrasive is used; the preparation or purchase of an abrasive of the type required is discussed in Chapter 8. A wax with a high softening point is also necessary to avoid transfer to the specimen during abrasion. On the other hand, the wax must not be too hard if the abrasive particles are to be held properly.

Method of Manufacture

Melt the wax and stir in the abrasive while maintaining heat so that the wax stays molten. Cast the mixture into a slab mold of the desired shape and allow it to solidify slowly. Attach the slab to a suitable base by means of an adhesive. Dress the working surface flat in a lathe, using an old carbide-tipped tool.

Method of Use

Rub the specimen on the working surface by hand, preferably using a unidirectional motion, the lap being used dry. The lap will clog rapidly and should be cleaned after one or two specimens have been abraded. Cleaning is carried out by swabbing vigorously with a pad of cotton wool moistened with alcohol until the old abrasion debris has been removed and fresh abrasive exposed. When the working surface becomes unduly irregular, it may be scraped flat by means of a steel straightedge after the surface layers have been softened by flooding the working surface with a strong solvent for the wax — e.g., petroleum ether or methyl ethyl ketone.

REFERENCES

1. L. Coes, "Abrasives", Springer-Verlag, New York, 1971.
2. T. O. Mulhearn and L. E. Samuels, *Wear*, 1962, *5*, 478.
3. A. J. Sedriks and T. O. Mulhearn, *Wear*, 1963, *6*, 457.
4. A. J. Sedriks and T. O. Mulhearn, *Wear*, 1964, *7*, 451.
5. E. J. Duwell and W. J. McDonald, *Wear*, 1961, *4*, 372.
6. T. C. Buttery and J. F. Archard, *Proc. Inst. Mech. Eng.*, 1970-71, *185*, 537.
7. J. Billingham and J. Lauridsen, *Wear*, 1974, *28*, 331.
8. M. M. Krushchov and M. A. Babichev, *Friction and Wear of Machinery*, 1956, *11*, 5, NEL Translation No. 831, National Engineering Laboratory, East Kilbride, Scotland.
9. R. C. D. Richardson, *Wear*, 1967, *10*, 291.
10. J. Larsen-Badse, *Trans. Met. Soc. AIME*, 1966, *236*, 1461.
11. J. M. Challen and P. L. B. Oxley, *Wear*, 1979, *53*, 229.
12. L. E. Samuels and E. D. Doyle, unpublished results.
13. M. A. Moore, *Wear*, 1974, *27*, 1.
14. G. K. Nathan and W. J. D. Jones, *Wear*, 1966, *9*, 300.
15. J. Larsen-Badse, *Wear*, 1968, *12*, 35.
16. J. Larsen-Basse and S. S. Sokoloski, *Wear*, 1975, *32*, 9.
17. B. W. E. Avient, J. Goddard and H. Wilman, *Proc. Roy. Soc.*, 1960, *258A*, 159.
18. B. W. E. Avient and H. Wilman, *Brit. J. Appl. Phys.*, 1962, *13*, 521.
19. S. W. Date and S. Malkin, *Wear*, 1976, *40*, 223.
20. G. W. Patterson and T. O. Mulhearn, *Wear*, 1968, *13*, 175.
21. L. E. Samuels, *J. Aust. Inst. Metals*, 1960, *5*, 63.
22. V. J. Haddrell, E. C. Sykes, and B. W. Mott, *J. Inst. Metals*, 1955-56, *84*, 112.

23. T. C. Jarrett, *Trans. Amer. Soc. Metals*, 1939, *27*, 758.
24. G. A. Ellinger and J. S. Acken, *ibid.*, 1939, *27*, 382.
25. K. Amberg, *Metal Progress*, 1940, *37*, 146.
26. M. Ferguson, *ibid.*, 1943, *43*, 743.
27. J. R. Vilella, *Metals and Alloys*, 1932, *3*, 205.
28. J. B. Cohen and J. H. Maker, *Metal Progress*, 1945, *47*, 508.
29. B. L. Johnson, *ibid.*, 1954, *66* (2), 111.
30. W. H. Bleeker, *Iron Age*, 1950, *165* (21), 71.
31. L. E. Samuels, *J. Inst. Metals*, 1952-53, *81*, 471.

CHAPTER 4
Machining With Abrasives: Surface Deformation

THE SEPARATION OF A CHIP during machining operations induces complex systems of plastic deformation in both the separating chip and the workpiece material. An inevitable consequence is that a layer plastically deformed during machining is left in the new surface that is produced. In general terms, the strains in this layer are very large at the surface and decrease approximately exponentially with depth.

This deformed layer becomes important in metallography when the plastic deformation changes the microstructure of the base material in a way that can be detected in the particular microscopical examination that is to be carried out. The layer is then an important potential source of false structures, or *preparation artifacts,* the avoidance of which is one of the primary objectives of a metallographic preparation sequence (see p. 3). Therefore, before attempting to study these features we need first to understand something of the circumstances under which manifestations of prior plastic deformation can be detected by optical microscopy.

DETECTION OF PLASTIC DEFORMATION BY OPTICAL MICROSCOPY

Cubic Metals Having High Stacking-Fault Energies

Malin and Hatherly[1] have established that the deformation of copper (a face-centered-cubic metal with a high stacking-fault energy) occurs in a series of overlapping stages during each of which a single operating mechanism predominates. At low strains (<10% reduction in cold rolling), deformation occurs by classic movement of dislocations (slip) on octahedral planes, but the dislocations relax after deformation to form the boundaries of equiaxed cells about 0.5 μm in diameter (background structure in Fig. 4.1b). No manifestations of deformation that occur by this mechanism can be revealed by optical microscopy.

In the strain range from 10 to 65% reduction, the slip process becomes progressively exhausted. Deformation tends to become concentrated inhomogeneously in bands, about 0.2 μm thick, aligned at about 35° to the trace of the compression plane (that is, the plane of maximum shear stress), although deviations of ±10° are common. The bands are confined to individual grains, and, usually, two systems form in each grain on {111} planes — but only in those grains in which the {111} plane happens to be aligned approximately parallel to the plane of maximum shear stress. These bands have been termed *microbands.* The mechanism of deformation within the bands has not been elucidated but probably involves some sort of slip process occurring on closely spaced planes. In any event, the structure found within the bands following deformation is one of plate-shape cells or subgrains with somewhat diffuse boundaries and with misorientations of only a few degrees between cells (Fig. 4.1b). Manifestations of the microbands can readily be seen by optical micros-

FIG. 4.1. Manifestations of plastic deformation observed in compressed polycrystalline copper by optical and transmission electron microscopy. *(Malin and Hatherly, Ref 1)*

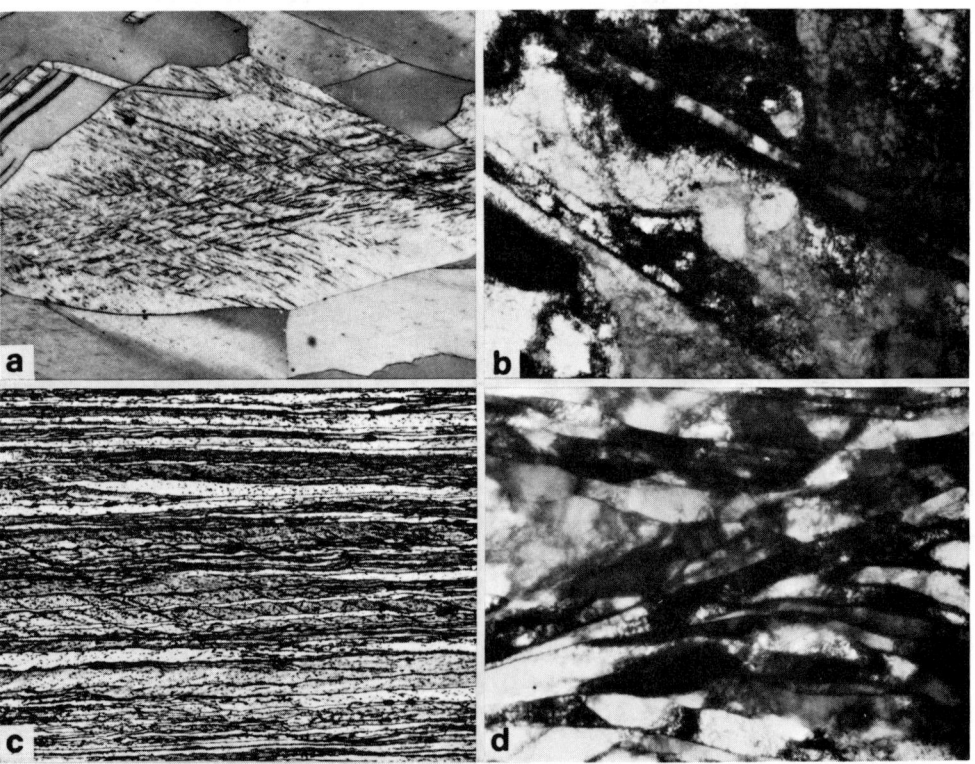

(a) Compressed to 10% reduction. Etched in the sodium thiosulfate reagent. Optical micrograph. Magnification, 500×. (b) Rolled to 10% reduction. Transmission electron micrograph. Magnification, 15,000×. (c) Rolled to 86% reduction. Etched in the sodium bisulfite reagent. Optical micrograph; 250×. (d) Rolled to 97% reduction. TEM; 20,000×. The features shown in (a) and (b) are microbands, and those shown in (c) and (d) are shear bands. The compression direction is vertical.

copy after etching in any of the reagents commonly used for this material,[2] although they are perhaps best seen after etching by the sodium thiosulfate method developed by Jacquet* (Fig. 4.1a); the markings so developed may be called *microband strain markings*.

The boundaries of the microbands sharpen with further deformation and the bands rotate progressively to become aligned parallel to the compression plane, until deformation can no longer be sustained by microband formation and a new deformation process is required. The new process involves the formation of *shear bands*, the morphology of which is dictated by considerations of macroscopic plasticity rather than crystallographic glide. Shear bands form on planes aligned at about 35° to the trace of the compression plane, and individual bands cross many grain boundaries without deviation. The structure found in these bands following deformation is one of slab-shape cells with lateral dimensions of 1 to 20 μm and a thickness of 0.02 to 0.1 μm (Fig. 4.1d); the cells have sharply defined boundaries. Manifestations of shear bands (*shear-band strain markings*) can be seen in copper by optical microscopy after etching by any of the methods commonly used for that metal (Fig. 4.1c).

Both microband and shear-band strain markings have been observed in several other metals,[3,4] and probably can be developed in any metal. For example, microbands identical in morphology to those shown in Fig. 4.1(a) have been observed in iron by optical microscopy, and investigation by transmission electron microscopy has shown that the bands have the same type of subgrain structure as that described above.[4]

Cubic Metals Having Low Stacking-Fault Energies

Many solid-solution alloys which have face-centered-cubic crystal structures are in this category, an archetype being 70:30 α brass. This discussion will concentrate on 70:30 brass because the alloy provides a convenient model for many investigations of the structures of abraded and polished surfaces. Deformation of this alloy precipitates a sequence of events somewhat different from that just described for copper.[5,6]

At small plastic strains, just beyond yield, deformation occurs by slip on {111} planes, but major relaxation does not occur and dislocations remain in arrays on the slip planes after deformation (Fig. 4.2b). Manifestations of these dislocation arrays can be developed through etching by several specific methods,[2,7] the most effective being Jacquet's sodium thiosulfate etch, which can be used in two ways (see Appendix 4-A). First, under high-sensitivity conditions, shaded dark lines (subsequently called *slip strain markings*) develop along the traces of {111} planes and on several such systems in most grains (Fig. 4.2a). It can be taken that these markings can be developed whenever any slip has occurred — i.e., when the elastic limit has been exceeded in a particular grain.[7,8] The markings extend only partly across grains after small strains (Fig. 4.2a), but extend completely across all grains after slightly larger strains (main body of the

---

*Details of this etching method, and those of others referred to in this chapter, are given in Appendix 4-A.

FIG. 4.2. Manifestations of plastic deformation observed in compressed 70:30 brass by optical and transmission electron microscopy.

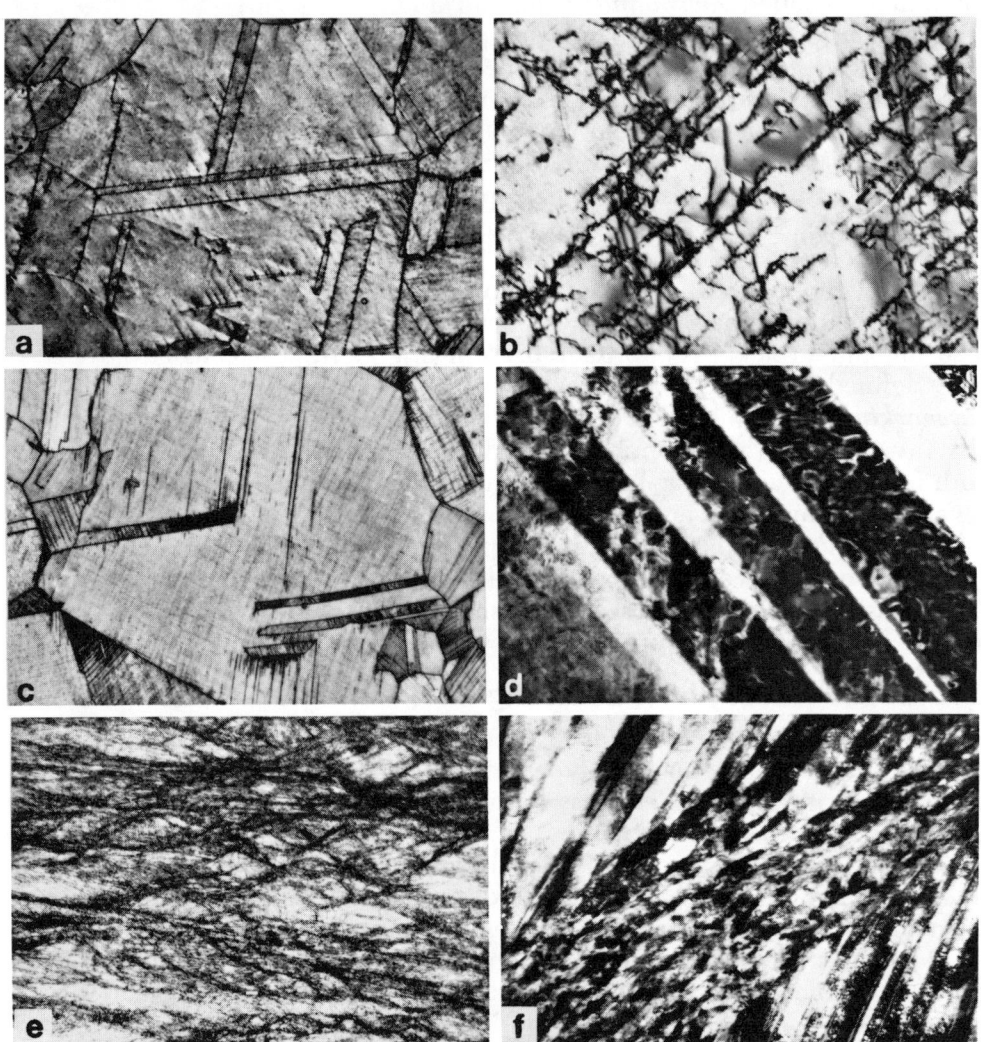

(a) Slip strain markings, revealed by the high-sensitivity sodium thiosulfate etch, in a specimen compressed just beyond yield. Optical micrograph. Magnification, 500×.
(b) Specimen similar to that shown in (a), exhibiting dislocation arrays on {111} planes. Transmission electron micrograph. Magnification, 20,000×. (c) Twin strain markings, among slip strain markings, revealed by the high-sensitivity sodium thiosulfate etch in a specimen compressed to 10% reduction. Optical micrograph; 500×.

(d) Specimen similar to that shown in (c); electron diffraction reveals that the diagonal bands have a twin orientation with respect to the parent grain. TEM; 50,000×.
(e) Shear-band strain markings revealed by an ammonium hydroxide – hydrogen peroxide etch in a specimen compressed to 80% reduction. Optical micrograph; 500×.
(f) Specimen similar to that shown in (e); the diagonal band that contains elongated subgrains is a shear band. TEM; 20,000×. In all cases, the compression direction is vertical.

grains illustrated in Fig. 4.2c). Secondly, under low-sensitivity conditions, the first manifestations of deformation are not observed until after a strain of about 0.1% reduction and then take the form of rows of etch pits developed along the traces of {111} planes (Fig. 4.3a). Two other etching reagents that can develop slip strain markings are the ammonium hydroxide – hydrogen peroxide reagent, which has about the same sensitivity as the high-sensitivity sodium thiosulfate etch, and the cupric ammonium chloride reagent, which develops the markings when the strain exceeds about 0.5% reduction.[2]

The second mode of deformation involves formation of thin twins clustered in bands (Fig. 4.2d). Manifestation of these twin bands (*twin strain markings*) can be developed by any of the etchants commonly used for the alloy. At low magnifications they appear as dark, structureless bands among the slip strain markings after etching by a method capable of developing slip strain markings (Fig. 4.2c), and as isolated structureless bands after etching in reagents which are not capable of developing slip strain markings (Fig. 4.3b). In the latter case, the markings are perhaps most clearly shown after etching in a ferric chloride reagent. The twin strain markings are certainly detectable after 5% reduction by compression, which can be taken to be the threshold strain for their observation.* They develop on the traces of a {111} plane, usually on a single system of {111} planes at this stage and mostly only in regions adjacent to twin and grain boundaries. After larger strains, the strain markings become more numerous and tend to extend completely across individual grains, to develop on two systems, and to rotate to become parallel to the compression plane (Fig. 4.3c).

The twin-band mode of deformation begins to exhaust at 40 to 50% reduction, at which point shear bands begin to form; the shear bands extend across many grains and are aligned at approximately 35° to the trace of the compression plane.[6] Manifestations of the shear bands can be developed through etching by any of the methods commonly used for brass (Fig. 4.2e). The structure of these bands is similar to that already described for shear bands in copper; that is, they are composed of slab-shape subgrains with lateral dimensions of 1 to 20 μm (Fig. 4.2f). The array of bands divides the structure into rhombohedral prisms of twinned material, the prisms subdividing with further strain until at 90% reduction more than 50 vol % of the structure consists of shear bands. The bands commonly are up to 1 μm thick, but occasionally may be 10 μm thick or more (Fig. 4.3d). Large strains can occur in individual shear bands (Fig. 4.3d) — certainly, natural strains of over 2.

Metastable Alloys Susceptible to Strain-Induced Transformations

Phase transformations are induced by plastic deformation in certain metastable alloys. Perhaps the most important example is the austenitic stainless steels, to which the present discussion will be confined, although

---

*The threshold strain for the observation of twin bands depends somewhat on the type of the strain, but here we need consider only compressive strains.

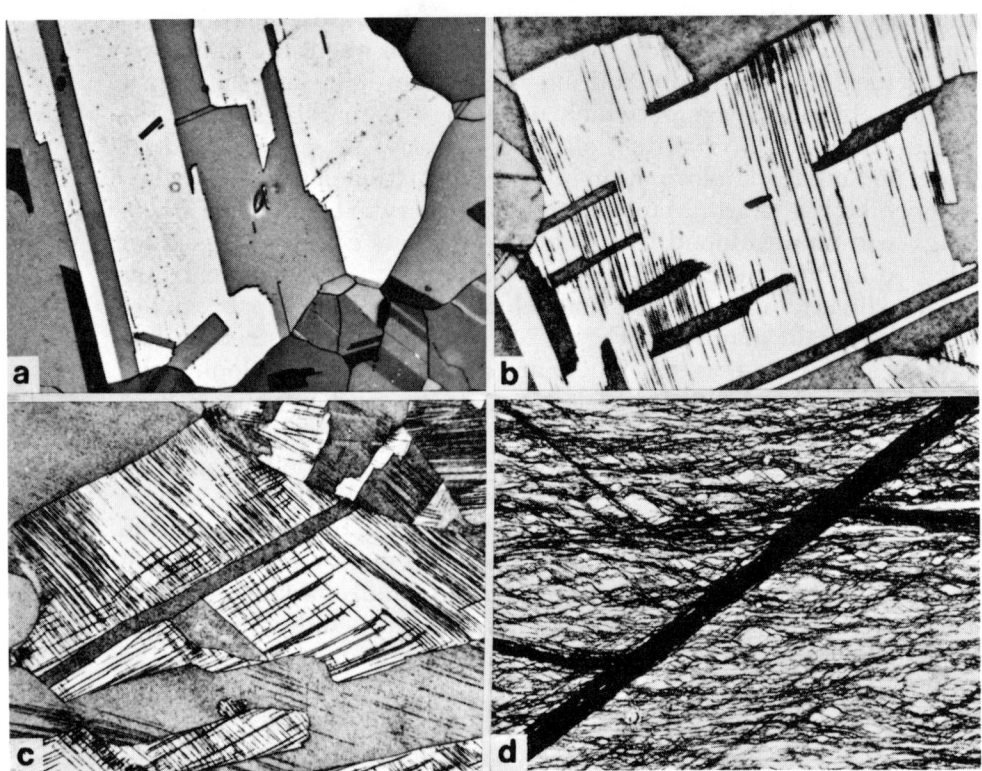

FIG. 4.3. Manifestations of plastic deformation observed in compressed 70:30 brass by optical microscopy.

(a) Slip strain markings, revealed by the low-sensitivity sodium thiosulfate etch, in a specimen compressed to 0.2% reduction. Magnification, 500×. (b) Twin strain marking, revealed by etching in a ferric chloride reagent, in a specimen compressed to 10% reduction. Magnification, 500×. (c) Same as (b), but compressed to 20% reduction. Magnification, 500×. (d) A particularly wide shear-band strain marking, revealed by etching in a ferric chloride reagent, in a specimen compressed to 90% reduction. Note that large shear displacements have occurred in the most prominent band shown in this field. Magnification, 500×. In all cases, the compression direction is vertical.

these alloys can also be taken to be representative. Several distinct martensite phases have been identified as the products of deformation of these steels, but the details of the transformations are in some dispute and are not of great importance in the present context. What is known is that the transformations are associated with faulting caused by the movement of slip dislocations during deformation, and that the transformation occurs in bands in which these dislocation movements are inhomogeneously concentrated. The bands in which the transformation occurs (*transformation strain markings*) can be seen by optical microscopy after etching in any of the reagents commonly used for these alloys.

Examples of transformation strain markings developed in an AISI type 304 austenitic stainless steel over a range of strains are given in Fig. 4.4. Morphologically, these markings bear some resemblance to the twin strain markings in 70:30 brass (cf. Fig. 4.3b and 4.4b), but they develop at somewhat smaller strains (<1% reduction by compression for the alloy illustrated) and generally develop on only one system in each grain even at high strains. The density of markings increases with increasing strain (Fig. 4.4c) until, at about 30% compression, the markings become so numerous that the structure is rather confused (Fig. 4.4d). Markings similar to the shear-band markings illustrated in Fig. 4.2(e) appear and again become progressively more clearly delineated at higher strains.

The magnitude of the strain at which these various events occur will vary with the different alloy types.

FIG. 4.4. Manifestations of plastic deformation observed in AISI type 304 austenitic stainless steel by optical microscopy.

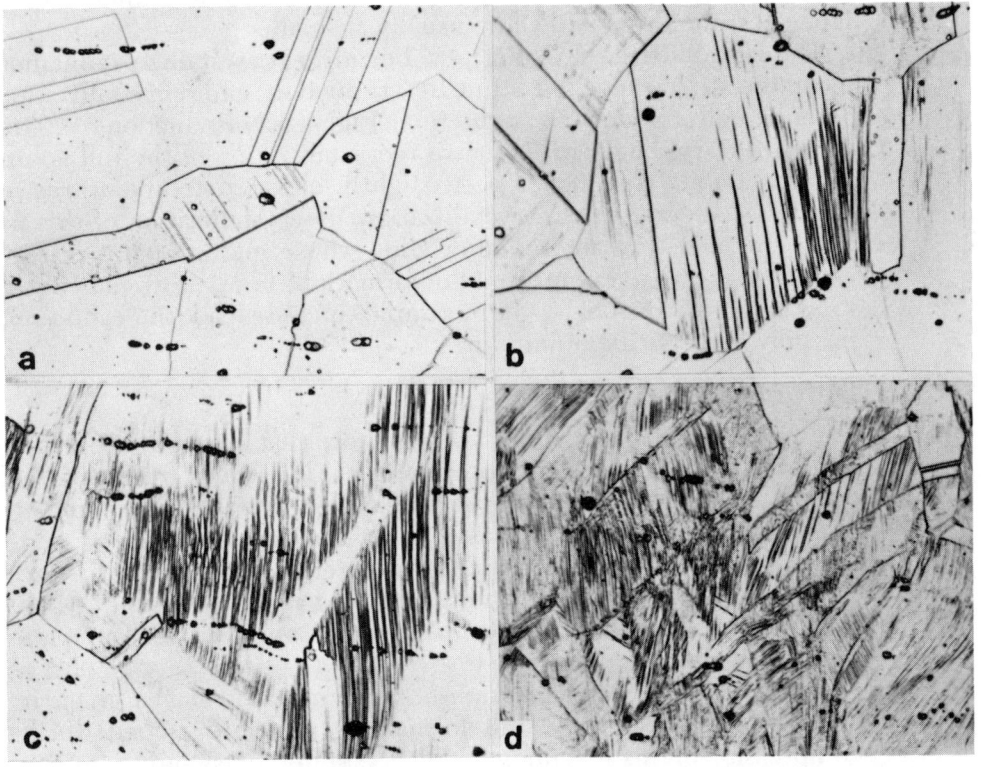

(a) 1% reduction. (b) 7.5% reduction. (c) 20% reduction. (d) 30% reduction. The strain markings shown in these micrographs were developed by transformation of the metastable alloy in bands along which dislocation movements were concentrated. All specimens were etched electrolytically in an oxalic acid solution. Magnification, 250×. The compression direction is vertical.

Massive Twinning and Recrystallization

The extent to which deformation can be accommodated by dislocation glide is limited in most polycrystalline noncubic metals because the number of slip systems that are available is limited. Deformation at higher strains consequently tends to be accommodated by the formation of *massive twins,* in which a comparatively large block of the grain shears into an orientation which is a mirror image of that of the parent crystal. Twins of this nature can always be detected by optical microscopy after etching in any of the reagents commonly used for the particular metal or alloy. It is usually also possible to detect them by examination in polarized light without etching.

Examples of massive twins produced by compression are given in Fig. 4.5, the material illustrated being polycrystalline zinc of commercial purity. Prominent twins were detected after strains as small as 1% reduction by compression (Fig. 4.5a), and the density of twinning increased markedly with further deformation up to about 7% reduction (Fig. 4.5b). The magnitude of the strain threshold for the development of twins depends, however, on a number of factors including the purity of the metal and the type of strain. It is also different in different metals. However, the threshold value usually is small.

The zinc illustrated in Fig. 4.5 began to recrystallize spontaneously when the strain exceeded about 10% reduction, and recrystallization was extensive after about 15% reduction (Fig. 4.5d). Reduction of 70 to 80% was necessary, however, to cause recrystallization of the full volume of the specimen. Heating to even slightly elevated temperatures would greatly accelerate this recrystallization. Recrystallization of this nature can be expected in any metal or alloy whose melting point is less than about twice the temperature of deformation, temperatures being measured in degrees Kelvin. Other examples include lead, tin, cadmium, and possibly high-purity aluminum.

Deformation Bands

The local variations in strain, such as those that arise between adjoining grains when a polycrystalline material is deformed, frequently are accommodated by coordinated bending of the crystal lattice of one or both of the grains. Broad bands consequently develop which are separated by narrow regions of marked lattice curvature. Bands of this nature are particularly likely to develop during straining in compression.[8] Manifestations of these *deformation bands* may be seen during examination by optical microscopy.

An example of a *deformation-band strain marking* is given in Fig. 4.6(a). A series of regular deformation bands developed during the compression of this specimen of polycrystalline brass and are visible after etching in the ferric chloride reagent; note that the bands coincide with bending of the annealing twins in the grain illustrated. A region at the boundary between a light-etching band and a dark-etching band in this grain is shown in Fig. 4.6(b), the specimen this time having been etched by a method that develops slip strain markings; the slip planes can be

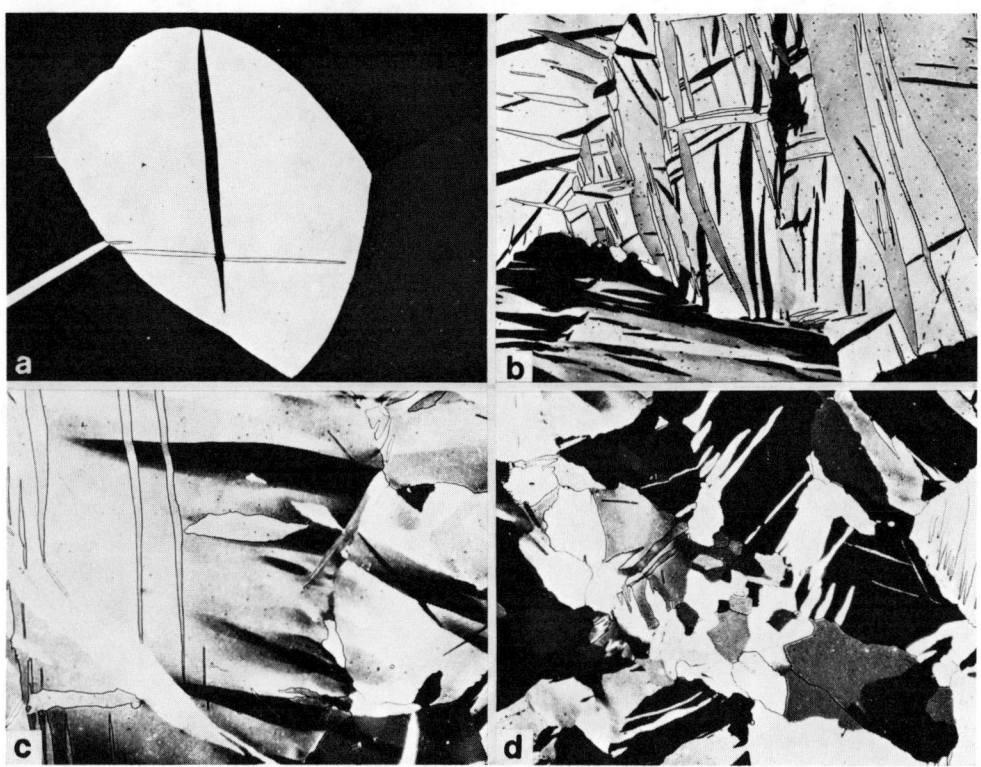

FIG. 4.5. Manifestations of plastic deformation observed in compressed zinc by optical microscopy.

(a) 1.2% reduction. **(b)** 7.8% reduction. **(c)** 10% reduction. **(d)** 17% reduction. Typical massive twins are present in all cases, as well as deformation bands in **(c)** and recrystallized grains in **(d)**. All specimens viewed in polarized light. Magnification, 250×. The compression direction is vertical.

seen to have bent in a coordinated way at the boundaries of the deformation bands. More characteristically, however, the deformation bands have a rather irregular outline (Fig. 4.6c).

Deformation bands of this nature have been observed in a wide range of metals. For example, diffuse deformation bands in compressed copper are visible in Fig. 4.6(d); several are also visible in the specimen of compressed zinc illustrated in Fig. 4.5(c). Twin strain markings or transformation strain markings when developed may be grouped to individual deformation bands, a typical example being illustrated in Fig. 4.4(c).

Lamellar Structures

Lamellar structures, whether eutectic or eutectoid in nature, frequently are of metallographic interest and often are composed of alternate plates of hard and soft phases. The hard phase, although brittle in bulk, may be

96 / Metallographic Polishing by Mechanical Methods

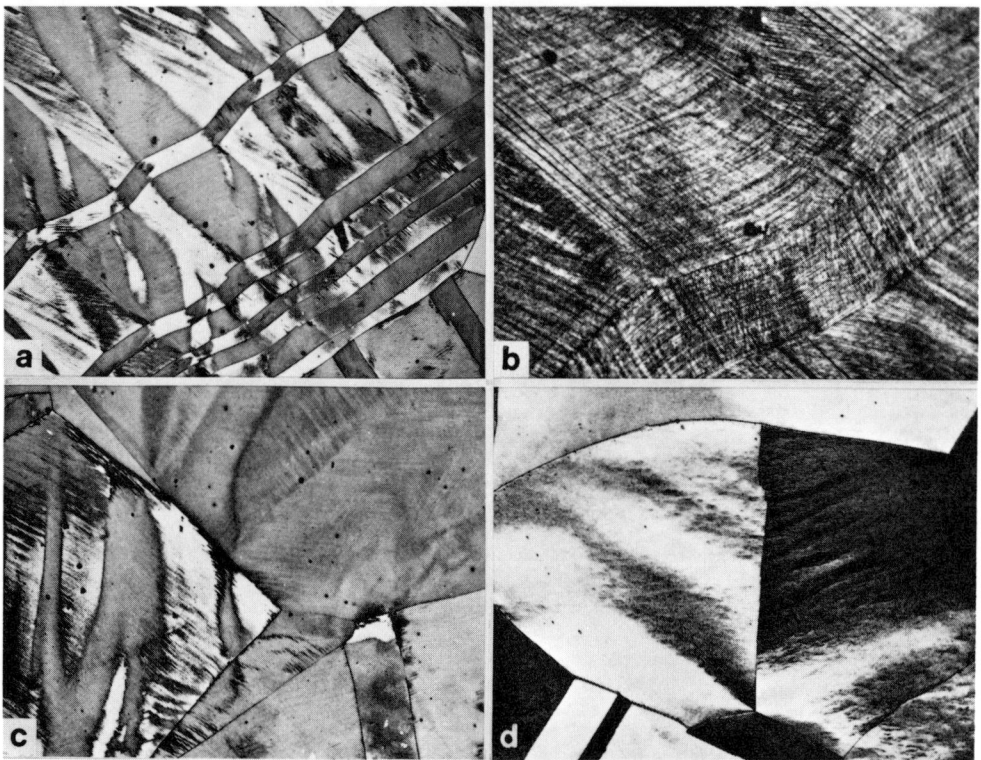

FIG. 4.6. Manifestations of deformation bands observed in compressed copper and brass by optical microscopy.

(a) Brass compressed to 12% reduction. Ferric chloride etch. Regular arrays of deformation bands, which are associated with bending of twin boundaries, are present in this field of view. Magnification, 100×. (b) Portion of the field of view shown in (a). High-sensitivity thiosulfate etch, which has developed slip-line etch markings. The slip-line markings bend at the boundaries of the deformation bands. Magnification, 500×. (c) Brass compressed to 12% reduction. Ferric chloride etch. Deformation bands that are irregular, but more typical than those in (a), are present in this field of view. Magnification, 100×. (d) Copper compressed to 10% reduction. Ferric chloride etch. Irregular deformation bands are present. Magnification, 100×. In all cases, the compression direction is vertical.

capable of deforming plastically when present in such a lamellar arrangement, and distortion of the lamellar morphology may then be recognizable.

Taking lamellar pearlite in steel as an example, the cementite plates bend in coordinated banded arrays, particularly during deformation in compression. The bands begin to become noticeable at about 20% reduction and are strongly developed by 40% reduction (Fig. 4.7a); closely spaced secondary kinks also develop in the cementite plates at higher strains. The coordinated bending is most likely to occur in plates that are

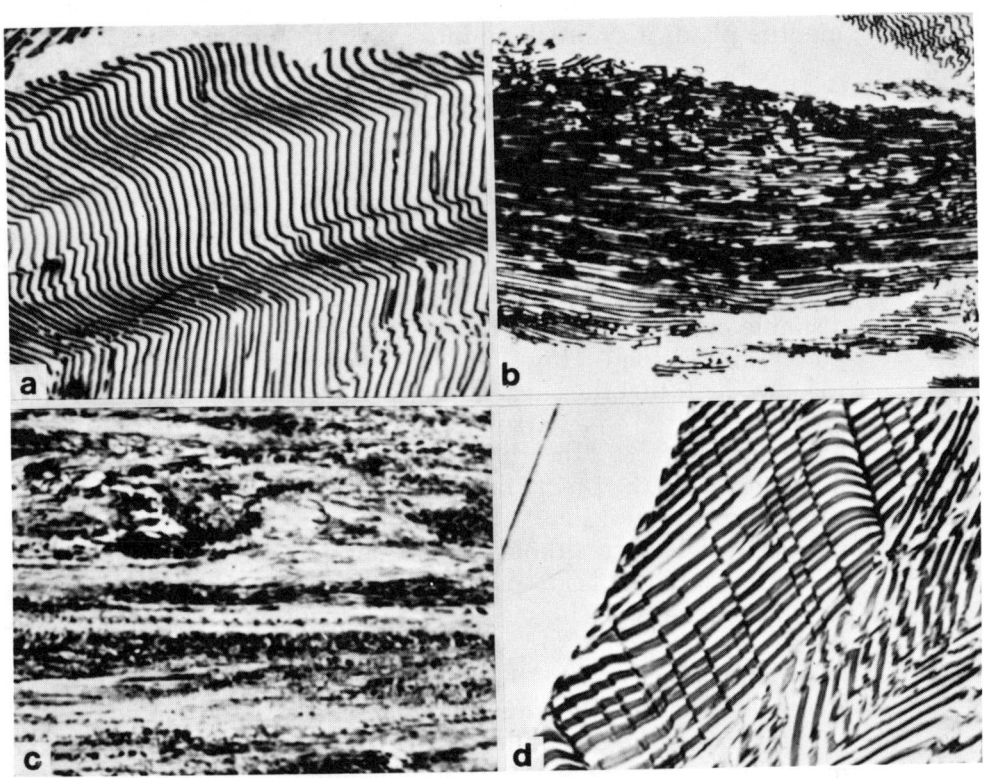

FIG. 4.7. Manifestations of deformation observed in lamellar pearlite in steel by optical microscopy.

(a) Compressed to 40% reduction. Magnification, 1500×. (b) Compressed to 60% reduction. Magnification, 1500×. (c) Drawn to 85% reduction. Magnification, 500×. (d) Necked region of a fractured tensile-test piece. Magnification, 1500×. All specimens etched in picral. The strain axis is vertical.

aligned approximately parallel to the compression axis, and is largely confined to those that are aligned within about 45° to the compression axis. The plates not favorably oriented for bending rotate to become aligned perpendicular to the compression axis, most being so aligned after about 60% reduction (Fig. 4.7b). A similar sequence of events has been observed in eutectics.[9]

The spacing and thickness of the cementite plates in pearlite are progressively reduced during some types of compressive deformation in proportion to the gross reduction, so that eventually the structure of the pearlite cannot be even partly resolved by optical microscopy (Fig. 4.7c). Examinations by transmission electron microscopy leave little doubt that the structure of both the white, completely unresolved areas and the mottled, partly resolved areas in Fig. 4.7(c) consist essentially of a lamellar arrangement of cementite and ferrite.[10,11] The cementite plates probably have been partly broken up, but they would still be continuous over considerable distances; longitudinally, they would be aligned parallel to the wire axis. However, under some other straining conditions, particularly those involving a component of tensile elongation, the cementite plates may break up into many short fragments (Fig. 4.7d).[12,13]

ORIGIN OF THE SURFACE DEFORMED LAYER ON ABRADED SURFACES

We saw in Chapter 3 that an abrasive machining device can be regarded as an array of vee-point machining tools, tools which have a range of rake angles but for which most of the rake angles are negative. The deformation processes involved in the formation of a machining chip by even a simple orthogonal tool are very complex and are far from being completely understood. However, for the present purposes, it is sufficient to know, first, that the shear strains involved in the separation of the chip are concentrated in a narrow zone extending from the point of the tool to the free surface of the chip. This is usually known as the shear line or, more correctly, the *shear zone* (Fig. 4.8a). A region adjacent to this shear zone, and extending into the workpiece in advance of the tool, is also plastically deformed although to a lesser degree. The strain in this region decreases until an elastic-plastic boundary is reached well in advance of the tool (Fig. 4.8a). The second feature of interest in the present context is illustrated in Fig. 4.8(b), which shows diagrammatically the streamline flow pattern followed by the material of the workpiece as it approaches the tool. Note that the streamlines separate slightly in advance of, and slightly above, the edge of the tool. Thus portions of the shear zone and most of the region of lesser strain flow into the new surface that is being created.

These features are illustrated in the section of a partly formed chip shown in Fig. 4.9. This chip was cut in 70:30 brass by an orthogonal tool with a rake angle that was negative but slightly less so than the critical angle under the particular circumstances. The section has been etched in a ferric chloride reagent which, it will be recalled, develops twin and shear-band strain markings but not slip strain markings. The shear zone (labeled) is the comparatively light-etching zone extending from the indentation made by the edge of the tool, and a portion of this zone can be seen to have flowed into the new surface as the chip separated. A zone containing twin strain markings is present ahead of the shear zone, and the boundary of the zone can be described as an isostrain line to which a strain value of about 5% compression can be ascribed. Much of this zone has found its way into the new surface. Similar effects are found to an even more marked extent in chips cut with tools that have more negative rake angles. They are also found in chips cut by tools with more positive rake angles,[15,16] although the shear zone may be too thin to be resolved in transverse sections by optical microscopy. The consequences of using tools with no flank clearance, as might occur with real abrasive points, are neglected; little is known about this.

FIG. 4.8. Diagrammatic illustration of some features of the formation of machining chips in orthogonal cutting. (Enahoro and Oxley, Ref 14)

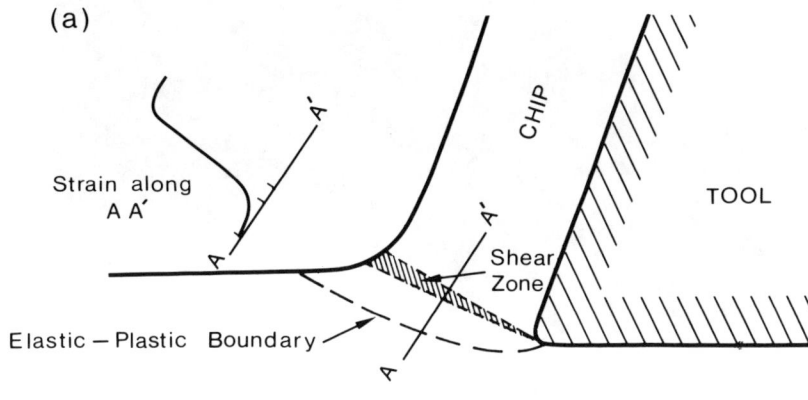

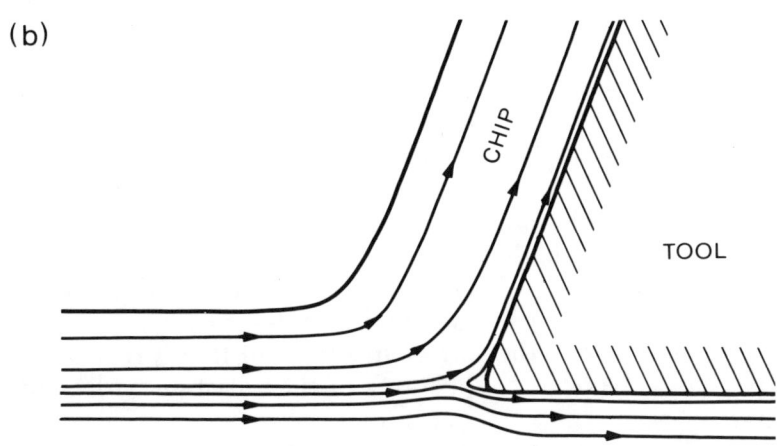

(a) Distribution of strains. (b) Streamline flow pattern in the workpiece material as it approaches the cutting edge of the tool.

STRUCTURE OF THE PLASTICALLY DEFORMED LAYER PRODUCED DURING ABRASIVE MACHINING

General Pattern of Deformation

The general pattern of the plastically deformed layer introduced into the surface of a metal by abrasive machining is most conveniently revealed by examining, by optical microscopy, sections of specimens of 70:30 brass.[7,17] As we saw earlier (p. 89), a number of manifestations of plas-

100 / Metallographic Polishing by Mechanical Methods

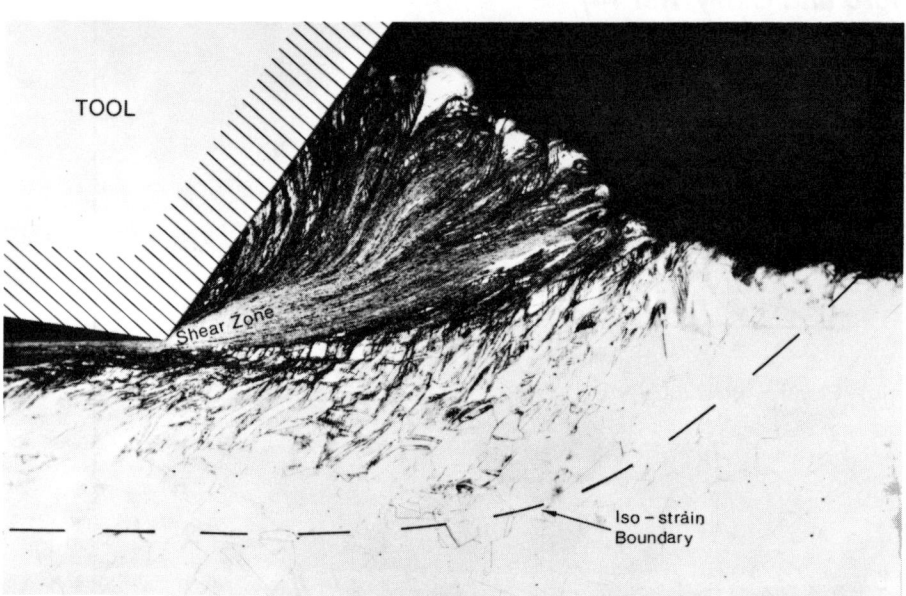

FIG. 4.9. Section of a chip cut in 70:30 brass by an orthogonal tool with a highly negative rake angle.

Etched in ferric chloride reagent. Magnification, 35×. The shear zone and one iso-strain line revealed by etching are identified.

tic deformation can be seen in this alloy by optical microscopy after etching by suitable methods. Examples of a section of a representative abraded surface that has been etched by several such methods is given in Fig. 4.10.* As might be expected, the deformation is distributed inhomogeneously across the surface in a manner that is related to the surface scratches. Noteworthy features of the variation with depth are as follows:

First, a dark-etching, structureless layer can be seen at the surface after etching by all of the methods used for this illustration (this layer is shown in more detail in Fig. 4.11). The layer contours the surface scratches, and its thickness is about the same as, or a little more than, the depth of the scratch with which it is associated. We shall call this layer the *fragmented layer* and we shall discuss its structure in more detail later.

---

*The section shown in Fig. 4.10, and in many subsequent illustrations, is a *taper section*. Details of the techniques of taper sectioning are given in Appendix 4-B. A taper ratio is quoted with each photomicrograph and this ratio indicates the geometric magnification perpendicular to the section line. The section is referred to as a *longitudinal taper section* when the section line is made parallel to the scratches. It is referred to as a *transverse taper section*, or simply as a *taper section*, when the section line is made perpendicular to the scratches.

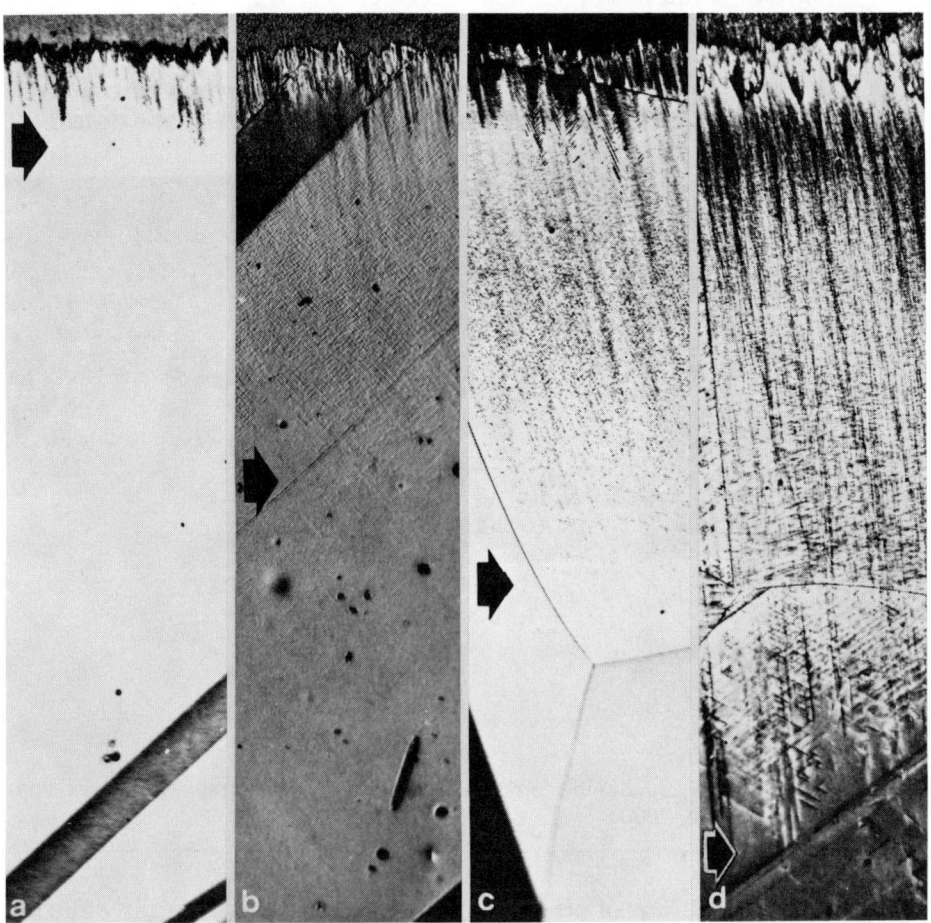

FIG. 4.10. Taper section of the surface of annealed polycrystalline 70:30 brass that has been abraded on 220-grade silicon carbide paper.

The section has been etched by several methods that have different threshold strains for revealing deformation, as follows. **(a)** Ferric chloride reagent (threshold strain: 5% compression). **(b)** Cupric ammonium chloride reagent (threshold strain: 0.5% compression). **(c)** Low-sensitivity thiosulfate etch (threshold strain: 0.1% compression). **(d)** High-sensitivity thiosulfate etch (threshold strain: elastic limit). In each case, the base of the layer in which the manifestations of deformation have been developed is indicated by an arrow. Taper ratio, 8.2. Magnification, 250×.

Secondly, a zone extends beneath the fragmented layer in which strain markings characteristic of comparatively small strains are developed. For example, the high-sensitivity thiosulfate etch develops the dark, shaded lines characteristic of slip strain markings (cf. Fig. 4.10d and 4.2a), and the base of the zone containing these markings can be identified as the elastic-plastic boundary. The layer so defined can be called the *deformed layer*. The base of this layer in more coarsely abraded surfaces is gently undulating in a manner not clearly related to individual surface scratches (as, for example, in Fig. 4.10d). For finely abraded surfaces it may undulate markedly in a manner that is apparently associated with individual surface scratches or groups of scratches (Fig. 4.12).

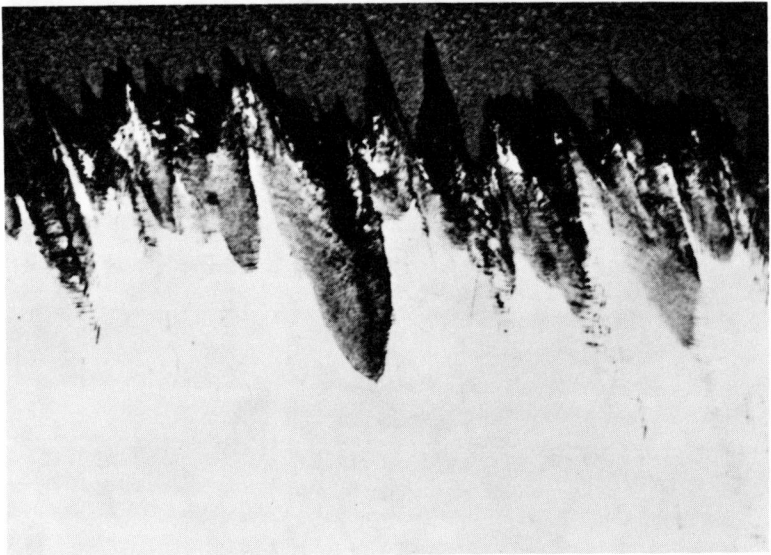

FIG. 4.11. Taper section of the same surface shown in Fig. 4.10(a), showing the fragmented layer in more detail.

Etched in the ferric chloride reagent. Taper ratio, 8.2. Magnification, 2000×.

The positions of other isostrain boundaries within the deformed layer can be established by suitable selection of etchants. For example, etching by the low-sensitivity sodium thiosulfate method develops arrays of etch pits recognizable as slip strain markings (Fig. 4.10c; cf. Fig. 4.3a), and the base of the layer containing these markings can be identified as the isostrain boundary for a strain level of approximately 0.1% compression. Likewise, a 0.5% strain boundary can be identified by using a cupric ammonium chloride reagent (Fig. 4.10b). Etching in a ferric chloride reagent, or any other reagent commonly used for brass, develops twin strain markings but not slip strain markings, and a strain value of about 5% compression can be ascribed to the boundary of the layer containing these markings (Fig. 4.10a). Thus an approximate indication of the strain gradient in the deformed layer can be developed (Fig. 4.13).

More detail of the layer developed by etching in the ferric chloride reagent is shown in Fig. 4.14. The layer contains closely spaced twin strain markings, most of them being on one crystallographic system aligned perpendicular to the original surface. The twin strain markings

FIG. 4.12. Taper section of the surface of annealed polycrystalline 70:30 brass that has been abraded on a fine lap.

The elastic-plastic boundary undulates markedly in a manner associated with individual surface scratches or groups of scratches. Etched by the high-sensitivity sodium thiosulfate method. Taper ratio, 9. Magnification, 500×.

are concentrated in rays extending beneath individual surface scratches, and diffuse dark-etching bands are associated with these rays; the bands can be interpreted as being deformation bands (p. 94). Thus the deformation at higher strain levels is always distributed inhomogeneously beneath individual scratch grooves, which might well be expected considering the variations in the characteristics of the abrasive points that produced the grooves.

The Fragmented Layer

Optical microscopy suggests that there is usually a comparatively sharp boundary between the fragmented layer and the rest of the deformed layer (Fig. 4.11 and 4.14), but gives little further information about the structure of the layer. This structure has, however, been successfully elucidated by transmission electron microscopy (TEM) of thin sections,[16,18,19] the interpretation of such sections being supplemented and confirmed by transmission electron diffraction.

The structure of the fragmented layer in an abraded surface of copper as determined by TEM methods is summarized diagrammatically in Fig. 4.15. The outermost layers (~150 nm deep) are composed of small, equiaxed grains (approx 30 nm in diameter) with some larger equiaxed

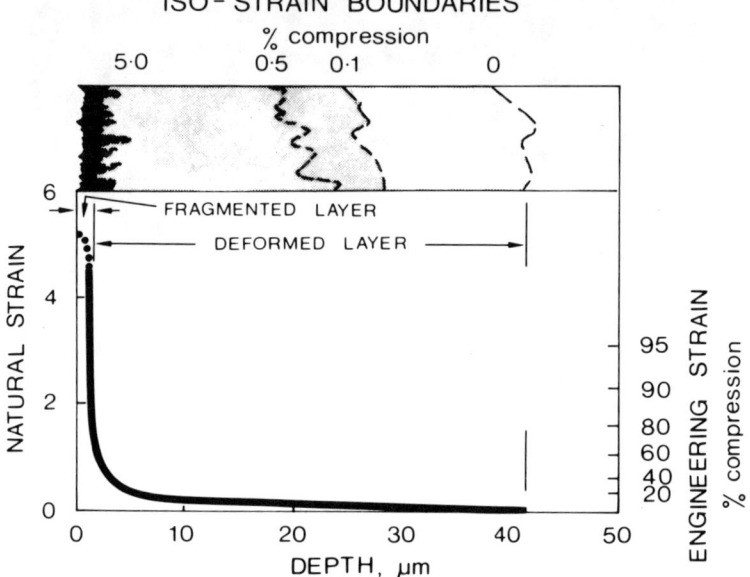

FIG. 4.13. Diagrammatic illustration (sketch at top) of the isostrain boundaries located by the etching methods used for the specimens in Fig. 4.10, and an approximate indication of the strain gradient in the deformed layer.

The estimate of the strain gradient in the fragmented layer was developed through investigation by transmission electron microscopy.

grains containing annealing twins, which appear to be recrystallized grains. These equiaxed structures shade with increasing depth into slab-shape cells (1 to 2 μm long and 0.1 to 0.2 μm thick) which have sharp boundaries and then into cells of similar morphology but with progressively more diffuse boundaries. Leaving aside for the moment the outermost equiaxed grains, this is the same sequence of structures, in reverse, as that observed in copper that has been cold rolled to increasingly large reductions (see p. 87); the structures of the outermost portions of the fragmented layer correspond to those found in shear bands, and the structures in the deeper portions correspond to those found in microbands. This is in agreement with the suggestion made earlier (p. 98) that the fragmented layer comes from the shear zone that is associated with the formation of a machining chip.

Comparison of the structures observed in the fragmented layer by transmission electron microscopy with those observed in the same material compressed to known amounts of strain allows an approximate estimate to be made of the strain gradient in the layer.[16] An estimate of this type is incorporated in Fig. 4.13, which emphasizes the very large magnitude of the strains in the fragmented layer.

FIG. 4.14. Detail of the specimen shown in Fig. 4.10(a).

Rays of twin strain markings and diffuse deformation-band strain markings extend preferentially beneath individual surface scratches. Etched in ferric chloride reagent. Taper ratio, 8.2. Magnification, 1000×.

The equiaxed subgrains present at the surface are not found in material deformed in bulk, but it seems that they result from recrystallization or relaxation of the shear-band or microband structures. The associated larger and obviously recrystallized grains almost certainly have resulted from thermal modification of these structures. Recrystallization of this nature can occur at room temperature in some materials, including copper, after large enough strains, but in the particular material used in the investigation summarized in Fig. 4.15 it is possible that the recrystallized grains resulted from heating to temperatures above ambient; if so, the temperatures would have been within the range 100 to 150°C. In any

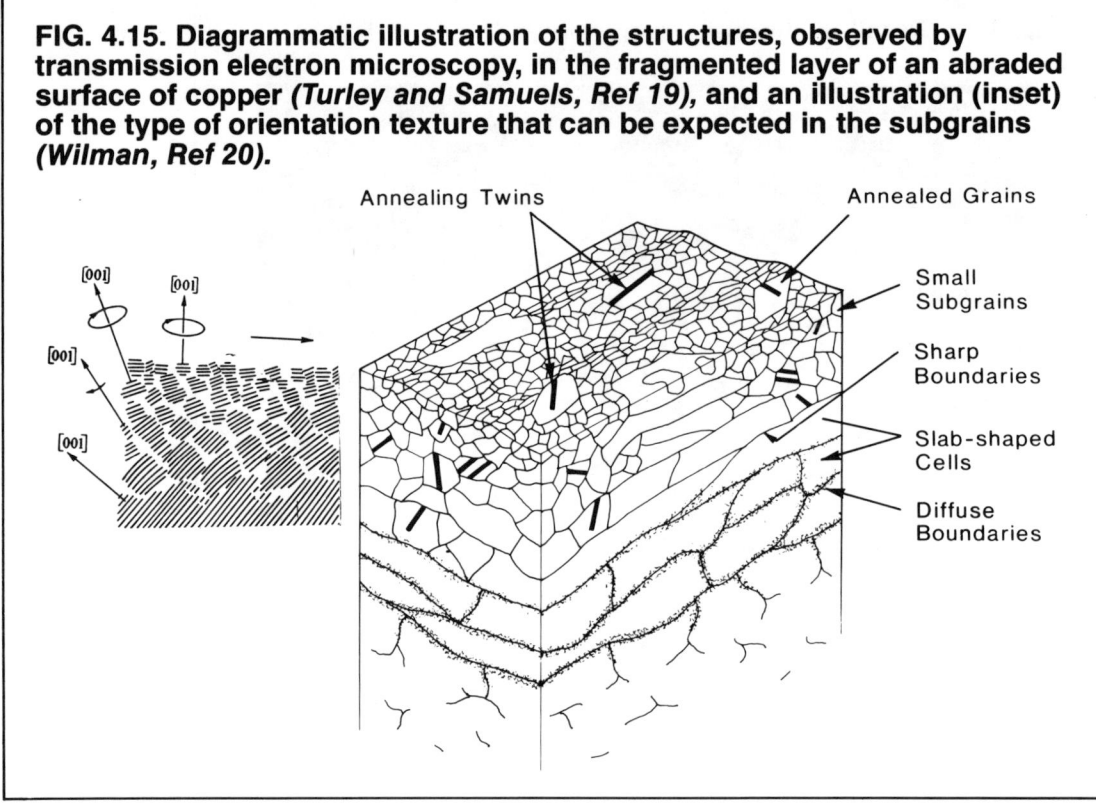

FIG. 4.15. Diagrammatic illustration of the structures, observed by transmission electron microscopy, in the fragmented layer of an abraded surface of copper *(Turley and Samuels, Ref 19)*, and an illustration (inset) of the type of orientation texture that can be expected in the subgrains *(Wilman, Ref 20)*.

event, the extent to which the deformation structures are modified thermally can be expected to be different in other metals and alloys. In general terms, they can be expected to have been more extensively modified in metals of lower melting points; for example, recrystallization is extensive enough in zinc and tin to be detectable by optical microscopy (see below). On the other hand, modification may not occur at all in metals of higher melting point, in which the slab-shape subgrains of the shear-band structures would emerge at the surface.

The fact that the outermost layers of abraded surfaces are composed of small subgrains has also been deduced from investigations by electron diffraction at grazing incidence.[20] This investigational technique also provides the additional information that the subgrains have a preferred orientation of a fiber texture type; that is, a particular crystallographic plane in each subgrain is aligned approximately parallel to the surface, but the orientation of such planes about their normal varies randomly over a range. The texture determined is only an average one for the surface, but the changes in the average texture with depth can be followed by removing successive layers by etching.

A characteristic pattern of textures, illustrated diagrammatically in the insert in Fig. 4.15, is found in abraded surfaces of all metals. These textures are similar to those developed by cold rolling in bulk, the variation in texture with depth corresponding to the variations that occur with decreasing amounts of cold work. However, the fiber axes are, with the exception of the outermost layers, inclined against the direction of motion of the abrasive points, an inclination which must relate to the direction of the resultant applied stress in abrasive cutting. A further important feature is that the textures developed are independent of the orientation of the parent crystal. Thus all traces of the underlying crystal structure, including the grain boundaries, are obliterated in the fragmented layer (this can be seen directly, for example, in Fig. 4.23a).

So far we have considered only the average structure of the surface layer. Accepting the general mechanism of abrasion outlined in Chapter 3, it should be possible to obtain some indication of the variations of the deformation structures to be expected beneath individual scratches by investigating surfaces machined orthogonally by tools with a range of rake angles. These studies indicate that the depths of both the fragmented layer and the deformed layer increase as the depth of cut is increased and as the rake angle of the cutting tool is made more negative.[15,16] However, the maximum strain at the surface is virtually independent of both the depth of cut and the rake angle;[16] it follows that the strain gradient is steeper the shallower the fragmented layer. The depth of the fragmented layer consequently can be expected to vary across an abraded surface in a manner that is related to, but not completely dependent on, the depth of the scratch with which it is associated. The maximum strain in the layer can, however, be expected to be approximately the same everywhere at the surface.

In a sense, the fragmented layer is merely a high-strain portion of the deformed layer. It is convenient, nevertheless, to distinguish it from the deformed layer, because the strains in the fragmented layer are so large and because there is a steep strain gradient at its base. The structure of the fragmented layer consequently is significantly different from that of the remainder of the deformed layer in all metals and alloys. "Fragmented" also is an adjective that needs to be interpreted with some care. The parent crystal has been fragmented only in the sense that it has been divided into a number of much smaller units. Each unit is still fully crystalline, and the boundaries between the units are of a normal type.

Translation Parallel to the Surface

Translation parallel to the surface, as well as compression normal to the surface, can be expected during chip formation by any machining process. This can be seen in its simplest form by observing the distortion of grids marked on the side face of the workpiece that occurs during orthogonal machining. The distortion of a vertically oriented set of such grids is illustrated in Fig. 4.16, which shows that the markers are drawn around the nose of the tool as it advances.[21] This can be expected also to occur during abrasive machining. However, the important factor here is the extent to which this occurs, because extensive plastic flow parallel to the surface might seriously obscure many types of structural features.

The order of the effects to be expected in a metallographic abrasion process is illustrated in Fig. 4.17(a), which shows a surface of a bismuth-silver eutectic alloy that has been abraded on a 600-grade silicon carbide

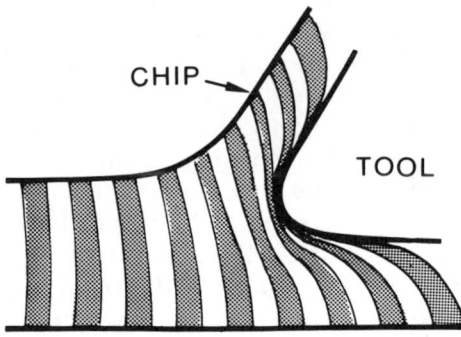

FIG. 4.16. Distortion, during orthogonal chip formation, of a vertical set of lines ruled on the side face of the workpiece. *(Oxley, Ref 21)*

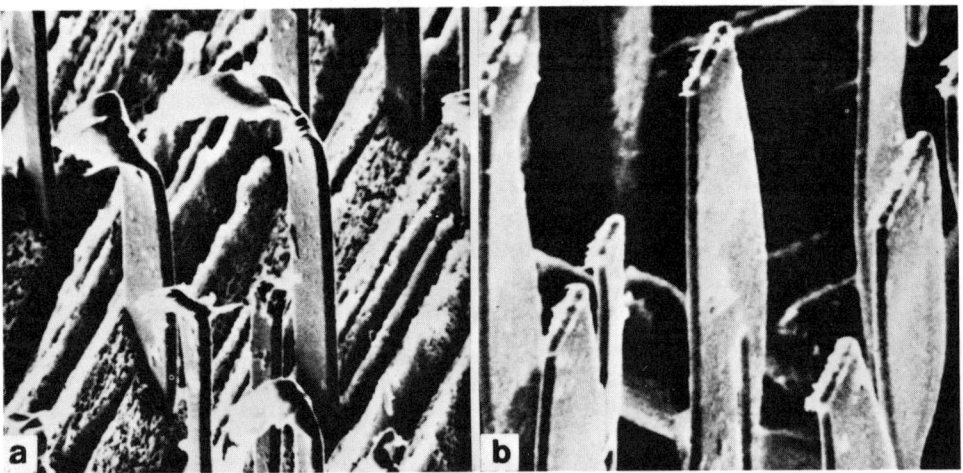

FIG. 4.17. Surfaces of a eutectic bismuth-silver alloy (a) abraded on 600-grade silicon carbide paper and (b) polished on 3-μm and 0.05-μm aluminum oxide abrasives. *(Kerr, Ref 22)*

The surfaces have been etched to dissolve the bismuth phase, which is the continuous phase of the eutectic, leaving the ribbons of the silver phase standing in relief. The extent to which these ribbons were flowed plastically parallel to the surface can be seen. Scanning electron micrographs. Magnification, 2200×.

paper.[22] The continuous phase of the eutectic (Bi) has been removed from the surface layers by preferential etching after abrasion, leaving ribbons of the silver phase of the eutectic standing in relief. These ribbons have certainly been bent in the direction of abrasion, but significantly so only to shallow depths. Another example is given in Fig. 4.18(a), which is a longitudinal taper section of a eutectoid steel that has been abraded on

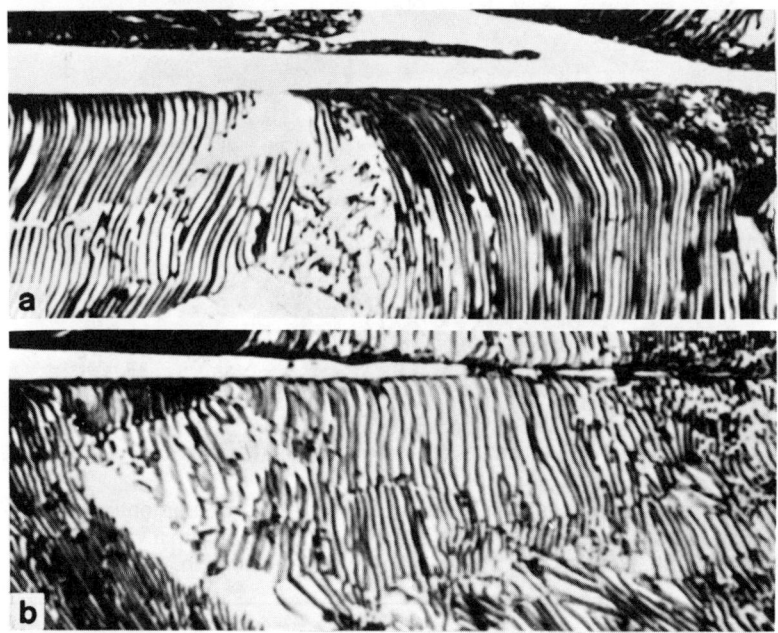

FIG. 4.18. Longitudinal taper sections of the surface of a eutectoid steel that has been abraded on a 100-grade aluminum oxide belt.

(a) A scratch where marked bending of the adjoining cementite plates of the pearlite has occurred. **(b)** A scratch where little or no bending of the cementite plates has occurred. Both etched in picral. Taper ratio, 10. Magnification, 2000×.

a 100-grade aluminum oxide belt. The plates of cementite adjacent to this scratch have been bent significantly in the direction of abrasion. However, very little translation has occurred adjacent to other scratches in the same surface (Fig. 4.18b). Presumably, the extent of the translation is determined by the rake angle of the abrasive point that made the groove. Another example of the magnitude of this effect is the distortion of the large cementite particle at a ground surface shown in Fig. 4.19; the distortion of this particle is spectacular, but extends only to a very shallow depth.

Thus the effects of plastic flow parallel to the surface may be of considerable importance when the characteristics of the as-abraded surface are under consideration, but are not of special consequence in metallographic practice. This is because the effects virtually are confined to the fragmented layer, in which other structural changes of at least equal consequence have occurred. The importance of these effects in metallographic practice seems often to be exaggerated.

The illustrations in Fig. 4.17 to 4.19 are, incidentally, examples of the plastic deformation of nominally brittle materials that can occur in the

FIG. 4.19. Section of the ground surface of a hardened hypereutectoid steel.

Considerable plastic flow has occurred in a large cementite particle (arrow) that was exposed at the surface, but only to a very shallow depth. Replica electron micrograph. Magnification, 12,800×.

fragmented layer. The strain system in the immediate vicinity of a contacting abrasive point can be expected to contain a large hydrostatic component which, as the classic work of Bridgeman[23] showed, permits deformation of nominally brittle materials.

DEPTH OF THE DEFORMED LAYER

There are three levels of deformation in an abraded surface that can be of interest in metallography, namely:

1. *Depth of the fragmented layer* (D_f), which can be taken to be approximately equal to the depth of the surface scratches
2. *Depth of deformation* (D_d), the maximum depth beneath the root of the surface scratches to the elastic-plastic boundary
3. *Depth of significant deformation* (D_s), the maximum depth beneath the root of the surface scratches of the deformation that will noticeably affect the observations to be made on the finished surface.[24]

The value of D_s is the most important one from the point of view of the design of a metallographic preparation sequence, but note that it will vary with the sensitivity of the examinational technique to be used to the effects of deformation. For example, D_s would be the same as D_d in 70:30 brass if the final section were to be etched by the high-sensitivity sodium thiosulfate method, but it would be defined by the 5% isostrain level if the section were to be etched in a ferric chloride reagent (or in any other common etchant for brass). For metals in which only the severest strains can be detected, D_s might have as small a value as that of D_f, but it would never be smaller. These matters will be discussed in more detail in Chapter 7.

The surface layer affected by a machining process is often referred to as the *damaged layer*. The term is a generic one which really should be applied only when the effects of the deformation are detrimental under the particular circumstances being considered. It is only in this sense that the term will be used here. In the context of metallographic surface preparation, the layer containing significant deformation, as defined above, can be taken to be the damaged layer.

Our purpose in the following sections will be to compare the depths of the plastically deformed layers produced by various machining and abrasive machining processes that are likely to be used in the preliminary stages of a metallographic preparation sequence; procedures that produce the minimum depth of deformation are, other things being equal, always to be preferred. For this purpose we shall use values of D_d and D_s for annealed polycrystalline 70:30 brass (hardness, 70 HV), the value of D_s being that for the layer in which strain markings are developed by a ferric chloride etchant. Results for abrasive-machining processes are listed in Table 4.1, and results for several conventional machining processes are presented in Table 4.2.

Noteworthy features for abrasive machining are as follows:

1. The depth of significant deformation (D_s) in this material can be anywhere between two and fifteen times the depth of the surface scratches.
2. The depth of the deformed layer (D_d) is commonly between 25 and 50 times the depth of the surface scratches.
3. There is no certain correlation between the depths of the deformed layers and the parameter on which the quality of the surface normally would be judged — namely, the depth of the surface scratches. A surface may have a superior surface finish and yet have a comparatively deep deformed layer (compare the data for diamond and aluminum oxide cutoff wheels in Table 4.1). The quality of surface finish depends on both the depth of indentation by individual abrasive points and the extent to which these points are coplanar. The depth of damage, on the other hand, is determined principally by the geometric configurations of the contacting abrasive points.
4. The effects of wear of the abrasive papers are illustrated in Fig. 4.20, which is representative of the behavior of all types and grades of abrasive paper.[17] The depths of both the surface scratches and the deformed layers decrease markedly early in the life of an abrasive paper, but the paper soon stabilizes, at least for materials which show group 1 and group 2 abrasion behavior (see p. 56). The values listed in Table 4.1 are for abrasive papers that have reached this stabilized condition.
5. The pressure applied to the specimen does not affect the *maximum* depth of deformation (Fig. 4.20), although this maximum depth is

TABLE 4.1. Depth of the Plastically Deformed Layer Produced on Annealed Polycrystalline 70:30 Brass by Abrasive-Machining Processes

Process	Abrasive Material	Abrasive Grade	Conditions	Cutting Fluid	Depth, μm Scratches	Significant deformation (D_s)(a)	Deformed layer (D_d)(a)
Cutoff wheels	Aluminum oxide	A 60 DE wheel	1500 m/min(b)	Proprietary oil-water emulsion	4	16	700
			10 m/min(b)	Proprietary oil-water emulsion	4	14	—
	Diamond		10 m/min(b)	Proprietary fluid	1	14	—
Surface grinding	Aluminum oxide	28A 46 KVBE wheel	0.001-in. feed, machine surface grinding	Proprietary oil-water emulsion	6	50	350
			0.0001-in. feed, machine surface grinding	Proprietary oil-water emulsion	4	30	150
	Aluminum oxide	38A 60 MVBE wheel	Hand grinding	None	15	40	170
Belt surfacing	Aluminum oxide	100 mesh	Specimen hand held	None	15	35	250
	Silicon carbide	80 grade	Specimen hand held	Water	10	45	240
		120 grade			5.5	25	190
		240 grade			3.5	15	95
		400 grade			1.2	5	60
Abrasion on abrasive papers	Emery	1/0 grade	Hand abrasion	Kerosine	1.8	7.5	45
		2/0 grade	Hand abrasion	Kerosine	1.0	7.0	38
		3/0 grade	Hand abrasion	Kerosine	0.4	4.0	30
		4/0 grade	Hand abrasion	Kerosine	0.3-1.0(c)	3-10(c)	20-50(c)
	Silicon carbide	220 grade	Hand abrasion	Water	2.0	7.5	77
		400 grade	Hand abrasion	Water	1.5	6.5	43
		600 grade	Hand abrasion	Water	0.8	5.0	22
Abrasion on fixed-abrasive laps	Aluminum oxide	10-20-μm grade	Hand abrasion	None	0.3	3	16

(a) Maximum depth beneath the roots of the surface scratches or machining marks. (b) Linear cutting speed. (c) The higher values result from clogging of the paper.

TABLE 4.2. Depth of the Plastically Deformed Layer Produced on Annealed Polycrystalline 70:30 Brass by Conventional Machining Processes

Process	Conditions	Cutting fluid	Machining marks	Depth, μm — Significant deformation (D_s)(a)	Deformed layer (D_t)(a)
Hand hacksaw	18 t.p.i., roll-set blade	None	100	55	750
Lathe turning	0.001-in. feed	Proprietary oil-water emulsion	1	15	150
Filing	Bastard cut	None	70	55	450
	Second cut	None	30	50	370

(a) Maximum depth beneath the roots of the surface machining marks.

attained at fewer positions with lighter pressures. According to the model of abrasion discussed in Chapter 3, the maximum depths of both of these parameters will be determined by the maximum load carried by, and the shape of, individual contacting abrasive points. The number of contacting points will increase with increasing applied pressure, and the distribution of load among the contacting points will change, but not for the points that have the worst combinations of load and rake angle.

6. The linear speed of the abrasive has no effect on the depth of deformation.
7. A paper which has clogged with debris (see p. 62) causes very severe damage to the surface, increasing particularly the depth of deformation (see the values for 4/0-grade emery paper in Table 4.1; see also Fig. 4.21). Clogging of waterproof aluminum oxide and silicon carbide papers can be easily prevented by flushing away the abrasion debris with a stream of water (see p. 62), but it is virtually impossible to do this with emery papers because they have to be used merely flooded with a liquid such as kerosine. This is a serious disadvantage of emery papers — serious enough to preclude their use in metallographic practice. Fine abrasive laps are susceptible to the same type of problem; consequently, a single track should not be used for extended periods, and certainly not after clogging has become obvious.
8. The depths listed in Table 4.1 are for one specific alloy, and a comparatively soft one at that. All that can be done to extrapolate these values to other materials is to assume that the depths of the deformed layers will be inversely proportional to the hardness of the specimen, although this is probably an excessively simple assumption.

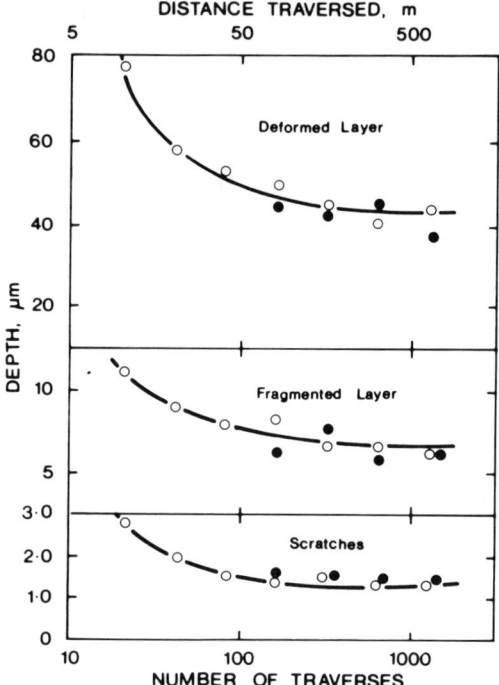

FIG. 4.20. Variation of the depth of surface scratches and surface deformed layers produced on annealed polycrystalline brass by abrasion on 400-grade silicon carbide paper flushed with water.

Closed symbols: light abrasion pressure. Open symbols: heavy abrasion pressure.

Table 4.2 gives some corresponding information on a few typical conventional machining processes that might be used to produce a preliminary flat surface. The depths of the surface irregularities and of the deformed layers are comparatively large, as is to be expected with such coarse machining processes. However, the ratios of the depth of the deformed layer to the depth of the machining marks can be comparatively small, because machining tools have geometrical configurations more appropriate than those of abrasive points.

METALLOGRAPHIC CHARACTERISTICS OF THE FRAGMENTED LAYER

The fragmented layer is one portion of the surface deformation that will always be detectable by optical microscopy. In single-phase metals or alloys, it will usually appear after etching as a sharply delineated but

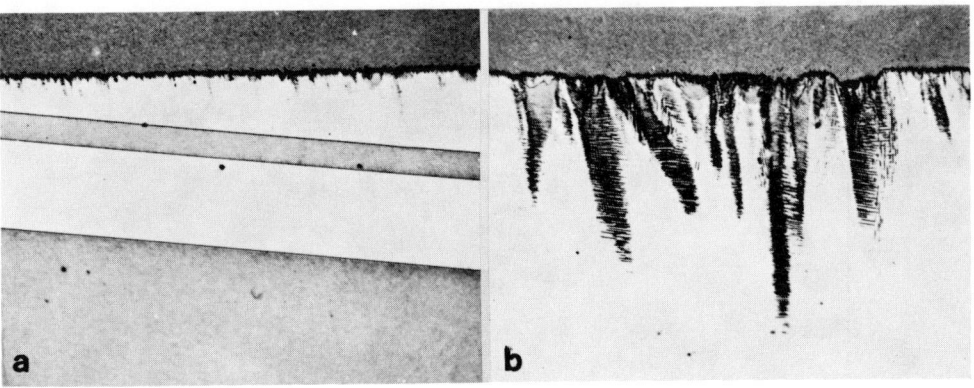

FIG. 4.21. Adjoining regions of a taper section of a brass surface abraded on 4/0-grade emery paper flooded with kerosine.

(a) Region where the deformed layer is characteristic of unclogged paper. (b) Adjoining region with a very deep deformed layer resulting from contact with a clogged area of the abrasive paper. Both etched in ferric chloride reagent. Taper ratio, 7.1. Magnification, 250×.

rather diffusely resolved layer; examples for 70:30 brass, for copper, and for an iron and a steel are presented in Fig. 4.11, 4.22 and 4.23, respectively.

The fragmented layer shown in Fig. 4.23(a) is characteristic of that seen in the ferritic phases of irons and steels. There is a suggestion of a resolvable structure in this case, and this mottling sometimes is interpreted as indicating the presence of a layer which has been heated to such a high temperature during abrasion that it has transformed to austenite (a *transformed* layer). Such an interpretation often is claimed to be supported by the observations of Agarwala and Wilman,[25] who investigated abraded surfaces of a single crystal of iron by reflection electron diffraction techniques. However, the sequence of structures that Agarwala and Wilman reported with increasing depth is so improbable metallurgically that considerable reservations must be held about their interpretations. We shall see later (p. 126) that austenitization of a surface layer of this thickness almost certainly cannot occur under the conditions of metallographic abrasion. The mottled appearance is more simply and reliably attributable to the effects of large plastic strains.

The fragmented layer (in a quench-hardened and tempered steel) shown in Fig. 4.23(b) is typical of those that frequently are referred to as the *white-etching layer* and that, again, are often assumed to be austenitized layers that have transformed to martensite during subsequent cooling. This conclusion rarely is based on any evidence other than the fact that both the fragmented layer and untempered martensite etch lightly and are comparatively hard — scarcely a positive identification. More positive identification has been sought by using reflection electron

FIG. 4.22. Taper section of surface of annealed copper abraded on 220-grade silicon carbide paper.

A diffuse-etching fragmented layer is present at the surface. Rays of microband strain markings are also present. The feature indicated by the arrow is probably an embedded abrasive particle. Etched in sodium thiosulfate reagent. Taper ratio, approx. 10. Magnification, 1500×.

and x-ray diffraction techniques, but here again, as will be discussed later, it is extremely difficult (if not impossible) to distinguish between the possible alternative structures.

Turley[26] has investigated the structure of the fragmented layer by the more certain techniques of transmission electron microscopy and transmission electron diffraction of thin sections cut through the layer. He established that the layer is composed of small subgrains of body-centered-cubic crystal structure; this is the structure to be expected in a layer of ferrite that has undergone severe plastic deformation. The small carbide particles initially present in the parent structure of the steel were absent in the white-etching layer, but Turley showed that this too could reasonably be attributed to the severe plastic deformation that the layer had undergone; carbon can be taken into solution at the sites of the many crystal defects that are present in the layer. Either the fine grain structure or the absence of small carbides, or a combination of the two, could account for the light-etching characteristics of the layer. A layer with this structure could also be expected to have high hardness, both because of its small grain size and because the dislocations present would be locked by carbon. For these reasons also, the layer might be expected to have some other unusual properties, but it is not possible to indulge in anything other than speculation on this point. The important points in the present context are that the layer clearly is not a reaustenitized layer, is not composed of untempered martensite, and is not an indicator of major surface heating during abrasion (see also p. 129).

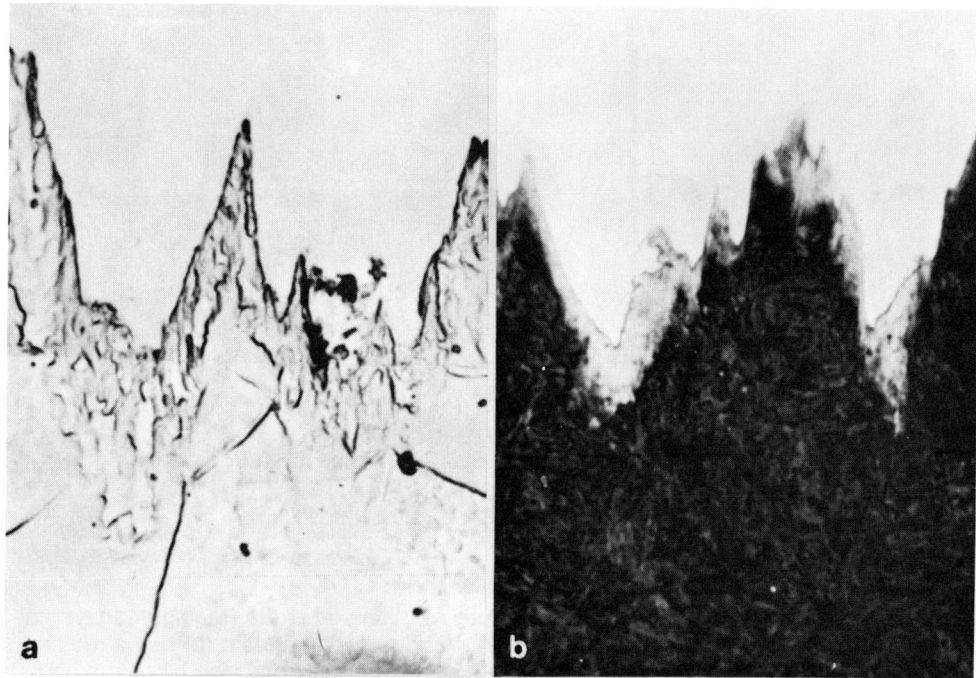

FIG. 4.23. Taper sections of the abraded surfaces of an iron and of a quench-hardened and tempered steel.

(a) Iron (hardness, 75 HV) abraded on 220-grade silicon carbide paper. The fragmented layer has a mottled appearance. Note that the grain boundaries of the parent structure do not extend into the fragmented layer. Etched in nital. Taper ratio, approx. 10. Magnification, 1500×. (b) Steel (hardness, 650 HV) abraded on 220-grade silicon carbide paper. The comparatively light-etching layer is the fragmented layer, which is comparatively shallow in this hard material. Etched in sodium bisulfite reagent. Taper ratio, 13.1. Magnification, 2000×.

The morphologies of the phases in multiphase alloys can be expected to be altered drastically within the fragmented layer, and the following examples are indicative of the possible extent of these alterations. The first example (Fig. 4.24) concerns lamellar pearlite, which can be taken to be generally representative of lamellar structures. A lamellar arrangement of phases often is not recognizable within the fragmented layer. The appearance may vary from one suggestive of distorted pearlite (Fig. 4.24a), to a general mottling (Fig. 4.24b) suggestive of cementite plates that have been broken up into small fragments, to a light-etching distorted appearance (Fig. 4.24b) suggestive of a lamellar structure that has been deformed in such a way that the lamellar spacing has become much smaller than the limiting resolution of optical microscopes (see p. 95).

FIG. 4.24. Taper sections of the surface of a pearlitic steel abraded on 80-grade silicon carbide paper.

These sections show the range of structures observed in the fragmented layer in pearlite. The pearlite in **(a)** appears merely to be distorted; that in **(b)** has a mottled appearance; and that in **(c)** appears light-etching, probably because it has not been resolved. Taper ratio, approx. 10. Magnification, 2000×.

The result is likely to be different, however, when the second brittle phase is large compared with the volume of material deformed during the passage of an abrasive grit. The matrix phase cannot then provide the constraint necessary to permit plastic deformation of the brittle phase. Nonmetallic inclusions, such as those found in steels, are an example. Some types of inclusions, such as silicates, are brittle, and the portions of these inclusions that become located in the fragmented layer fracture into small pieces (Fig. 4.25a). The disintegration of the inclusions can be severe enough to result in parts of the inclusions falling out of the surface when they intersect the surface (left-hand inclusion in Fig. 4.25a). On the other hand, some inclusions, such as manganese sulfides, are sufficiently ductile merely to be bent in the direction of abrasion (Fig. 4.25b). Graphite flakes in gray cast iron, at the other extreme again, are particularly susceptible to damage of the type under consideration. The graphite may be lost completely from its enclosing cavity in the fragmented layer and the cavity may then close up completely (Fig. 4.26a).[27] In this event, the graphite flakes appear to be much thinner than their true width when observed on the abraded surface (Fig. 4.26b; cf. Fig. 4.26f).

A similar situation arises when a cavity or its equivalent already exists. One example is a shrinkage cavity in a casting. Such cavities may collapse

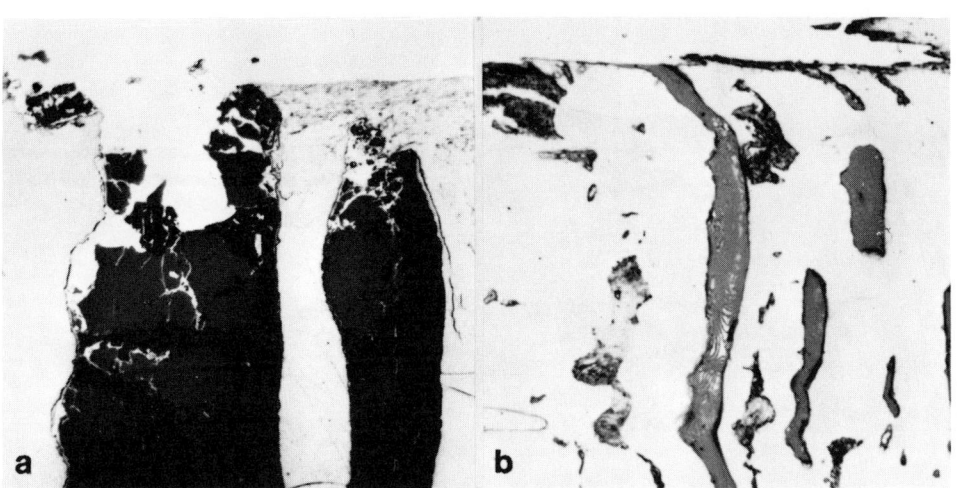

FIG. 4.25. Longitudinal taper sections showing abrasion damage of nonmetallic inclusions in iron and steel. **(a)** Fragmentation of brittle silicate inclusions in wrought iron. Taper ratio, 10.5. Magnification, 500×. **(b)** Distortion of a manganese sulfide inclusion in steel. Taper ratio, 10. Magnification, 1000×.

when they enter the fragmented zone, rather as a balloon deflates (Fig. 4.27),[28] in which event an open cavity is replaced by a filamentary feature entirely different in appearance. Cavities filled with a very soft material (e.g., lead particles in leaded copper alloys) behave in the same way (see Fig. 8.13).[29]

The effects described in the immediately preceding paragraphs can all reasonably be explained in terms of the large compressive strains developed in the fragmented layer. It can on these grounds be expected that the extent to which they develop will be determined by the ratio of the volume of the microstructural feature concerned to the volume of material deformed during the passage of an abrasive point.

An example of the scaling effect described above is given in Fig. 4.26. All of the graphite flakes of this particular size were removed from their cavities, and the cavities collapsed in the fragmented layer produced by abrasion on 220-grade paper (Fig. 4.26a). This also occurred with some of the flakes during abrasion on 600-grade paper, but other flakes were merely removed from their cavities (Fig. 4.26c). The flakes consequently appeared as fine lines on surfaces abraded on 220-grade paper (Fig. 4.26b) but as lines of varying width after abrasion on 600-grade paper (Fig. 4.26d). Thinner flakes would all have collapsed during abrasion on 600-grade paper. On the other hand, none of the flakes in the iron under consideration were damaged during abrasion on a fine lap (Fig. 4.26e), and the flakes then appeared in their correct width on the abraded surface (Fig. 4.26f). In fact, the flakes stand slightly above the surface after abrasion on this lap (Fig. 4.26c). This can be attributed to greater elastic re-

120 / Metallographic Polishing by Mechanical Methods

FIG. 4.26. Taper sections of abraded surfaces of a gray cast iron (left column), and the appearance of the same surfaces before sectioning (right column).

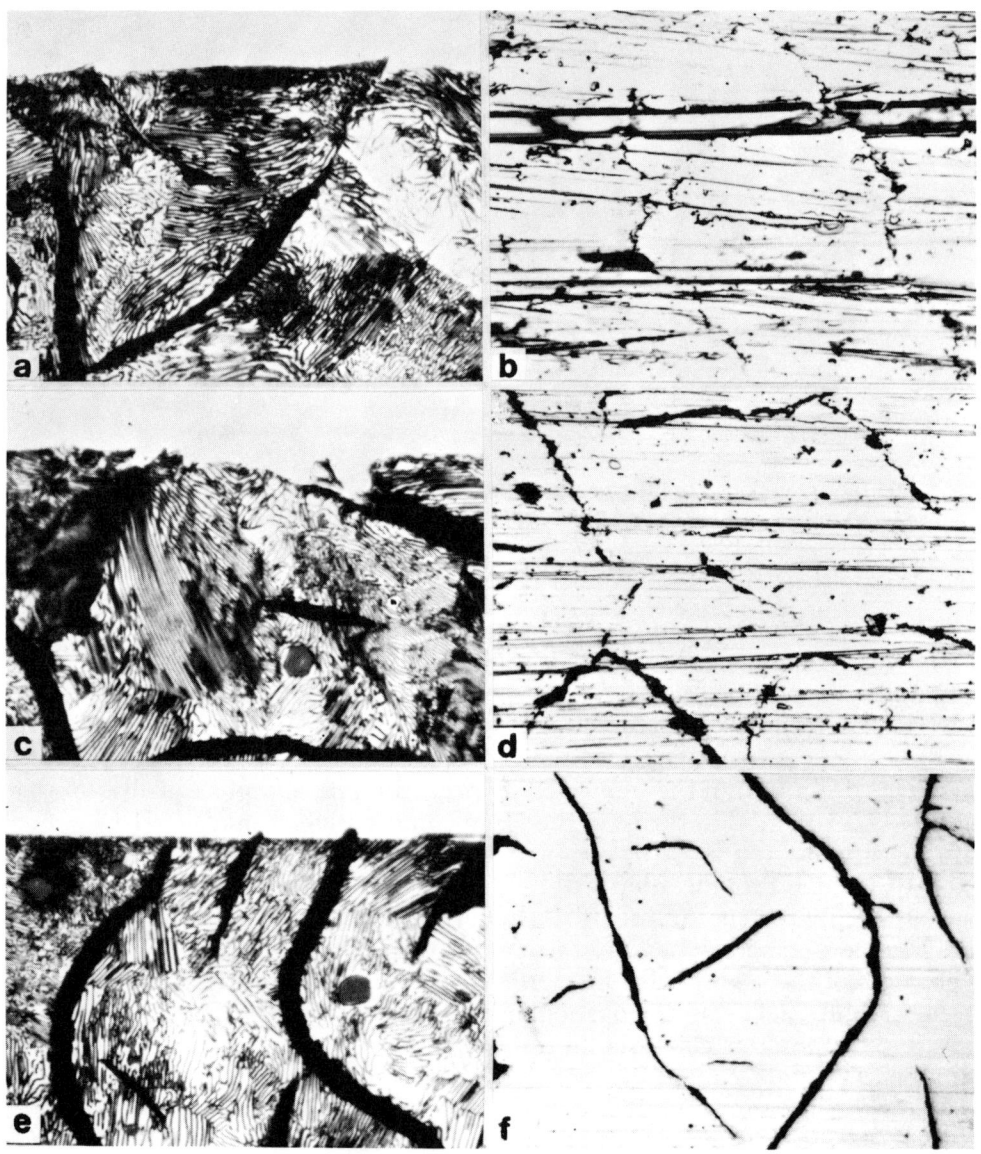

(a) and (b) Abraded on 220-grade silicon carbide paper. **(c) and (d)** Abraded on 600-grade silicon carbide paper. **(e) and (f)** Abraded on an aluminum oxide – wax lap. Sections etched in picral. Taper ratio, approximately 10. Magnification, 1500×. Surfaces unetched. Magnification, 500×.

covery in the graphite than in the metallic matrix. Graphite flakes are single crystals whose basal planes are oriented parallel to the long axis of the flake, and graphite has an extremely high elastic modulus in this crystallographic direction.

The reduction in apparent width of features such as those illustrated in Fig. 4.26 and 4.27 commonly is ascribed to smearing of a burr across the mouth of the feature. Although, as discussed on pp. 106 and 198, burrs do form during abrasion and during polishing, their magnitude both laterally and in depth is much too small to account for the phenomenon that actually occurs. The phenomenon of collapse by plastic compression is in many respects a more serious one, particularly as the structure is distorted to a comparatively great depth.

MODIFICATIONS OF THE STRUCTURE OF THE DEFORMED LAYER DUE TO STRAIN-INDUCED TRANSFORMATIONS

The modifications of structure induced by general deformation of a metastable alloy were discussed in connection with Fig. 4.4 (p. 93). The resultant effects on the microstructure of the surface deformed layers pro-

FIG. 4.27. Longitudinal taper section showing abrasion damage of shrinkage cavities in a cast bronze.

A normal cavity is indicated by arrow A. The cavity indicated by arrow B, which is now located in the abrasion-fragmented layer, has collapsed. Etched in ferric chloride reagent. Taper ratio, 8.4. Magnification, 500×.

duced during abrasive machining are illustrated in Fig. 4.28,[30] the alloy being the same as that shown in Fig. 4.4.

The outer fragmented layer is again clearly discernible (Fig. 4.28a), and examination by electron diffraction methods shows that this layer has a high content of transformation products.[31] Below this is a layer in which transformation has occurred in characteristic laths (cf. Fig. 4.28a and b),

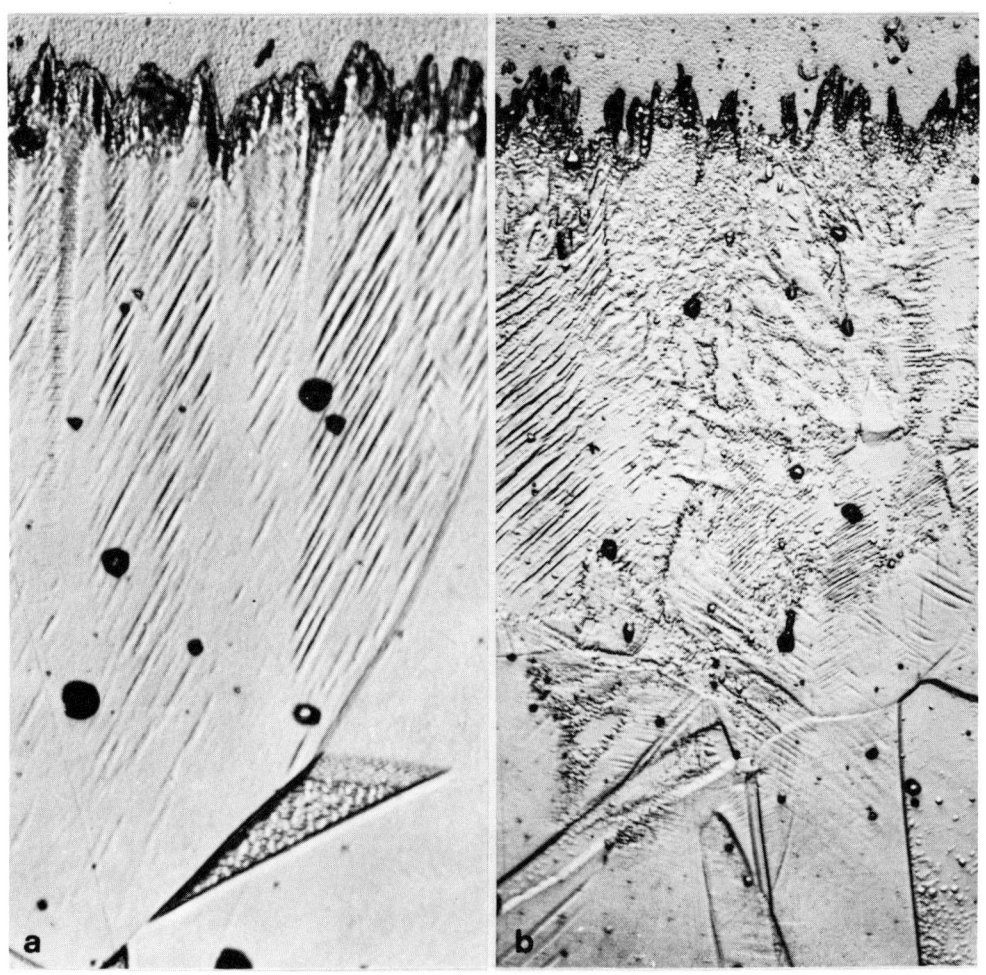

FIG. 4.28. Taper sections of abrasive-machined surfaces of an austenitic stainless steel.

(a) Abraded on 220-grade silicon carbide paper. Etched in oxalic acid. Taper ratio, 10.1. Magnification, 1000×. (b) Machine surface ground with a large downfeed per pass. Etched in oxalic acid. Taper ratio, 10.7. Magnification, 500×. The manifestations of deformation shown in these sections are typical of those resulting from strain-induced transformations, which have been partly suppressed in (b) due to surface heating during grinding.

which tend to be grouped in bands associated with individual surface scratches. The appearances and distributions of these strain markings are similar to those of the twin strain markings in abraded brass surfaces (cf. Fig. 4.28a and 4.14), but the strain level at the bases of the markings is only about 1% compression instead of the 5% compression for brass. The base of the lath-containing layer is the isostrain boundary at which the abrasion deformation can be regarded as being significant in the metallography of these alloys. The depths of significant damage produced by various surface-preparation methods are listed in Table 4.3.

TABLE 4.3. Depth of the Deformed Layer Produced on a Type 304 Austenitic Stainless Steel (18/8 Type) by Machining, Grinding and Abrasion Processes *(Samuels and Wallwork, Ref 30)*

Process	Abrasive Material	Abrasive Grade	Conditions	Cutting fluid	Depth of significant deformation(a), μm
Surface grinding	Aluminum oxide	38A 46 KVBE wheel	0.001-in. feed; machine surface grinding	Proprietary oil-water emulsion	35
	Aluminum oxide	38A 60 MVBE	Hand grinding	None	43
Lathe turning	...	...	0.001-in. feed	Proprietary oil-water emulsion	45
Belt surfacing	Aluminum oxide	100 mesh	Specimen hand held	None	12
Abrasion on abrasive papers	Emery	1/0 grade	Hand abrasion	Kerosine	5
		2/0 grade	Hand abrasion	Kerosine	4
		3/0 grade	Hand abrasion	Kerosine	4
	Silicon carbide	220 grade	Hand abrasion	Water	6
		400 grade	Hand abrasion	Water	2.5
		600 grade	Hand abrasion	Water	2.2

(a) Maximum depth beneath roots of surface scratches or machining marks.

MODIFICATIONS OF THE STRUCTURE OF THE DEFORMED LAYER DUE TO MASSIVE TWINNING

Massive twins can be expected to be present in the deformed layer produced during abrasion of noncubic metals, for which the formation of twins is a normal feature of the deformation process (see p. 94 and Fig. 4.5). Examples are given in Fig. 4.29,[32] the material being the same as that illustrated in Fig. 4.5. The boundary of the twin-containing layer bears little relationship to the surface topography for coarsely finished surfaces (Fig. 4.29a), but the twins increasingly tend to be grouped in rays extending beneath groups of scratches or individual scratches with increasing fineness of finish (Fig. 4.29b and c). Deformation bands may

124 / **Metallographic Polishing by Mechanical Methods**

FIG. 4.29. Taper sections of abrasive-machined surfaces of polycrystalline zinc, viewed in polarized light.

(a) Machine ground. Taper ratio, 16.2. Magnification, 100×. (b) Abraded on 220-grade silicon carbide paper. Taper ratio, 13.6. Magnification, 250×. (c) Abraded on 600-grade silicon carbide paper. Taper ratio, 7.7. Magnification, 250×.

also be present, particularly in more finely finished surfaces, in which they extend beyond the twin-containing layer (Fig. 4.29c).

The boundary of the twin-containing layer sets one value on the depth of significant damage in these materials. The twin-containing layer is comparatively deep (Table 4.4) because the threshold strain for twinning

TABLE 4.4 Depth of the Deformed Layer Produced on Polycrystalline Zinc by Machining, Grinding and Abrasion Processes

Process	Abrasive Material	Abrasive Grade	Conditions	Cutting fluid	Depth, μm Recrystallized layer	Depth, μm Twinned layer
Surface grinding	Aluminum oxide	38A 46 KVBE wheel	0.001-in. feed; machine surface grinding	Proprietary oil-water emulsion	25	125
	Aluminum oxide	38A 60 MVBE wheel	Hand grinding	None	45	130
Lathe turning	...	...	0.001-in. feed	Proprietary oil-water emulsion	30	135
Belt surfacing	Aluminum oxide	100 mesh	Specimen hand held	None	35	85
Filing			Second cut	None	30	130
			Fine cut	None	20	70
	Emery	1/0 grade	Hand abrasion	Kerosine	6	40
		2/0 grade	Hand abrasion	Kerosine	5	25
		3/0 grade	Hand abrasion	Kerosine	4	24
	Silicon carbide	220 grade	Hand abrasion	Water	7.5	45
		400 grade	Hand abrasion	Water	5	25
		600 grade	Hand abrasion	Water	3	15
Abrasion on fixed-abrasive laps	Aluminum oxide	10-20 μm	Hand abrasion	None	5	15

(a) Maximum depth beneath the roots of surface scratches or machining marks.

is comparatively small (see p. 94). It will be deeper at comparable hardnesses in those metals which have lower threshold strains for twinning.

In some circumstances, the kink bands may define D_s. In this event, the values of D_s will be larger than those listed in Table 4.4.

MODIFICATIONS OF THE STRUCTURE OF THE DEFORMED LAYER DUE TO RECRYSTALLIZATION

We have seen earlier (p. 94) that spontaneous recrystallization can occur at room temperature in metals of comparatively low melting point (e.g., lead, tin, and zinc) after a certain level of deformation. A consequence is recrystallization of the fragmented layer and of some portions of the deformed layers in abraded surfaces of these metals (Fig. 4.29). The recrystallized layer tends to extend preferentially along the twin-containing rays, particularly in more finely finished surfaces (Fig. 4.29c), but its depth always is less than that of the twin-containing layer (Table 4.4).

The grain size of the recrystallized material is comparatively small (Fig. 4.29), and this grain size is related to the abrasive process (it will be larger for processes that tend to produce some surface heating) and not to the grain size of the parent metal. Consequently, it is different from the grain size of the parent metal, except by coincidence.

Recrystallization of the deformed layer occurs in any metal when an abraded surface is subsequently heated to an elevated temperature appropriate to the parent metal, in the same way that the parent metal would recrystallize during annealing after having been cold worked.[7]

MODIFICATIONS OF THE SURFACE STRUCTURE RESULTING FROM HEATING OF THE SURFACE

As already mentioned, it is commonly supposed that high transient temperatures are attained at local areas of contact whenever the surface of one solid is rubbed against that of another. This supposition probably is based largely on extrapolations of the work of Bowden and his colleagues, the validity of which will be considered later (see p. 145). At best, however, the concepts that Bowden developed were concerned only with small volumes of material and short periods of time. What we are going to consider now is the rather different question of whether a layer of finite thickness (thick enough to be detected by optical microscopy) can be heated to a sufficiently high temperature for a sufficiently long time to change its microstructure recognizably. These changes could take the form either of changes to the basic structure of the deformed layer or of introduction of new structures due to phase transformations in the parent metal. The most important change of the second type depends on whether or not a sufficiently high temperature ($>725°C$) can be reached to cause austenitization of the surface layers of steels: in this event, the austenitized layer could be expected to be quenched rapidly enough by the unheated bulk of the workpiece to transform to martensite during subsequent cooling.

Blok and Jaeger have made a theoretical analysis of the rise in temperature that might occur under the boundary conditions of a single flat point rubbing against a semi-infinite solid (see Archard[33] for a simplified formulation of the theory), but it is uncertain whether these treatments can reasonably be applied to the more complicated conditions that exist during abrasion. Even if they can, not enough is known about the values of important parameters (e.g., the actual coefficient of friction, the area of contact of individual abrasive points, and the load on each point) to permit meaningful calculations of transient temperatures to be made. Rabinowicz[34] has suggested that an order-of-magnitude estimate of the transient temperature can be made from the following simple formula: $\theta_m \simeq v/2$, where θ_m is the transient temperature in °F and v is the sliding speed in ft/min. He states that experience in friction experiments indicates that θ_m is estimated realistically to within a factor of 2 or 3, which is no worse than the uncertainties associated with the more complex theoretically based calculations. Rabinowicz's formula emphasizes an important point that also emerges from the full theoretical analyses — namely, that sliding speed is a major variable. It is not reasonable, therefore, to extrapolate without taking sliding speed into consideration, and this is often done.

A number of attempts have been made to measure directly the temperatures attained in the surface layers of steels during grinding, the most successful being those in which a fine thermocouple is mounted in the workpiece at a point slightly beneath the surface to be ground. Grinding is continued in successive passes until the thermocouple junction is exposed, a temperature pulse being recorded during each pass.[35,36] The temperature gradient in the surface, as well as the maximum temperature attained, can then be deduced when the thermocouple junction is finally exposed. These experiments establish directly that temperatures in excess of 800°C can be attained under certain circumstances, this temperature certainly being high enough to cause some austenitization of the surface layer even in a short period of time. On the other hand, temperatures of this magnitude were measured only during comparatively severe grinding conditions, the maximum temperature recorded under some quite normal grinding conditions being as low as 100°C. Factors which favor low temperatures are small downfeed per pass, the use of a grinding fluid as a coolant, the selection of an appropriate type of wheel, and good wheel-dressing practice.[35]

On the other hand, thermocouple experiments do not necessarily measure the real maximum temperature. The volume heated might be too small for the temperature rise to be sensed fully. It could be, therefore, that high temperatures are attained in thin surface layers even under those grinding conditions for which the thermocouple experiments record comparatively modest temperature rises. The only technique so far devised that improves on this situation is to seek evidence of phase transformations which would indicate that a certain critical temperature had been exceeded. Most work of this nature has been carried out on carbon or medium-alloy steels in which, as previously mentioned, evidence of an austenite transformation is sought. We have already discussed the difficulty of positively establishing by optical microscopy the presence of a martensitic layer on an abraded or ground surface (p. 117). Reflection x-ray and electron diffraction techniques would seem to be capable of providing more positive evidence, but there are great difficulties here, too. The possible alternative structures are a body-centered-cubic structure (ferrite) which is heavily deformed or a body-centered-tetragonal structure (martensite) which has a c/a ratio not much larger than one and which also is heavily strained. It is virtually impossible to distinguish with certainty between the two under most circumstances.

The more reliable technique developed by Sedriks and Craig[37] depends on the special transformation characteristics of an 18%Ni-Co-Mo maraging steel, the thermal diffusivity of which is not very dissimilar from that of general structural steels. A metastable γ' phase can be detected unambiguously in this alloy by reflection x-ray diffraction methods. Its presence can be taken to indicate that a surface temperature of about 650°C has been exceeded in the surface layers, provided that the affected layer is at least 0.1 to 0.5 μm thick — which is within the limits of what can be regarded as being significant in the present context.

Sedriks and Craig established that the critical temperature was attained in specimens abraded *dry* on silicon carbide belts at high abrasion velocities, as in belt surfacing, but that abrasion had to be continued for a definite period of time before evidence of a heated layer was obtained. The time required increased significantly as abrasion speed decreased (from about 8 s at 1300 cm/s to 5 min at 700 cm/s) and also decreased slightly as the load applied to the specimen decreased. The attainment of high surface temperatures, as determined by this technique, was always associated with the production of a shower of sparks during abrasion. The most important observation, however, was that no evidence of the transformation was detected even under the most severe abrasion conditions if the abrasive belt was kept flooded with water. There seems, therefore, to be no possibility whatsoever that surface heating of this magnitude will occur during good metallographic abrasion practices, in which comparatively low speeds and copious supplies of coolant are used. Certainly, showers of sparks are never produced under these circumstances.

The structure of the fragmented layer in copper, as elucidated by transmission electron microscopy and summarized in Fig. 4.15, is also of relevance here. The surface concerned was abraded on a silicon carbide paper at low speed in the presence of a flowing stream of water. The presence of recrystallized grains in the fragmented layer probably indicates that some general heating of the surface did occur, but only to temperatures from 100 to 150°C.[19]

We consequently have now reached the position where we can state that there are some abrasive-machining processes that definitively cause sufficient surface heating to produce austenitized layers on the surfaces of steels (these processes include grinding with heavy passes and dry belt surfacing) and that there are others that definitively do not do so (these processes include metallographic abrasion at comparatively low loads and speeds in the presence of copious amounts of liquid). We need to be able to recognize by optical microscopy, if this is at all possible, which of the two situations applies. We have already seen that the observation of a white-etching layer at the surface is not sufficiently definitive for this purpose. More positive and reliable evidence can be obtained, at least in quenched-hardened steels that have been tempered only at low temperatures, by looking for the presence of a comparatively dark-etching tempered layer immediately beneath the white-etching layer. Such a layer must be present if the white-etching layer has been austenitized.

A clear example of a positive identification of this nature is given in Fig. 4.30(a). An extensive dark-etching (tempered) layer is contiguous with the white-etching surface layer. The latter consequently can safely be assumed to be a martensitic layer, a conclusion supported by the fact that the grinding conditions used in preparing this specimen were similar to those used in one of the experiments of Littman and Wulff[35] in which temperatures of the order of 800°C were measured. The martensitic nature of a layer with these characteristics has been confirmed by Turley[26] using transmission electron microscopy. He found, incidentally, that two layers with different structures are present in the white-etching layer. The outermost layer is composed of a martensite with an unusual structure which probably has resulted from plastic deformation taking place concurrently with both the heating and the cooling transformations. The deeper layer has a structure similar to that formed during normal quench hardening, although its substructure is much finer than for normal quench hardening.

The section illustrated in Fig. 4.30(b) is of a surface of the same steel illustrated in Fig. 4.30(a) but which has been ground under much gentler conditions, conditions under which only modest rises in temperature were detected in the Littman and Wulff experiments. Beneath a few of the scratches — and not necessarily the deepest scratches — are dark-etching layers contiguous with the white-etching layer: one such scratch is arrowed at the center of Fig. 4.30(b). The adjacent white-etching layer in this case may well have been austenitized. Beneath most of the surface scratches, however, there are no such dark-etching layers, and the white-etching layer almost certainly is a plastically deformed fragmented layer (see p. 115). It seems that the characteristics of the contacting abrasive point must have an influence on the temperature attained as well as on more obvious parameters such as traversing speed and depth of interaction. Surfaces treated *dry* on metallographic belt surfacers often contain similar mixtures of the two types of scratches, but surfaces abraded with copious supplies of liquid flowing over the belt, and hand-abraded surfaces, never do.

Note that the tempered layer, when present, is comparatively deep. Also, it must be possible for a tempered layer to form without an austenitized layer if the surface heating is less severe. Another possible con-

FIG. 4.30. Taper sections of abrasive-machined surfaces of a hardened but untempered carbon steel.

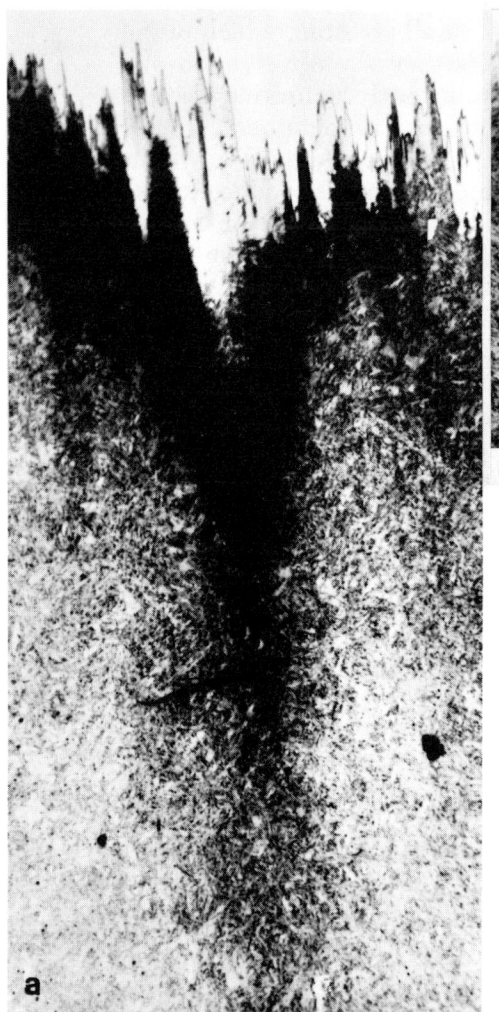
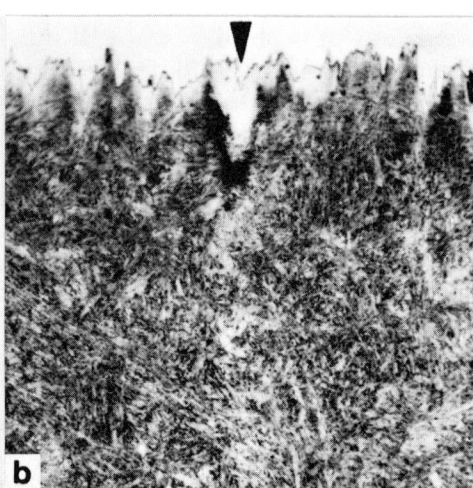

(a) Machine surface ground with a downfeed of 0.001 in. Taper ratio, 10.9. (b) Same as (a) but with a downfeed of 0.0001 in. Taper ratio, 11.2. Both etched in nital. Magnification, 2000×. A white-etching surface layer is continuous across both surfaces. A contiguous dark-etching (tempered) layer is present beneath all surface scratches in (a), but beneath only one scratch (arrow) in (b).

sequence of modest surface heating is a reduction in the volume fraction of retained austenite in quench-hardened steels.[38] Changes in the volume fraction of retained austenite usually are too small to be detected by optical microscopy, but the effect could be of importance when the surface is to be examined by more sensitive techniques.

Our attention so far has been confined to ferritic steels, which are the materials that have been most studied because of the practical importance of the effect in these materials. Similar effects are to be expected in any metal or alloy in which appropriate structural changes can occur. For example, the strain-induced transformation in austenitic steels that we discussed earlier does not proceed above a certain characteristic temperature (about 200°C for a type 304 austenitic steel). The outer portions of the damaged layer in ground surfaces of an austenitic steel consequently may under appropriate conditions contain deformed austenite as well as the normal laths of transformation product.[31] This has a considerable influence on the microstructure of the damaged layer (cf. Fig. 4.28a and b).

It might also be expected that surface heating during abrasive machining would cause recrystallization of the plastically deformed layer being formed at the same time. No evidence has yet been obtained on the circumstances under which this occurs. A related effect has been observed, however, in metals (such as zinc) in which recrystallization would have occurred anyway: the grain size of the recrystallized layer is larger after those treatments which are most likely to have caused surface heating,[32] and the recrystallized layer occupies a greater proportion of the damaged layer (Table 4.4).

EMBEDDED ABRASIVE

We have seen that the points of contacting abrasive particles may fracture during even the gentlest abrasion operation, releasing fragments of abrasive (Fig. 3.11a). These fragments tumble between the specimen and abrasive paper, producing a track of irregular indentations (Fig. 4.31a), and some may eventually jam and embed deeply into the specimen surface (Fig. 4.31b).[39]

The presence of embedded abrasive particles is not always directly obvious when an abraded surface is examined by optical microscopy. The presence of the embedded particles can be established positively by area scanning for the metallic (or semimetallic) element of the abrasive with an electron probe microanalyzer, which discerns the resultant chemical inhomogeneity in the surface (Fig. 4.32b). Williams[40] has also employed autoradiographic techniques using pre-irradiated abrasives to establish positively the presence of embedded particles in lapped surfaces of hardened steel, but this is a much more difficult technique. Embedded particles are also observed not too uncommonly in sections of abraded surfaces (Fig. 4.31c and d, and Fig. 4.22).

The particles of abrasive clearly are too deeply embedded to be removed by any simple washing or cleaning process (Fig. 4.31c and d). It is possible to remove them, however, by etching the surface rather deeply and then wiping it well or treating it with an ultrasonic cleaner. The sites of the abrasive fragments that are thereby released can be recognized microscopically as small irregular cavities (Fig. 4.32f). The presence of any unreleased fragments can then also be recognized during a

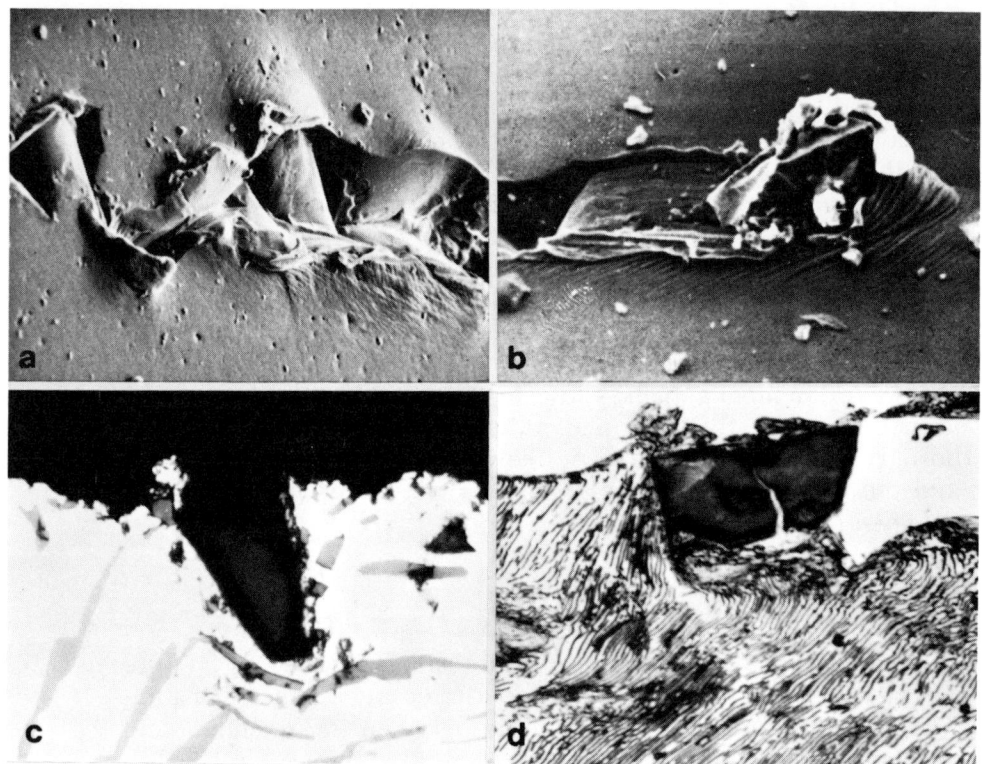

FIG. 4.31. Embedding of abrasive during treatment on 220-grade silicon carbide paper.

(a) A polished surface of aluminum abraded for a single stroke, showing a track produced by a tumbling abrasive fragment. Scanning electron micrograph. Magnification, 220×. (b) Termination of a track similar to that shown in (a); an abrasive fragment has embedded deeply into the surface. Scanning electron micrograph. Magnification, 220×.
(c) Section of an abraded surface of an aluminum-silicon alloy, showing a deeply embedded abrasive fragment. Magnification, 1000×. (d) Taper section of an abraded surface of an annealed eutectoid steel. An abrasive fragment similar to that shown in (b) has embedded in the surface. Taper ratio, 10. Magnification, 1000×.

microscopical examination. Embedded particles are exposed when the abraded surface is subsequently polished (Fig. 4.32c). Continued polishing eventually removes the particles, but note that the abrasive fragments are then released into the polishing pad and could affect its subsequent performance.

The concentration of embedded abrasive in the specimen illustrated in Fig. 4.32(a) to (c) is very high, and concentrations of this magnitude are obtained only with very soft materials such as annealed tin, lead, zinc, magnesium, and aluminum of comparatively high purity. The problem is reduced in magnitude but can still be significant with slightly harder

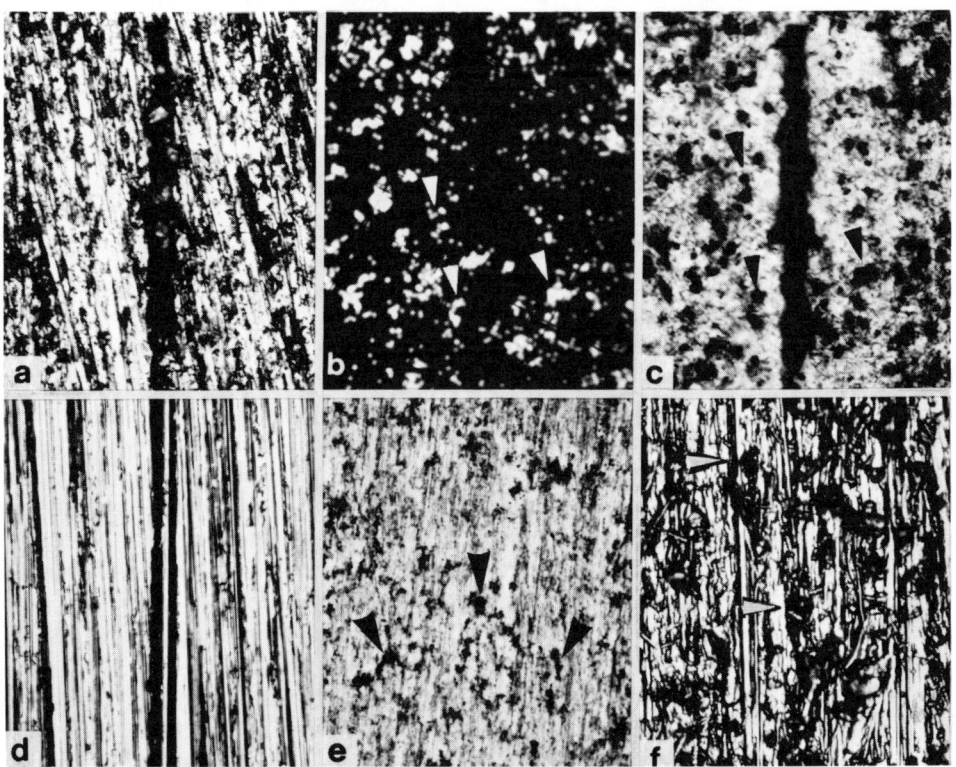

FIG. 4.32. Embedding of abrasive fragments in high-purity aluminum abraded on 600-grade silicon carbide paper. Magnification, 250×.

(a) to (c) Abraded using water as a lubricant: **(a)** Optical micrograph. **(b)** Distribution of silicon in the field shown in **(a)** as indicated by scanning in an electron probe microanalyzer; arrows point to silicon concentrations indicative of embedded abrasive fragments. **(c)** Same field shown in **(a)** after rough polishing; arrows indicate embedded fragments corresponding to the silicon concentrations indicated by arrows in **(b)**.
(d) Abraded on a wax-loaded abrasive paper. **(e)** Abraded using water as a lubricant and then abraded on a fixed-abrasive lap. Arrows indicate abrasive fragments from the 600-grade paper which have been pushed deeper into the surface by treatment on the lap. **(f)** Abraded using water as a lubricant; etched in a 20% NaOH solution at 70°C. Arrows indicate cavities left by released abrasive fragments.

materials such as annealed copper and iron. Some embedding undoubtedly also occurs with still harder materials, but this is not a serious problem in metallography. When embedding is severe, the surface finish of the specimen and the deterioration of the abrasive paper and the abrasion rate are affected (see p. 60). As a rough guide, the existence of a problem can be suspected whenever a rough, irregular surface such as that shown in Fig. 4.32(a) is obtained, but is not likely when a regular finish such as that shown in Fig. 4.32(d) is obtained.

The tendency for abrasive particles to embed is a function of the type of abrasive paper used. Waterproof alumina papers are, for example, much more satisfactory in this respect than corresponding silicon carbide papers; emery papers seem to be even less satisfactory. Moreover, the finer grades of paper (e.g., P1200 grade) cause more severe embedding than coarser grades. The abrasive particles in the finer grades tend to have a thin needle-like shape, and are more likely to fracture in use. More importantly, embedding is much more severe, by at least an order of magnitude, when the papers are used dry than when they are flushed with a stream of water during use. The stream of water presumably washes away many of the fragments of abrasive before they can embed in the surface of the specimen. The problem can be virtually eliminated by loading the surface of the abrasive paper with candlewax or any similar wax before use; this treatment is effective even if the papers are used dry. For example, no evidence of embedded abrasive was found when the surface shown in Fig. 4.32(d) was examined in an electron probe microanalyzer. It is possible that the fragments of abrasive, instead of embedding in the specimen surface, are held by the wax that fills the spaces between the contacting abrasive particles and so are effectively removed from the system.

Laps of the type in which the abrasive particles are held in a matrix of wax (see p. 76) do not cause significant embedding of abrasive, possibly for the reason just discussed. These laps may also be effective in removing abrasive particles which were embedded during preceding treatments on abrasive papers; but this is not so for the softest specimen materials, which are the very ones for which the problem is most serious (see Fig. 4.32e).

This discussion has been concerned only with abrasion on coated abrasive products. Embedding of abrasive can be much more severe, and occur with hard workpiece materials, during grinding with vitreous-bonded wheels and laps.[41] It is particularly likely when the working surface of the wheel or lap has recently been dressed.

APPENDIX 4-A
Etching Methods

Etchants are identified by an abbreviated form of their chemical composition. Water, wherever listed, should be taken to be distilled water, or water of equivalent purity. All reagents should be taken to be analytical grades, and acids to be concentrated.

Ammonium Hydroxide – Hydrogen Peroxide

Composition:

Ammonium hydroxide	1 vol
Hydrogen peroxide (3%)	2 vol
Water	2 vol

Method of use:

Apply by swabbing at room temperature.

Uses:

Develops all types of strain markings, including slip strain markings, in copper alloys.

Remarks:

The swabbing necessary for effective etching inevitably scratches the surface of the specimen.

Cupric Ammonium Chloride

Composition:

Cupric ammonium chloride	10 g
Water	100 ml
Ammonium hydroxide (conc)	Add drop by drop until the solution is alkaline

Method of use:

Immersion at room temperature.

Uses:

Develops all types of strain markings, including slip strain markings, in copper alloys.

Ferric Chloride

Composition:

Ferric chloride	5 g
Hydrochloric acid	10 ml
Water	100 ml

Method of use:

Immersion at room temperature.

Uses:

General etchant for copper and copper alloys. Develops strain markings with the exception of slip strain markings.

Nital (3%)

Composition:

Nitric acid	3 ml
Water	100 ml

Method of use:

Immersion at room temperature.

Uses:

General etchant for steel. Develops, in particular, the structures of ferritic phases.

Oxalic Acid

Composition:

Oxalic acid	10 g
Water	100 ml

Method of use:

Etch electrolytically, using specimen as anode and using a cathode of stainless steel or platinum located at a distance of 2-3 cm. Apply 1-6 V.

Uses:

General etchant for austenitic steels.

Picral

Composition:

Picric acid	4 g
Ethyl alcohol	100 ml

Method of use:

Immersion at room temperature.

Uses:

General etchant for steels, particularly to reveal carbides.

Sodium Bisulfite

Composition:

Sodium bisulfite	20 g
Water	100 ml

Method of use:

1. To ensure even wetting of the specimen by the primary etchant, it is often desirable first to etch briefly in nital.
2. Immerse at room temperature for 10-25 s.
3. Wash and dry, taking particular care not to disturb the surface film deposited by the etchant.

Uses:

Enhanced contrast in and between ferrite grains. Produces good distinction between lightly tempered martensite and ferrite.

Sodium Thiosulfate

Composition:

	High Sensitivity	Low Sensitivity
Sodium thiosulfate	0.25 g	0.50 g
Water	100 ml	100 ml
Sodium chloride	15-25 ppm	...

Method of use:

Etch electrolytically, using a copper anode and an applied potential of 6 V and the following current densities:

High sensitivity: 3.5 A/dm^2
Low sensitivity: 1.0 A/dm^2

Uses:

Develops strain markings, including slip strain markings, in copper alloys.

Remarks:

1. If the specimen is electrolytically polished, the specimen must be removed from the bath with the potential still applied; otherwise, an interfering film of copper will deposit on the surface. Moreover, an electrolytically polished surface must first be etched for 20 s in a solution containing 100 g of phosphoric acid per litre, then immersed for 20 s in water, and then placed immediately in the etching solution; the purpose of this treatment is to remove an interfering film of a copper phosphate that forms during electrolytic polishing.

2. This etching method was devised and developed by Jacquet[7] but the additional precautions included above, which were elucidated by Samuels[2] and by Manion and Mulhearn,[42] must be followed if reliable results are to be obtained.

APPENDIX 4-B
Taper Sectioning

Surface layers or profiles are usually investigated by means of a section made perpendicular to the surface, in which case the layer or profile is seen with a magnification in depth equal to the lateral magnification employed. If, however, the section is cut at an oblique angle to the surface, then the layer or profile is magnified by the geometrical effect of the *taper section*. This principle is illustrated in Fig. 4.33. The geometrical magnification, called the *taper ratio*, is defined by cosec τ. For the technique to be worthwhile, the taper ratio needs to be about 10, for which $\tau = 5°\ 43'$. Because the surface has to be sectioned at such a small angle and because good retention of the edge of interest is of primary importance, the section edge must be protected in some way during preparation of the taper-section surface (see p. 300); the most effective way of doing this is to cover the surface with an electrodeposit before cutting the taper section.

The taper section may be cut at the required angle in a number of ways, but only two general principles are involved (Fig. 4.34), namely: (a) the surface of the specimen is laid on a preformed plastic wedge of the required angle before mounting (Fig. 4.34 a)[43] or (b) the specimen is laid on a wedge block of the required angle during machining of the taper section (Fig. 4.34b).[44] Alternatively, a mounted specimen may be clamped in a block in which a hole has been machined at the required angle from the normal (Fig. 4.34c).[45] It is best to ensure that an obtuse angle is left in the specimen.

FIG. 4.33. Model of a taper section of a surface containing two grooves (A and B) with an associated surface layer of uniform thickness.

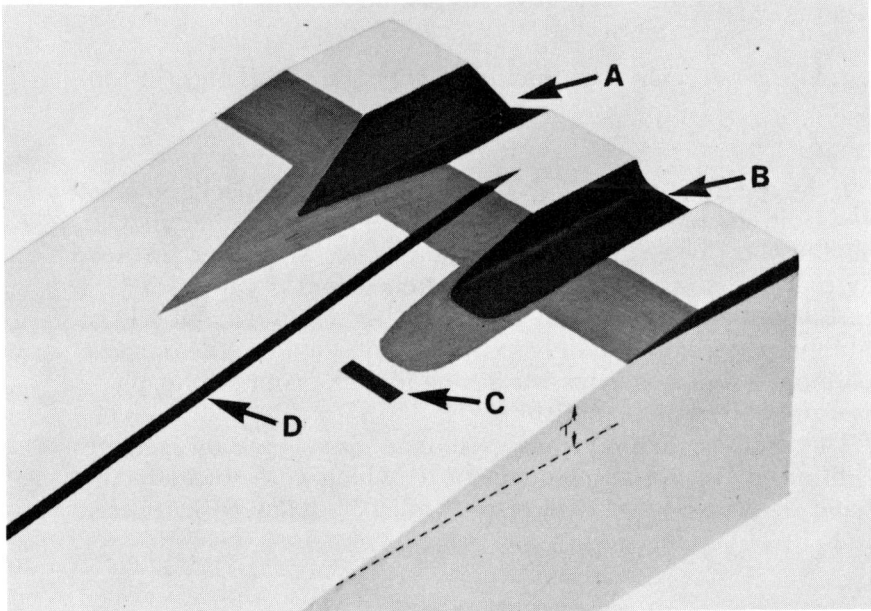

The feature labeled C is aligned normal to the original surface. The feature labeled D is a groove in the taper-section surface.

FIG. 4.34. Sketches illustrating methods of cutting a taper section.

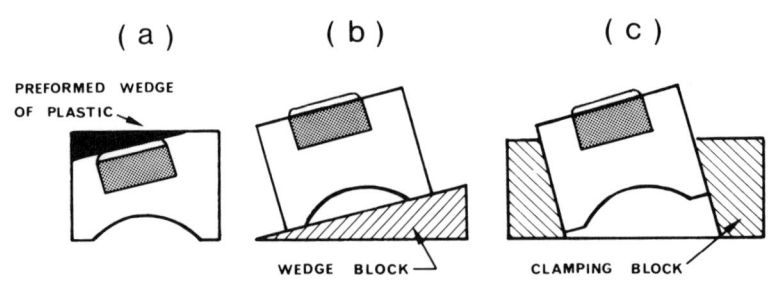

The section surface is machined horizontally to a depth that exposes the feature of interest.

The first difficulty with taper sectioning is that, with a large taper ratio, a small variation in the taper angle results in a large variation in the taper ratio; for example, for a taper ratio of about 10 a change of only 0° 3′ in the taper angle changes the taper ratio by 1%. Even if the taper section could originally be cut with the desired accuracy, and this is by no means certain with any method, it is most unlikely that this accuracy would be retained in the final surface after it had been prepared metallographically — i.e., the final surface will not generally be parallel to the surface that was machined. Reliable depth measurements therefore require that the taper ratio of the section that is finally examined be known.

The most satisfactory method of doing this is, before electroplating, to set a length of fine wire parallel to the surface to be sectioned, and aligned approximately perpendicular to the intended section line. The surface is then electroplated and sectioned in the usual way, and the wire appears in section as an ellipse. The ratio of the major axis (measured perpendicular to the section line) to the minor axis of the ellipse gives the taper ratio of the section.[45]

Some care is necessary in the interpretation of taper sections. An accurate representation of any surface feature is obtained only if the feature is uniform in thickness over the length that is sectioned. Moreover, only features that are parallel to the surface are magnified geometrically by the amount of the taper ratio (A and B in Fig. 4.33); at the other extreme, features that are perpendicular to the surface are not magnified at all (C in Fig. 4.33). Thus considerable distortion of boundaries and interfaces that are at all irregular occurs; this is illustrated for two simple cases (A and B) in Fig. 4.33. It is often of help to think of a taper section as a series of closely spaced steps, each step representing a level parallel to the original surface. Finally, the contours of any irregularities in the final surface, such as polishing scratches, will also introduce irregularities into the contour of the section line (D in Fig. 4.33). These features will be magnified by the taper ratio but in the reverse direction to those in the original surface.

REFERENCES

1. A. S. Malin and M. Hatherly, *Metal Science*, 1979, *13*, 463.
2. L. E. Samuels, *J. Inst. Metals*, 1954-55, *83*, 359.
3. L. E. Samuels, "Optical Microscopy of Carbon Steels", American Society for Metals, Cleveland, 1980.
4. R. L. Aghan and J. Nutting, *Metal Science*, 1980, *14*, 233.
5. D. M. Turley, *J. Inst. Metals*, 1969, *97*, 237.
6. M. Hatherly and A. S. Malin, *Metals Tech.*, 1979, *6*, 308.
7. P. A. Jacquet, *Compt. Rend*, 1949, *228*, 1027; *Rev. Mét.*, 1950, *47*, 255.
8. L. E. Samuels and M. Hatherly, *J. Inst. Metals*, 1955-56, *84*, 84.
9. B. J. Shaw, *Acta Met.*, 1967, *15*, 1169.
10. M. A. P. Dewey and G. W. Briers, *J. Iron Steel Inst.*, 1966, *204*, 102.
11. J. D. Embury and R. M. Fisher, *Acta Met.*, 1966, *14*, 147.
12. L. E. Miller and G. C. Smith, *J. Iron Steel Inst.*, 1970, *208*, 998.
13. A. R. Rosenfield, E. Votava and G. T. Hahn, *Trans. Amer. Soc. Metals*, 1968, *61*, 807.

14. H. E. Enahoro and P. L. B. Oxley, *J. Mech. Eng. Sci.*, 1966, *8*, 36.
15. D. M. Turley, *J. Inst. Metals*, 1968, *96*, 82.
16. D. M. Turley, *J. Inst. Metals*, 1971, *99*, 271.
17. L. E. Samuels, *J. Inst. Metals*, 1956-57, *85*, 51.
18. D. M. Turley and L. E. Samuels, *J. Aust. Inst. Metals*, 1972, *72*, 114.
19. D. M. Turley and L. E. Samuels, *Metallography*, 1981, *14*, 275.
20. H. Wilman, *Wear*, 1969, *14*, 249.
21. P. L. B. Oxley, *Machine Tool Design and Research*, 1961, *1*, 89.
22. H. W. Kerr, *Metallography*, 1972, *5*, 363.
23. P. W. Bridgeman, "Studies in Large Plastic Flow and Fracture", McGraw-Hill, New York, 1952.
24. J. W. Richardson, "Optical Microscopy for the Materials Sciences", Marcel Dekker, New York, 1971.
25. R. P. Agarwala and H. Wilman, *J. Iron Steel Inst.*, 1955, *179*, 124.
26. D. M. Turley, *Materials Sci. and Eng.*, 1975, *19*, 79.
27. L. E. Samuels and J. V. Craig, *J. Iron Steel Inst.*, 1965, *203*, 75.
28. L. E. Samuels, *J. Aust. Inst. Metals*, 1959, *4*, 1.
29. V. J. Manners, J. V. Craig and F. H. Scott, *J. Inst. Metals*, 1967, *95*, 173.
30. L. E. Samuels and G. R. Wallwork, *J. Iron Steel Inst.*, 1957, *186*, 211.
31. J. Wulff, *Trans. Amer. Inst. Min. Met. Eng.*, 1941, *145*, 295.
32. L. E. Samuels and G. R. Wallwork, *J. Inst. Metals*, 1957-58, *86*, 43.
33. J. F. Archard, *Wear*, 1958-59, *2*, 438.
34. E. C. Rabinowicz, "Friction and Wear of Materials", Wiley, New York, 1965.
35. W. E. Littman and J. Wulff, *Trans. Amer. Soc. Metals*, 1955, *47*, 692.
36. K. Takazawa, *Jap. Soc. Precision Eng.*, 1966, *2*, 14.
37. A. J. Sedriks and J. V. Craig, *J. Iron Steel Inst.*, 1965, *203*, 268.
38. K. L. Beu and D. P. Koistinen, *Trans. Amer. Soc. Metals*, 1956, *48*, 213.
39. R. W. Johnson, *Wear*, 1968, *12*, 213.
40. K. J. Williams, "Proceedings of the Conference on Lubrication and Wear", Institution of Mechanical Engineers, London, 1957, p. 602.
41. D. M. Turley and P. A. Ewing, "Proceedings of the Conference on Lubrication, Friction and Wear", Institution of Engineers, Australia, Melbourne, 1980, Paper No. 181.
42. S. A. Manion and T. O. Mulhearn, *Metallography*, 1971, *4*, 551.
43. A. J. W. Moore, *Metallurgia*, 1948, *38*, 71.
44. E. C. W. Perryman, *Metal Industry*, 1950, *79*, 23.
45. L. E. Samuels, *Metallurgia*, 1955, *51*, 161.

CHAPTER 5
Polishing With Abrasives: Principles

WE ESTABLISHED IN CHAPTER 4 that an abraded surface contains a layer in which the microstructure has been altered in a manner that can be detected by optical microscopy. It is apparent that this layer, if left in the final surface in part or in whole, could result in the appearance of false structures, or *preparation artifacts*. It follows that the first, and perhaps the most important, objective of the subsequent polishing stages of preparation must be to remove this abrasion-damage layer. Otherwise, the second criterion listed on p. 3 for a satisfactorily prepared surface will not be met. In this chapter, therefore, we shall consider whether mechanical polishing methods remove material from the surface and, if so, how the maximum rate of material removal can be achieved.

The first criterion listed on p. 3 for an acceptable finish-polished surface was that "surface layers which might obscure structural features must not be present". This requirement was listed because in the past it has been thought that a layer of amorphous-like material, known as the *Beilby layer,* is smeared over the surface during mechanical polishing and that this layer indeed obscures structural features. We shall therefore need to consider first the mechanisms by which mechanical polishing occurs and whether this obscuring layer really does form.

But before doing so, we need to clarify what we mean when we say that a surface is "polished". A polished surface is commonly taken to be one that reflects light brightly, like a mirror. This is called *regular* or *specular* reflection as opposed to the *diffuse* reflection obtained from a rough surface such as an abraded surface. There is, however, no sharp physical distinction between specular and diffuse reflection when light is incident on anything other than an atomically flat surface. Some light is always reflected specularly and some diffusely, the proportion of specular reflection decreasing with increasing roughness. For light in the visible region and for surfaces for which the roughness is comparable to the wavelength of the incident light, the specular reflectance depends on the probability distribution of the roughness heights while the diffuse reflectance depends on distribution of slopes[1] — that is, on the fine detail of the topography of the surface.

"Polish" consequently is a matter of degree, and basically all that is required in metallography is a surface that has a high component of specular reflection. A precise definition of what constitutes a metallographically "polished" surface consequently is not possible because, as the following discussion will show, the roughnesses of polished and abraded surfaces differ only in degree. Any definition of polishing must therefore necessarily be an arbitrary one and, moreover, one which needs to be qualified by an adjective such as "poor", "good", "coarse" or "fine". Nevertheless, it is still worthwhile in practice to make a distinction between abrasion and polishing because, although the difference is only one of degree, this difference is a considerable one.

Abraded surfaces are produced by one distinct type of process, which was discussed in Chapter 4. A polished surface, accepting that it is merely a surface of high specular reflectance, can on the other hand be produced by several different types of mechanical processes — processes which differ radically in both mechanism and effect. We shall be concerned in the present chapter basically with *standard metallographic polishing* processes — namely, those in which the abrasive is suspended in a nominally inert liquid and supported on an elastic backing, such as a cloth pad, the specimen being rubbed against this pad in such a way that it makes positive contact. Processes in which chemical action is made, either deliberately or unintentionally, to play a predominant role by incorporating an etching reagent in the abrasive slurry will be considered briefly, and only when specifically mentioned. Cruder processes, such as those used in industrial practice or those involving burnishing (rubbing one smooth solid against a softer one), will not be considered on the grounds that they are not appropriate methods of preparing surfaces for metallographic examination.

MECHANISMS OF POLISHING: EXISTENCE OF THE BEILBY LAYER

The Beilby Theory of Polishing

The original view of polishing held, for example, by Hooke, Newton and Herschel was the obvious one that the fine abrasive particles cut away the asperities in an abraded surface, replacing them with a set of finer ones. Doubts about this simple view arose early in the present century and culminated in a proposal put forward by Sir George Beilby[2] that polishing occurs by material being smeared over the surface to fill in the pre-existing depressions. He proposed that the rough surface becomes covered by this means with a layer of material the surface of which is smooth, rather like the frosting on a cake. Beilby thought that the smeared layer had "passed through the liquid state, and had solidified under the influence of surface tension". He also thought that it was amorphous, but later workers modified this view slightly. Thompson,[3] for example, when reviewing the electron diffraction controversy mentioned later, suggested that the smeared layer was probably composed of "more or less amorphous crystallites so minute that this definition (amorphous) can be used".

Beilby did not propose any specific mechanism for the smearing process, but a very plausible one was subsequently advanced by Bowden and Hughes.[4] Their proposal was based on the premise that very high transient temperatures are attained at the points where the abrasive rubs against the asperities in a rough surface, the asperities being heated to the melting point or at least to a temperature close to the melting point. The heated region would thus melt or become very plastic, and so readily be translated into adjoining depressions by the shearing forces imposed by the abrasive points (Fig. 5.1). It was further supposed that, due to rapid chilling, the transported material developed an amorphous-like structure, thus building up the Beilby layer. The development of an amorphous structure would these days be regarded as being feasible if the heated material had actually melted, but not if it had merely become plastic.

These concepts became well entrenched in the minds of scientists and technologists. The evidence on which they are based has been shown on a number of occasions to have been falsely interpreted, but the concepts have nevertheless continued to be resurrected on the basis of new evidence, which in turn has been shown to have been falsely interpreted. The question of the validity of the theory is one of such fundamental importance to the subject of this book that it is necessary to review these waves of evidence.

Beilby's Original Microscopical Evidence

Beilby's direct evidence for his theory was based on the examination of metal surfaces by optical microscopy in reflected bright-field illumina-

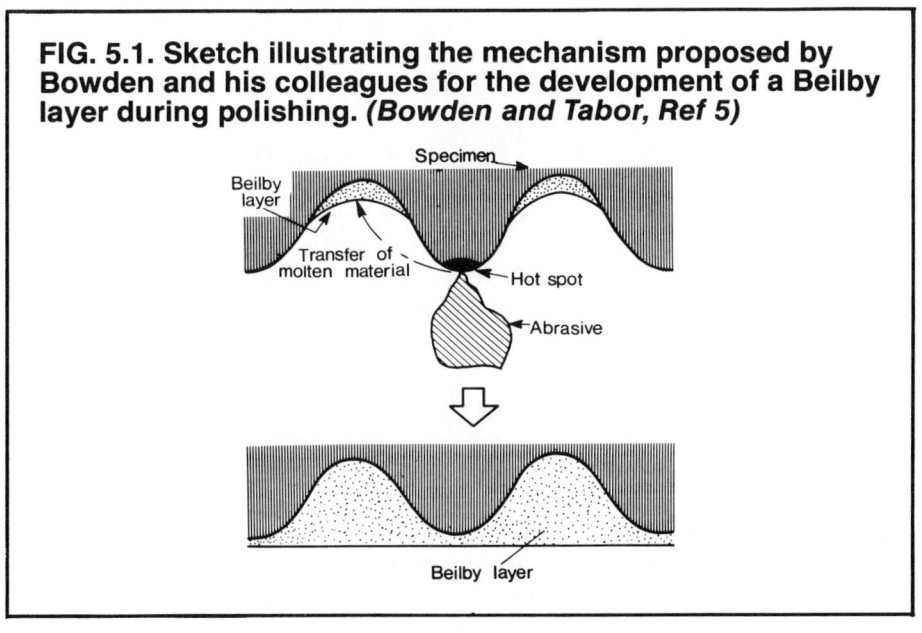

FIG. 5.1. Sketch illustrating the mechanism proposed by Bowden and his colleagues for the development of a Beilby layer during polishing. *(Bowden and Tabor, Ref 5)*

tion. It is now realized that this technique has severe limitations for investigation of minor surface irregularities, and much more sensitive techniques have become available. The application of these new techniques indicates that, without exception, Beilby's interpretations, perceptive though they were at the time, were not sound.

The first observation that concerned Beilby and his contemporaries was that topographical irregularities could be seen in abraded surfaces but not in polished surfaces. Polished surfaces appeared to have a glassy appearance, and this was perhaps a prime motivator for the search for a new mechanism of polishing and the basis for the concept that polished surfaces are covered by a glassy layer. Modern methods of examination show, however, that fine grooves (*polishing scratches*) are present in all surfaces polished by the methods being considered (Fig. 5.2). Indeed, when a series of progressively finer abraded and polished surfaces is examined by any sensitive modern technique, it is apparent that the scratches in the surfaces finished by the two methods differ only in degree and not in kind. As we noted earlier, the surfaces can be expected to become progressively more specularly reflecting as the roughness heights and the slopes of the scratch grooves decrease.

Beilby[2] also advanced, as direct evidence of surface flow, observations made in an experiment in which a set of unidirectional scratches was established on a surface; the surface was then polished in a direction perpendicular to the abrasion scratches until bright flats developed between them. He observed irregularities at the edges of these flats which he thought were material being flowed from the flats into the grooves. This experiment is repeated in Fig. 5.3. Most of the abrasion scratches that remain after partial polishing appear to have sharp edges when viewed by optical microscopy (scratch at bottom in Fig. 5.3a), as would be expected if the crests had simply been cut away. In some cases, however, material from the flats on the crests appears to have extended into the grooves from the direction of polishing (scratch at center in Fig. 5.3a), and this is the type of observation to which Beilby referred. But in some other cases the crest material seems to have extended from the other side of the abrasion groove, or even from both sides (scratch at top in Fig. 5.3a), which is scarcely in accord with the precepts of the Beilby theory. Examination of these grooves by scanning electron microscopy shows unequivocally that the features indicated by the small arrows in Fig. 5.3(a) are not material that has been transported into the abrasion grooves. Instead, the evidence suggests that the surface has been reduced as a plane during polishing and that this plane sometimes has intersected irregularities in the abrasion groove, such as the one indicated by the arrow in Fig. 5.3(b). A burr does sometimes extend from a polishing scratch into the entry side of an abrasion groove (small arrow in Fig. 5.3c), but this burr is less than 0.1 µm long even for this exceptionally deep scratch and would not be resolved by optical microscopy.

A most important piece of indirect evidence advanced by Beilby, and which could be described as the keystone of this theory, concerned a phenomenon that is of considerable importance in metallographic prac-

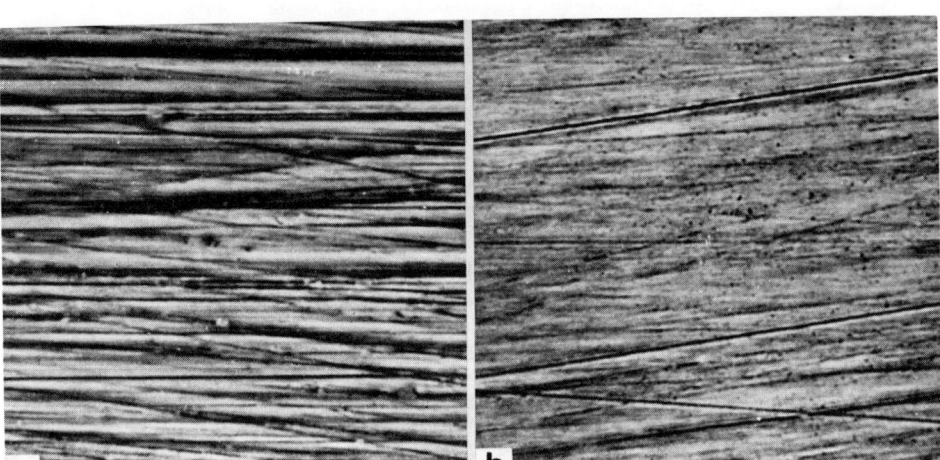

FIG. 5.2. Scratches on mechanically polished surfaces of a single crystal of silver, as revealed by phase-contrast illumination.

(a) Polished on 1-μm-grade diamond. Magnification, 2000×. **(b)** Polished on 0.1-μm-grade aluminum oxide. Magnification, 2000×. Both of these surfaces appear to be almost structureless when examined under bright-field illumination.

tice. He noticed that a set of scratches that had been just obliterated during a subsequent polishing stage seemed to reappear when the polished surface was etched. Beilby's explanation was that the scratches were filled by the Beilby layer during polishing (Fig. 5.1); he then proposed that this layer is chemically more reactive than the crystalline base material and hence is preferentially dissolved out of the scratches during etching (Fig. 5.4). It is now quite clear that this explanation, however neat, is not correct.[6] The phenomenon can be simply and completely explained in terms of etching of the deformed layer associated with the pre-existing scratches, a deformed layer which is not completely removed during polishing even though the scratches themselves have been removed (Fig. 5.4). We shall discuss this phenomenon in detail later (see p. 237).

Evidence of the Attainment of High Transient Temperatures

The Bowden mechanism of polishing is based on the premise that high transient temperatures are attained at asperities when abrasive particles rub past them. This premise was based on two pieces of evidence, one direct and one indirect.

The direct evidence was obtained by pressing a metal specimen against a rotating flat disk of glass, the area of contact being observed through the back surface of the glass.[7] Several tiny stars of light appeared at the interface between the rubbing surfaces, the position of the stars changing

146 / Metallographic Polishing by Mechanical Methods

FIG. 5.3. One of Beilby's experiments repeated.

A set of unidirectional abrasion scratches was established in a brass surface, and the surface was then polished on 1-μm diamond in a direction approximately perpendicular to the abrasion scratches (large arrows). Polishing was carried out for only a brief period, and thus the abrasion scratches have been only partly obliterated.

(a) Optical micrograph of an area containing representative scratches. Magnification, 250×. **(b)** and **(c)** Scanning electron micrographs of opposite edges of a typical scratch. Magnification, 6400×.

from instant to instant. The stars became brighter when the speed or the applied load was increased but were unaffected by the application of a liquid or an abrasive slurry to the rubbing surface of the disk. The light emitted by the stars was analyzed by the use of infrared-sensitive photocells from the output of which, after suitable calibration, the temperature of the emitting spots could be estimated. In general, the temperature increased with load (after a critical load had been exceeded) and

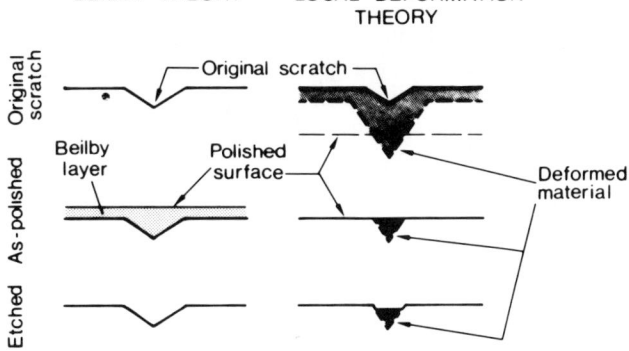

FIG. 5.4. Sketches illustrating the mechanism proposed by Beilby to explain the phenomenon of the apparent reappearance of scratches during etching (left) and the correct mechanism (right).

For a discussion of the correct mechanism (local deformation theory), see p. 237.

with rubbing speed and, at least in the case of low-melting-point metals, was limited by the melting point of the metal. Nevertheless, very severe conditions were required in rubbing steel on glass before temperatures exceeding 500°C were registered, and hot-spot rise times of up to 4 ms were then recorded. The conclusion drawn was that the stars of light were emitted from transient hot spots developed at asperities in the surface of the slider.

However, Heighway and Taylor[8] later repeated these experiments using more sophisticated equipment to analyze the characteristics of the light emitted by the stars. They paid particular attention to the rise time, and produced compelling evidence to show that the hot spots were not produced in the metal surface at all, but arose from recovery from deformation of small pieces of glass which had previously been sheared from the surface of the glass disk and then embedded in the surface of the metal slider.

The indirect evidence, advanced by Bowden and Hughes,[4] was based on a comprehensive series of experiments which was interpreted to indicate that it was the relative melting points that determined whether a particular powder polished a particular metal, and not the relative hardnesses. This was taken for many years as compelling evidence that a surface had to be melted, or at least heated to a temperature close to its melting point, before it could be polished. However, Rabinowicz[9] has pointed out that a detailed analysis shows that Bowden and Hughes' experimental results were not as conclusive as they seemed at the time. The correlation between melting temperature and ability to be polished does not hold for metals of high melting point (Fig. 5.5). From this fact it can be concluded that the Bowden and Hughes' experiments did not in fact rule out relative hardness, or abrasion-type cutting, as the main factor in polishing. A rise of only 100 to 200°C (and this is possible: see p. 197) would soften metals of low melting point sufficiently to reduce their hardness below that of the polishing powder. Metals of high melting point, on the other hand, would not in this event be softened significantly, and the decisive factor would then be the relative hardnesses of the unsoftened metal and the polishing powder.

Indirect evidence that high temperatures are *not* attained during metallographic polishing is found when the fine structures of polished surfaces are observed by transmission electron microscopy. This evidence, which is discussed on p. 197, indicates that, at most, temperatures from 100 to 150°C are attained. Certainly, melting could not occur with most metals.

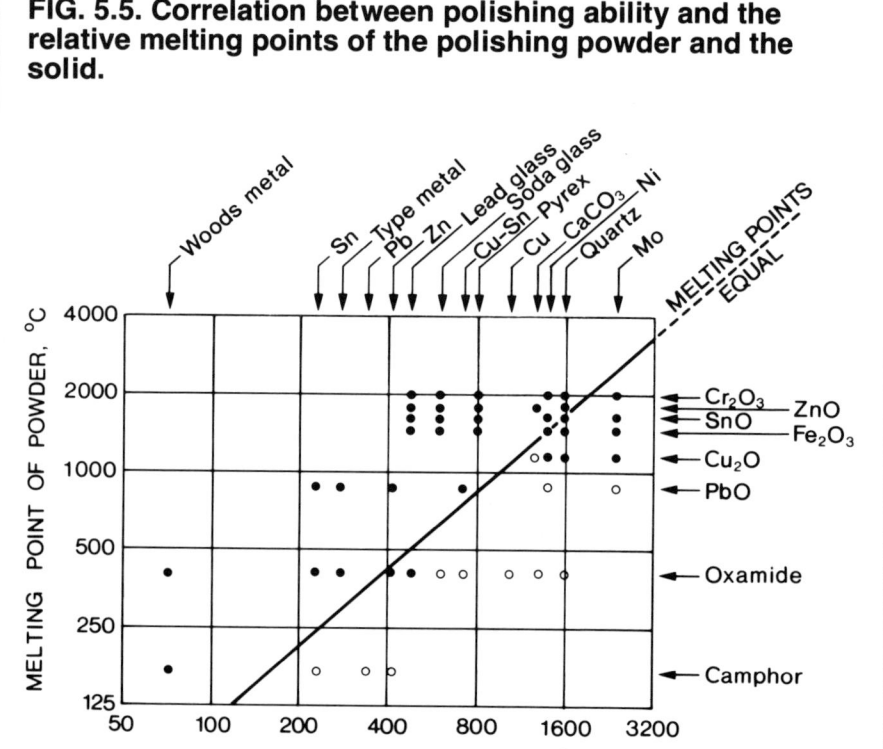

FIG. 5.5. Correlation between polishing ability and the relative melting points of the polishing powder and the solid.

The black dots signify that polishing does occur. In the combinations investigated in the original experiments of Bowden and Hughes,[4] black dots were found only above the line of equal melting points. Subsequent experiments by Rabinowicz[9] produced the combinations plotted above in which black dots are also found below this line.

Continuity of Overgrowth of Deposits

It was known at the time Beilby carried out his work that a crystal of one type grown on the surface of another will grow with the same orientation as the substrate crystal on which it nucleates, provided that there is an appropriate relationship between the crystal structures of the two. This phenomenon is known as *oriented overgrowth* or as *epitaxy*. It is to be expected that the presence of a Beilby layer in any of its proposed manifestations would prevent oriented overgrowth from occurring.

Beilby[2] carried out overgrowth experiments on polished surfaces of calcite and found that crystals of sodium nitrate evaporated from an aqueous solution did show oriented overgrowth on these surfaces. He supposed that the "crystalline influence" of the calcite extended through the Beilby layer, which modern views on the range of atomic forces would say is not possible. However, the question is not relevant here because it is not claimed any longer that calcite polishes by the Beilby mechanism.

Somewhat later,[10-12] observations made during the examination by optical microscopy of sections of electrodeposits grown on polished surfaces seemed to indicate that the deposits were not epitaxial. Numerous small grains were present in electrodeposits grown on as-polished surfaces, grains which seemed to bear no relationship to the underlying grain structure of the substrate (as in Fig. 5.6a). On the other hand, the grains in the deposits grown on the same surfaces after they had been heavily etched clearly were each related to a parent grain in the substrate (as in Fig. 5.6c). It was assumed that the deposits on the polished surface had not grown epitaxially because of the presence of a Beilby layer, which was removed by etching.

A detailed investigation has established, however, that this assumption is not valid.[13] Each crystallite of an electrodeposit is in fact nucleated epitaxially, but the orientations of the crystallites vary over a range because the orientations of the regions on the substrate at which they are nucleated vary over a range. This is due to plastic deformation of the surface during polishing, a concept which will be discussed later (p. 195). With coarse polishes, where the range of orientations of the crystallites in the polished surface is large, the range of orientations in the deposit crystallites is so large that no continuity of overgrowth is noticeable on the scale on which the phenomenon is seen by optical microscopy (Fig. 5.6a). However, the continuity of overgrowth of the crystallites can be seen in sections examined by transmission electron microscopy (Fig. 5.7).[14]

With finer polishing, where the range of orientations in the polished surface is smaller, the range of orientations in the deposit crystals is smaller, and some indications of a relationship between the grains of the substrate and those of the deposit can be seen by optical microscopy, although the grains in the deposit contain a number of subgrains of slightly different orientations (Fig. 5.6b). The continuity of overgrowth becomes more obvious with increasing fineness of polish (Fig. 5.6c), with the deposit grains containing fewer and fewer subgrains until none can be seen and continuity of overgrowth appears to be perfect even in optical

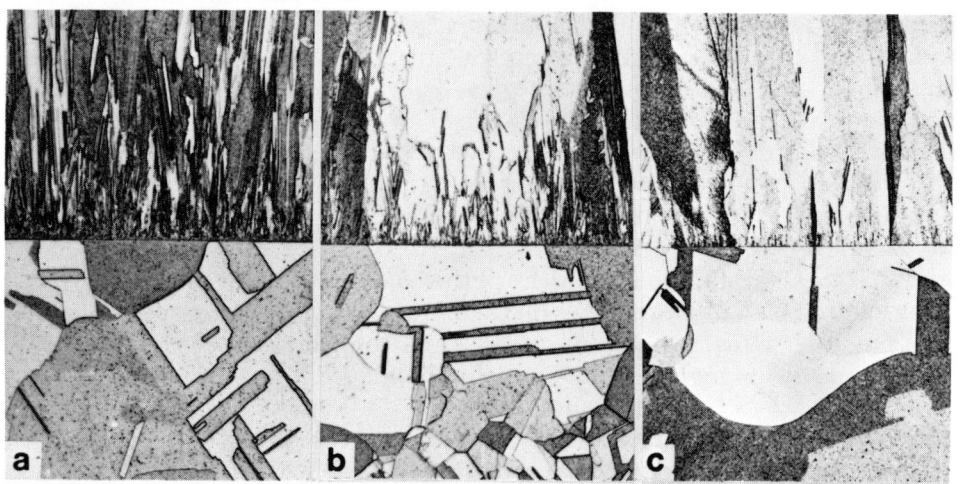

FIG. 5.6. Sections of electrodeposits of copper grown on mechanically polished copper surfaces.

(a) Polished on 1-μm-grade diamond abrasive. No continuity of overgrowth is apparent. (b) Polished on 2-μm-grade aluminum oxide. Some continuity of overgrowth is apparent, but many additional subgrains have developed in the deposit grains. (c) Polished on magnesium oxide. Almost full continuity of overgrowth is apparent. All sections were etched in ferric chloride reagent. Magnification, 250×.

microscopy. The fineness of polish necessary to achieve this condition varies with the metal. Less perfect polishes are needed for metals of low melting point (e.g., tin), in which relaxation of the misorientations introduced by polishing might be expected.

The implication of the more modern overgrowth experiments is that a Beilby layer is not present on the as-polished surface. Complicated assumptions would have to be made (such as that the Beilby layer is dissolved in the solution used to form the deposit or that the layer is discontinuous and the deposit crystallites are nucleated only at the discontinuities where the crystalline substrate is exposed) if the results were to be explained in terms of the Beilby theory.

Reflection Electron Diffraction

Perhaps the biggest boost to the Beilby theory followed the discovery in the early 1930's that electrons could be diffracted by a crystal lattice, if a beam of electrons was aligned at grazing incidence and the refracted beam detected in a reflection mode. Because electrons can penetrate only a few atomic layers into a metal under these circumstances, the technique seemed to be ideal for investigating surface layers as thin as the Beilby layer.

It was found[15-17] that polished metal surfaces gave rise to extremely diffuse diffraction patterns instead of the sharp spot patterns to be expected of crystalline materials (see Ref 3 for a review of these experi-

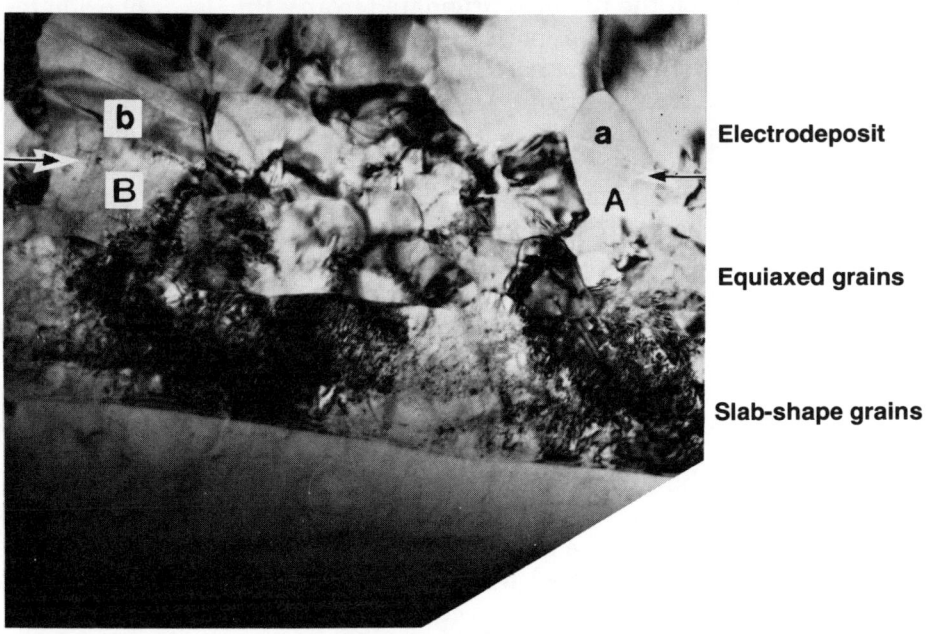

FIG. 5.7. Transmission electron micrograph of a perpendicular section through an electrodeposit (top) grown on a copper surface that was polished on 6-μm-grade diamond abrasive.

The interface between deposit and substrate can be discerned (between arrows). Grains in the deposit **(a and b)** which have grown epitaxially on substrate subgrains **(A and B)** are identified; the orientations of these pairs of grains, as established by electron diffraction, are the same. The layer of small equiaxed subgrains at the surface of the substrate was produced by the polishing process. Beneath this layer can be seen a layer of slab-shape subgrains with sharp boundaries (see p. 196). Note, however, that there is no layer of abnormal structure at the interface. Magnification, 70,000×.

ments). Two diffuse halos usually could be seen in the diffraction patterns of polished surfaces (somewhat as in Fig. 5.9a), and patterns of the same type were obtained from the surface of a liquid metal (mercury). Moreover, sharp diffraction patterns were obtained when a thin layer was removed from a polished surface by etching. Here, it was said, was uncontrovertible evidence of the presence of a layer of liquid-like material on polished surfaces, just as Beilby had proposed. By following the changes in the patterns that occurred as successively thicker layers were removed from the surface, it was estimated that the Beilby layer was 5 to 10 nm thick on metallographically polished surfaces.[18,19]

But it soon emerged that this evidence was not so uncontrovertible after all. The difficulty was that the very shallowness of penetration of the electrons makes it uncertain what particular surface layer has given rise to a particular diffraction pattern. Moreover, vacuum technology was difficult and rudimentary at the time and the diffraction apparatus would have contained much organic material, particularly from sealing waxes and greases. Some investigators consequently maintained that it was a layer of carbon deposited on the polished surface while it was in the diffraction apparatus that was responsible for the halo patterns. Others thought that the patterns originated from a thin layer of oxide formed on the surface during preparation. One subject that was not discussed at the time, but which is apparent when the original papers are consulted, is that the surfaces examined were polished by very abusive methods. Typical examples were rubbing at high speed against dry chamois leather or burnishing with a jeweler's agate.

Samuels and Sanders[20] were the first to investigate by this technique surfaces that had been polished by reasonable metallographic methods. They investigated single crystals of silver, on which no oxide should form, and used diffraction apparatus designed to reduce contamination of the experimental surface to the minimum. The surfaces investigated contained unidirectional polishing scratches, so that the electron beam could be made incident either parallel or perpendicular to the length of the scratches. Typical examples of the topography of the surfaces examined are given in Fig. 5.2.

Sharp diffraction patterns characteristic of fully crystalline material were obtained in these experiments when the electron beam was incident perpendicular to the length of the scratches (Fig. 5.8), but not the sharp spot patterns expected from a single crystal. Instead, continuous ring patterns were obtained from coarsely polished surfaces (Fig. 5.8a), which indicated that the surface layers were composed of small subgrains randomly oriented parallel to the surface; this interpretation is in full agreement with the concepts of the deformation structures in polished surfaces developed in Chapter 6. The diffraction rings became progressively less continuous (i.e., they became arced; see Fig. 5.8b and c) with increasing fineness of polish, indicating that the surface subgrains were less randomly oriented but still had orientations extending over a range. The location of the arcs also indicates that a pronounced preferred orientation texture was present, with a {011} plane approximately parallel to the surface and a <100> direction approximately parallel to the direction of the polishing scratches; this is one of the three orientation textures that have been observed on abraded surfaces of silver.[21] The finest polishes gave patterns clearly recognizable as grating patterns from single crystals in which the spots were diffused (Fig. 5.8c) to an extent indicative of lattice misorientations of only about ±5°.

The intensity and contrast of the diffraction patterns decreased progressively when the electron beam was rotated away from the direction perpendicular to the scratches, and were at a minimum when the beam was parallel to the scratches. In the latter position, the patterns were virtually featureless with the finest polishes (Fig. 5.9a); they contained barely discernible haloes which were, however, less pronounced than those classically associated with the Beilby layer. Similar featureless patterns were obtained in finely polished surfaces containing randomly oriented scratches. The absence of sharp diffraction maxima in these pat-

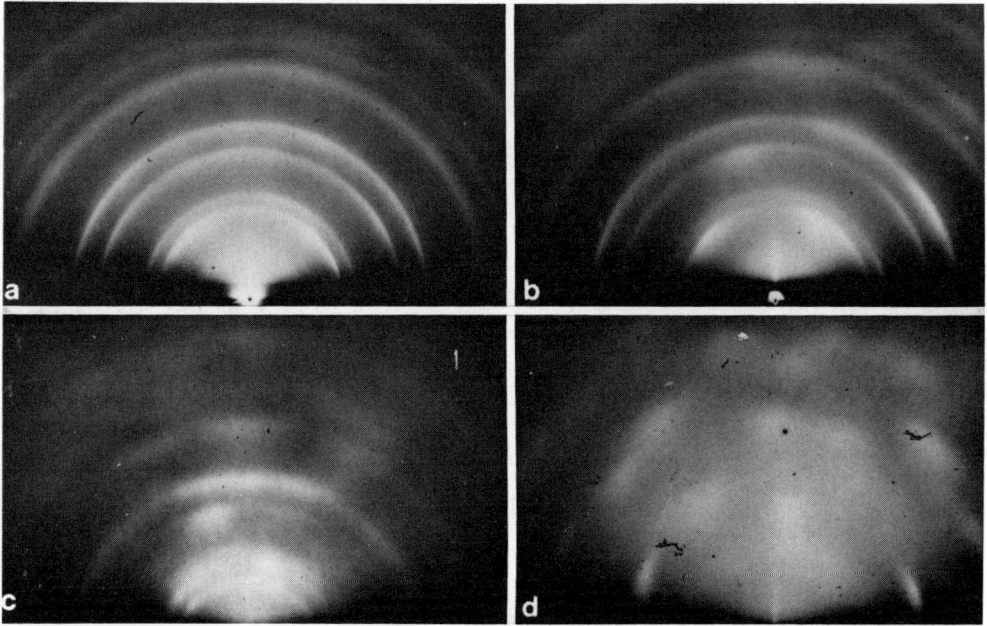

FIG. 5.8. Reflection electron diffraction patterns from polished surfaces of a single crystal of silver.

Polished on: **(a)** 1-μm diamond; **(b)** 0.3-μm α aluminum oxide; **(c)** 0.1-μm γ aluminum oxide; **(d)** magnesium oxide. In all instances, the electron beam was incident perpendicular to unidirectional polishing scratches. All the patterns are characteristic of fully crystalline material. The continuous rings in **(a)** indicate that the surface is composed of small, randomly oriented subgrains. The rings are progressively reduced to arcs in **(b)** to **(d)**, indicating progressive reduction in the misorientations of these subgrains.

terns was shown by Samuels and Sanders[20] to be due to internal reflection of the diffracted rays at the emergent surface when the surface is sufficiently flat in the direction of incidence of the electrons. More definite patterns are obtained when the electron beam is incident in the perpendicular direction, because diffraction then occurs through the inclined peaks of the polishing scratches. Faint halo patterns from surface contaminants, such as carbon, might be expected to appear when the primary pattern has been suppressed by internal reflection.

Although this work gave one possible explanation for the halo patterns classically obtained from polished surfaces, it is not necessarily the only one. The important point is that this work showed that the reflection electron diffraction issue is not a relevant one so far as the structure of metallographically polished surfaces is concerned. It shows that surfaces polished by such methods indubitably are crystalline in the normal sense, but that polishing has introduced textured lattice misorientation of the type to be expected from plastic deformation of the same sort that occurs during abrasion.

Removal of Material During Polishing

One simple matter that was not considered during early discussions of the Beilby theory was that of whether or not material is removed at a significant rate during polishing. The Beilby-Bowden mechanism illustrated in Fig. 5.1 proposes that material is merely moved from one point on the surface to another, and this implies that there is no net loss of material from the surface. This is just not so for metallographic polishing. Material is removed at a significant rate for as long as polishing is continued in the presence of an adequate supply of abrasive, a subject which will be discussed in detail in later sections of this chapter.

It is also worth noting that the peak-to-trough distance in the sketch in Fig. 5.1 is from 0.1 to 1.0 μm in a real metallographic situation. Irregularities of this depth clearly cannot be filled, as proposed in the lower sketch in Fig. 5.1, by a layer estimated to be less than 10 nm thick.

Direct Observations

Thin sections cut perpendicular to surfaces of copper polished by a number of standard metallographic methods have been examined by transmission electron microscopy.[14] The resolution of this method of investigation is good enough for it to be capable of detecting a layer of the supposed thickness of the Beilby layer. Yet no such layer is observed (see, for example, Fig. 5.7).

Assessment of the Beilby Theory

None of the evidence that has been advanced through the years to support the concepts of the Beilby theory can still be said to be valid, and in fact there is much direct and indirect evidence that an amorphous-like layer is not present on polished metal surfaces — at least not on those polished by standard metallographic methods. It follows that there is no reason to suppose that a smearing mechanism of the type proposed by Beilby, and later by Bowden, does occur.

MECHANICAL CUTTING MECHANISMS OF POLISHING

There can be no doubt that specularly reflecting surfaces *can* be produced by mechanical cutting processes. They are produced routinely by lathe turning and milling, albeit only in special low-vibration machines using diamond tools with carefully prepared cutting edges (see, for example, Fig. 10.7, p. 346). The question at issue here is whether cutting mechanisms operate during metallographic polishing processes. There is both direct and indirect evidence to support such a hypothesis.

The indirect evidence suggests that metallographic abrasion and polishing operations differ in degree and not in kind; this implies that the particles of polishing abrasives function as vee-point machining tools some of which remove material by cutting chips and some of which merely form grooves by plowing. The evidence is as follows:

FIG. 5.9. Reflection electron diffraction patterns from the surfaces of a single crystal of silver polished on magnesium oxide.

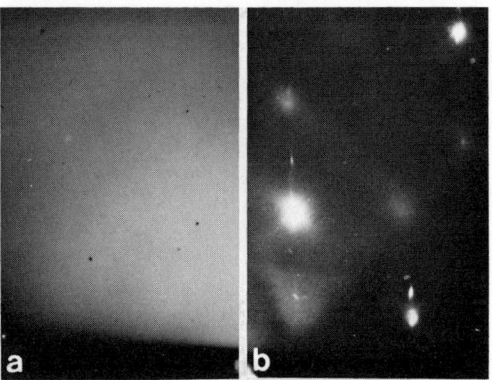

(a) Electron beam parallel to unidirectional polishing scratches (cf. Fig. 5.8d). (b) Surface etched after polishing. Electron beam perpendicular to unidirectional polishing scratches (cf. Fig. 5.8d). In each case, only half of the diffraction pattern is shown. The point of incidence of the direct electron beam is at center bottom in each of these diffraction patterns.

1. Material is removed continuously during both abrasion and polishing, the removal rate being less for polishing than for abrasion. The polishing rates achieved are discussed later in the present chapter; those achieved in abrasion were discussed in Chapter 3.
2. The topography of both types of surfaces is composed of approximately vee-shape grooves, the groove height and side slope decreasing with increasing fineness of polish.[20]
3. Plastically deformed layers that have generally similar characteristics are produced by both processes, these layers being shallower the finer the polish (see Chapters 4 and 6).

The direct evidence has been obtained from observations on surfaces which have first been very finely polished and then polished on the pad under investigation but for a very short stroke.[22] Ribbons of metal with the distinct characteristics of machining chips can be seen to be associated with some of the new polishing scratches (Fig. 5.10a and b). Flaky fins are also often developed along one edge of the grooves and such fins are likely to become detached by subsequent traverses on the polishing pad, a phenomenon representing a second and minor mechanical mechanism of material removal. Furthermore, the bulk of the metallic debris recovered from a polishing pad always is composed of long ribbons of material which have the morphology of machining and abrasive chips (Fig. 5.11 and 5.27a). The ribbons can be shown by transmission electron diffraction to have the same composition as the specimen material, and to

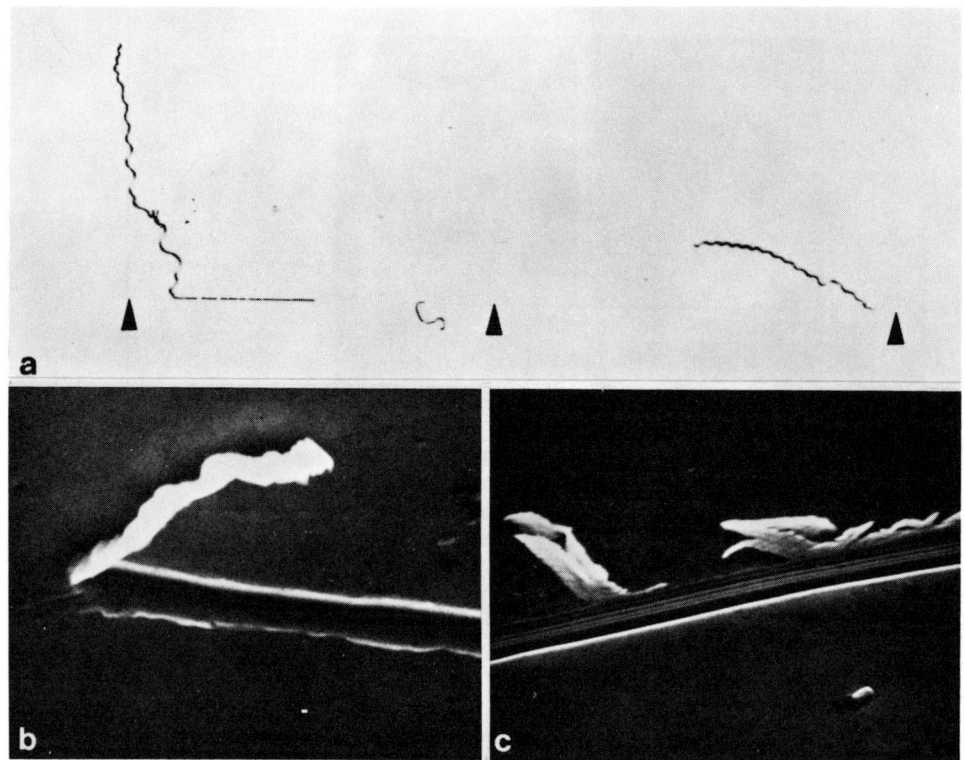

FIG. 5.10. Surfaces of annealed high-purity copper polished on a 6-μm diamond abrasive pad.

The surface was very finely polished and then was polished on the diamond abrasive in the normal manner but for only a short stroke from left to right. **(a)** A series of chips attached to a polishing scratch (arrows). Optical micrograph. Magnification, 750×. **(b)** A chip attached to the termination of a polishing scratch. Scanning electron micrograph. Magnification, 7000×. **(c)** A fin formed at the edge of a polishing scratch. Side fins form when the side face of a vee-point tool has negative clearance. Scanning electron micrograph. Magnification, 7000×.

be fully crystalline but with a very small grain size (i.e., the material in the ribbons has been severely deformed). In all these respects, the ribbons are identical to definite chips cut by grinding at very shallow depths of cut. The widths of polishing chips, however, are orders of magnitude smaller than of abrasion chips (Fig. 5.11) and the diameters of the abrasive grits that produced them (cf. Fig. 5.11c and 5.16).

Consequently, it is reasonable to suppose that the same general concepts discussed in Chapter 3 for abrasion apply equally well to polishing. We can expect, for example, that the force applied to an individual abrasive particle will determine the depth and width of the scratch that it produces, that this force must be small to produce the shallow scratches found in polished surfaces, and that the rake angle of the contacting point

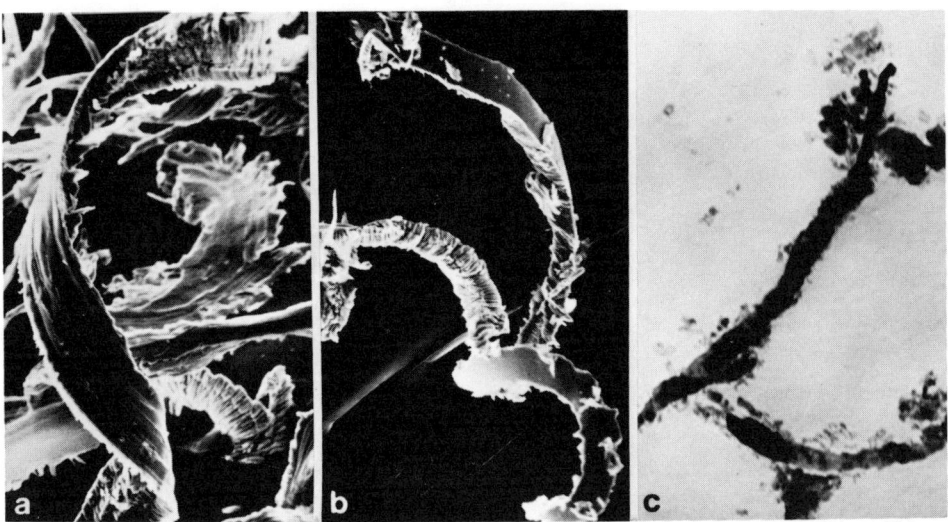

FIG. 5.11. Debris produced by subjecting copper to a range of abrasive-machining processes.

(a) Machine grinding. Scanning electron micrograph. Magnification, 1500×. (b) Abrasion on 400-grade silicon carbide paper. Scanning electron micrograph. Magnification, 1500×. (c) Polished on 1-μm diamond. Transmission electron micrograph. Magnification, 60,000×. The chips in (a) and (b) are identical to those produced by lathe turning except that they are more irregular; they have large aspect ratios, and are smooth on one side surface and serrated on the other. The main particles in (c) also have large aspect ratios and serrated structures. Note, however, that some small, irregular particles are also present in (c), such as might result from the side fins illustrated in Fig. 5.10(c).

of the abrasive particle will determine whether the point cuts a chip or plows a groove. The model has not been developed even semiquantitatively, but qualitatively it would seem that the differences in degree between abrasion and polishing can be explained in terms of differences in the manner in which the abrasive particles are supported and hence the loads that can be applied to them. The real difficulty lies in explaining how the abrasive particles are held sufficiently well for them to indent into the specimen and then stay in position long enough to produce scratches and polishing chips.

One possibility is based on the observation that abrasive particles are found embedded in the fibers of used and washed polishing cloths (Fig. 5.12). These embedded particles would be held against the surface of the specimen by the elasticity of the fibers, which would act as cantilever springs (Fig. 5.13) — very light springs for napped cloths and somewhat heavier springs for napless cloths. This is in contrast to abrasion, where comparatively large loads can be applied to each contacting abrasive point and where the points are held fairly rigidly in space. Although there is direct evidence that diamond grits do embed in the fibers of metallo-

FIG. 5.12. Synthetic suede cloth that has been used for polishing with a 6-μm diamond abrasive.

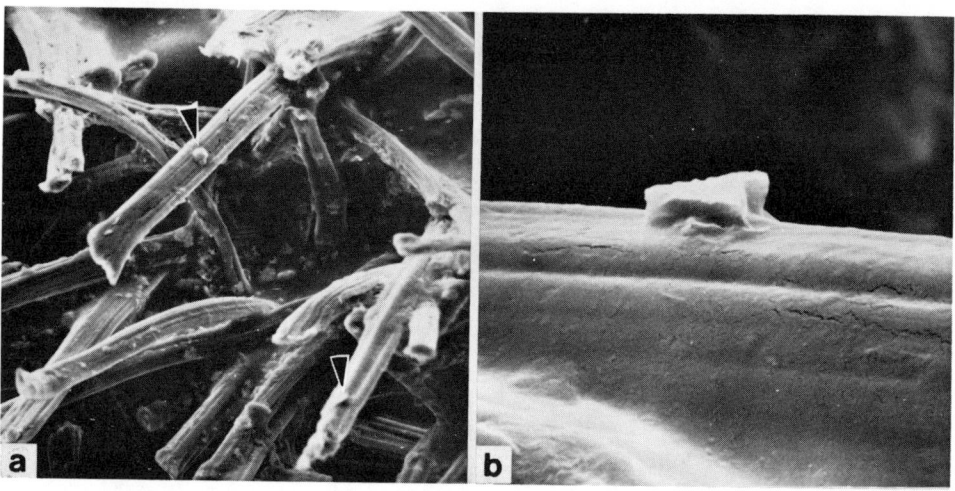

(a) General view; arrows indicate diamond particles embedded in the fibers of the polishing cloth. Magnification, 210×. **(b)** Detailed view of a diamond particle embedded in the cloth fiber. Magnification, 2000×. Both scanning electron micrographs. The cloth has been cleaned to remove the residuals of the carrier paste.

FIG. 5.13. Diagrammatic illustration of a possible mechanical mechanism of polishing, showing how an abrasive particle could cut a chip when it became embedded in a fiber of the polishing cloth.

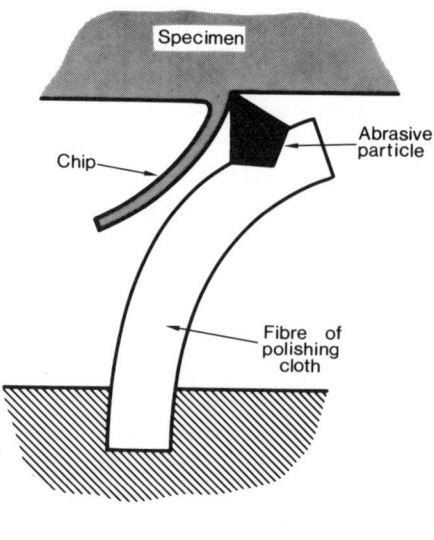

graphic polishing cloths (e.g., Fig. 5.12), very few particles are found to be embedded in this way. However, very few embedded particles are needed, because experiments of the type described on p. 36 indicate that the number of contacting particles in a typical metallographic polishing process is only on the order of $10/cm^2$, which is at least one order of magnitude smaller than that for abrasion (Table 3.2, p. 42).

Another possibility that exists when diamond abrasive is applied in a carrier paste is that the paste holds the particles to the cloth fibers. This certainly does happen (Fig. 5.14a and b), although it might be thought that the paste would scarcely hold the particles firmly enough; nevertheless, several characteristics discussed later (pp. 172 and 184) suggest that this is probably the most important mechanism. For example, the particular carrier paste being considered adheres in lumps to the cloth fibers when kerosine is used as polishing fluid, as in Fig. 5.14(a), but not when water is used, as in Fig. 5.14(b). The paste disperses when water is used as the polishing fluid and, although some abrasive particles embed in the cloth fibers, comparatively small polishing rates are obtained (Fig. 5.28).

A third possibility is that the abrasive particles are held mechanically between contacting or tangled fibers. For example, abrasive particles can be seen in Fig. 5.14(d) to be held between the filaments of nylon, but it is difficult to judge how significant this mechanism might be in practice.

The same principles as those illustrated in Fig. 5.13 would apply to any of these mechanisms. Moreover, they all imply that only a minute fraction of the many abrasive particles applied to a polishing pad ever contact the specimen surface in a useful manner, and that several factors determine how large this fraction is.

CHEMICAL-MECHANICAL MECHANISMS OF POLISHING

Chemicals known to be capable of etching the specimen material are sometimes added deliberately to polishing pads (see p. 270), and sometimes even an etching solution is used without any abrasive. It is self-evident that a chemical mechanism of material removal must operate under these circumstances, but the chemical mechanism could be accompanied by a mechanical mechanism (i.e., a *chemical-mechanical mechanism* could operate). Surfaces polished in this way usually show indications, when examined critically enough, of having been etched, but the etching is much more uniform than it would have been if the specimen had only been immersed in the particular active solution. It is reasonable to suppose that the role of the abrasive and/or the fibers of the polishing cloth under these circumstances is to remove continuously from the surface any protective reaction products of etching, so making the material removal more uniform and more rapid than would be achieved by etching alone. Removal of material by chemical dissolution might even occur when the liquid present would not normally have acted as an etchant. For example, oxygen will always be present in solution in

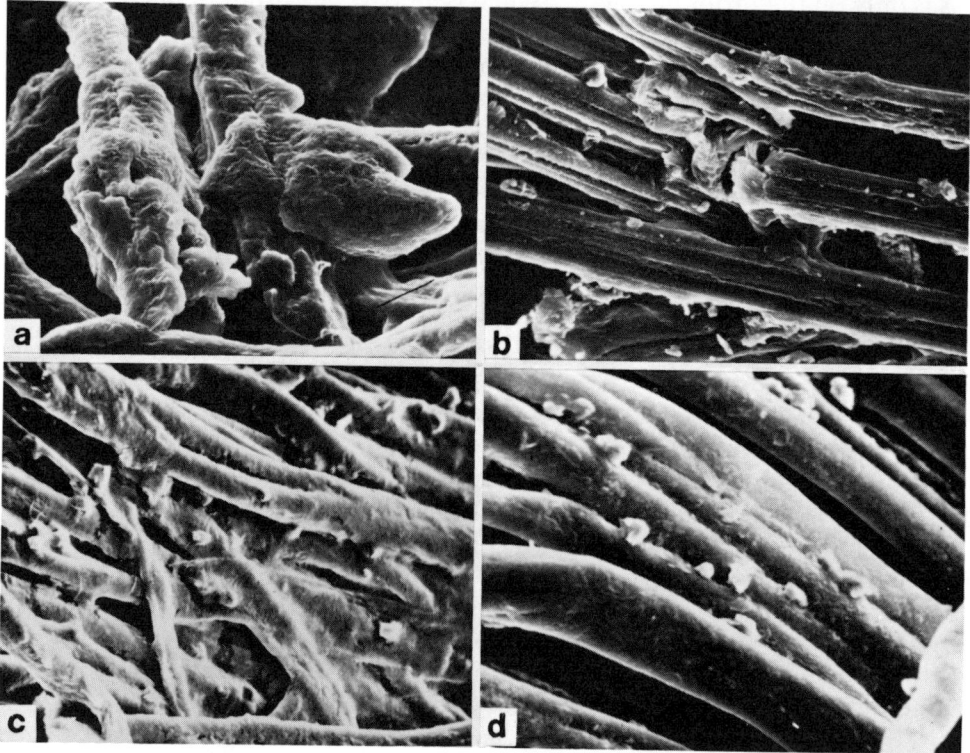

FIG. 5.14. Scanning electron micrographs of polishing cloths charged with 6-μm-grade diamond abrasive in a carrier paste and used for an extensive period of time.

(a) Synthetic suede cloth used with kerosine as the polishing fluid. Magnification, 500×. (b) Synthetic suede cloth used with water as the polishing fluid. Magnification, 500×. (c) Cotton drill cloth used with kerosine as the polishing fluid. Magnification, 360×. (d) Nylon cloth used with kerosine as the polishing fluid. Magnification, 350×.

the polishing fluid and could cause material to be removed continuously by oxidation with metals that normally are protected by an oxide layer if the protective layer is removed continuously. The skid-polishing techniques discussed later (p. 267) probably are examples of this; they produce neither scratches nor surface deformed layers and so can scarcely have occurred by a mechanical chip-formation process. Moreover, when examined at high magnifications by electron microscopy, skid-polished surfaces are seen to contain numerous small etch pits, and this is suggestive of chemical dissolution.

If chemical-mechanical mechanisms are possible without the deliberate addition of an etchant, the possibility then arises that this type of mechanism might make a contribution to all mechanical polishing processes. We would then have to recognize two limiting cases: (*a*) mechanical polishing in which material removal occurs entirely by chip removal, and (b) chemical-mechanical polishing in which material removal occurs entirely by chemical dissolution accelerated by mechanical action. The question for a particular practical polishing operation would then be in what proportion the two mechanisms contribute to material removal. There is no real evidence on which an assessment of this type can be made. The probability is, however, that the vast majority of standard metallographic polishing processes operate entirely, or nearly entirely, by a mechanical mechanism. It is only the finest final polishing processes to which chemical-mechanical mechanisms are likely to make a significant contribution.

CHEMICAL AND ELECTROCHEMICAL MECHANISMS OF POLISHING

Specularly reflecting surfaces can be produced by immersion in specific chemical solutions with (*electrolytic polishing*) or without (*chemical polishing*) a current being applied. The mechanisms and metallographic applications of these polishing processes have been thoroughly reviewed by Tegart.[23] The general concept is that either a relatively thick layer of a viscous liquid or a thin film of a solid reaction product is caused to form on the surface, with the result that high spots in the original surface are dissolved more quickly than the low spots. The surface consequently is smoothed until it becomes highly specularly reflecting.

This is a distinctive polishing mechanism that does not involve the action of abrasives at all. It is not the purpose of this book to discuss processes operating by this mechanism, although some chemical polishing processes will be referred to when they can be used to put a finishing touch on surfaces that have already been prepared substantially by mechanical means.

VARIABLES IN METALLOGRAPHIC POLISHING

We shall see in Chapter 8 that metallographic polishing processes need to be divided into two categories: (a) *rough polishing*, used to remove the abrasion damage while establishing a preliminary polish; and (b) *final polishing*, used to produce a final surface suitable for microscopical examination. We shall be concerned in this chapter only with rough-polishing processes, and we shall seek to establish those parameters that determine polishing rate and hence the methods by which maximum polishing rates can be achieved. This is important because, as also established in Chapter 8, the first rough-polishing stage of a preparation sequence must be designed to achieve maximum polishing rate consistent with polishing of adequate quality.

The number of possible combinations of abrasive, polishing fluid, polishing cloth, load and speed that have been or might be used in metallographic polishing is very large. It is not possible to consider them all here, so attention will be confined to a selection of those that are considered to be the most appropriate. However, the same principles of analysis can be applied to any other selection of variables. Indeed, the same type of analysis should be applied before any particular selection is adopted in practice; the choice should not be based on subjective impressions.

Type of Abrasive

The major options for the abrasive that might be used for rough polishing are listed in Table 5.1, together with an indication of the polishing rates achieved relative to that for 6-μm-grade diamond abrasive. The figures pertain to a particular specimen material and a particular set of polishing conditions, but nevertheless are generally representative.

The highest polishing rates are achieved with the coarser grades of conventional abrasives (silicon carbide and aluminum oxide), but at the expense of polishing quality as judged by such features as rounding of edges (Fig. 5.15a), relief between constituents (Fig. 5.15c), and retention of nonmetallic constituents (Fig. 5.15e and g). A matte finish may also be produced on soft metals with these abrasives (note the matrix regions in Fig. 5.15b). If these conventional abrasives must be used, they should at least be used in closely sized grades. A 600-mesh grade corresponds to a nominal particle diameter of about 15 μm but, as supplied commercially, this grade can have particles ranging from 5 to 40 μm in diameter. The few large particles do not greatly increase the polishing rate but do greatly increase the maximum depth of the surface scratches and the deformed layers; the fine particles do not contribute proportionately to the polishing rate. Grades with particle sizes of about 10 to 20 μm are available commercially and are about optimum for rough polishing. Alternatively, a grade of satisfactory size can be produced by levigation in the laboratory, the procedure developed by Rodda[24] being suitable (see Appendix 5-A); the fraction which has a settling rate of 2.5 cm in 20 s has a size range of about 10 to 20 μm.

Nearly the same polishing rates obtained with coarse conventional abrasives can be achieved with diamond abrasives of much finer grades (Table 5.1), and the quality of polish is improved considerably by all of the criteria just mentioned (Fig. 5.15). Moreover, adequate polishing rates and finishes can be achieved with hard and refractory metals which could not be polished at all with conventional abrasives. The advantages of diamond abrasives in these respects are so overwhelming that they are recommended for rough polishing unless there is a good and specific reason to the contrary. Further attention will be confined to them.

A factor which perhaps inhibits the use of diamond abrasives is high initial cost. This factor is more apparent than real because abrasive operating costs can be kept low. In any event, they are usually small compared with labor costs. However, the abrasives need to be used under optimum conditions if costs are to be kept within acceptable levels, and much of the following discussion is relevant to this issue.

Graded diamond abrasive powders are available commercially in three types, namely:

1. *Monocrystal natural:* Natural or mined diamonds reduced to a powder by controlled crushing, the powder then being graded.[25] The

FIG. 5.15. Comparison of the results obtained by rough polishing on 10-to-20-μm-grade aluminum oxide abrasive (left column) and on 6-μm-grade diamond abrasive (right column).

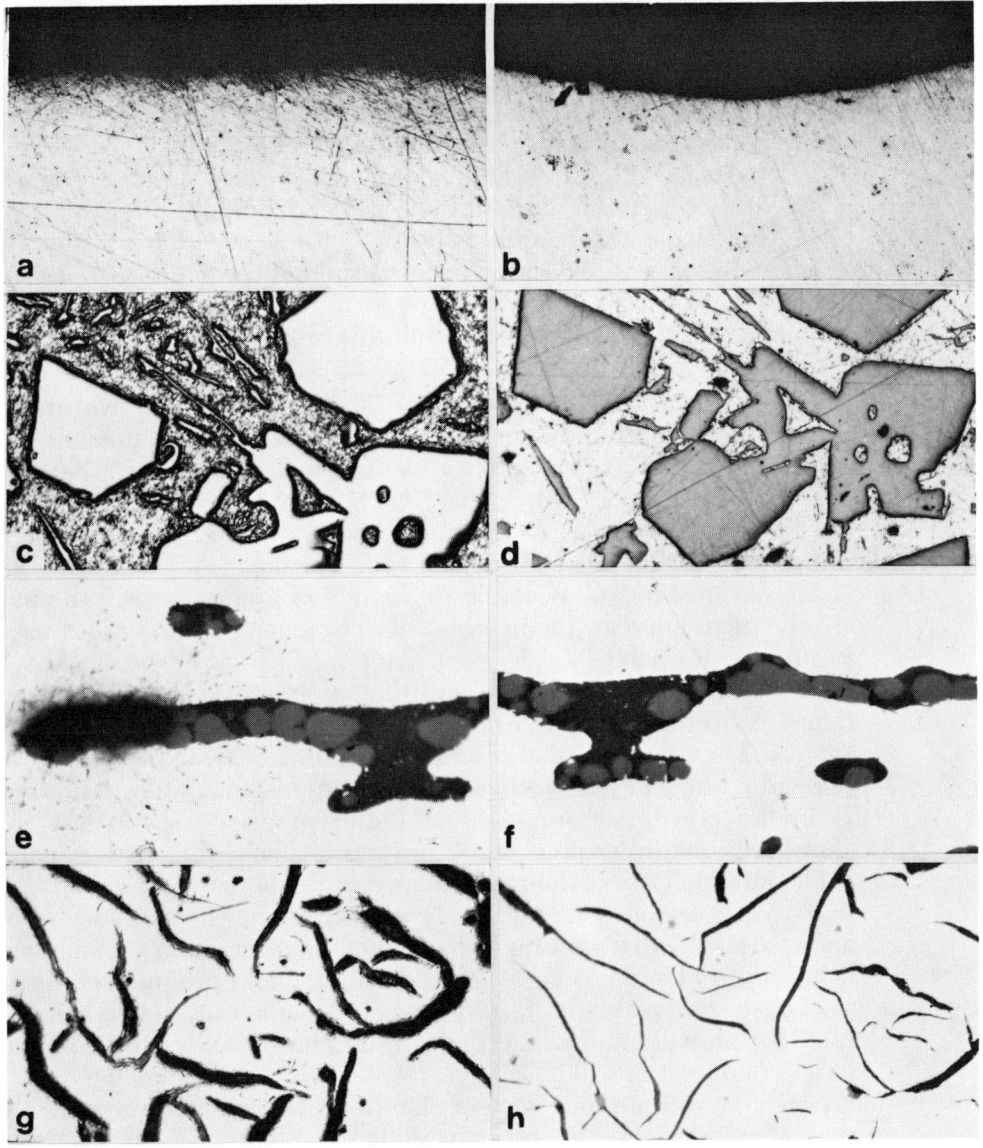

(a) and **(b)** Edge of a hardened steel. Magnification, 100×. **(c)** and **(d)** Aluminum – 13% silicon alloy. Magnification, 500×. **(e)** and **(f)** Silicate inclusions in wrought iron. Magnification, 500×. **(g)** and **(h)** Flake graphite in cast iron. Magnification, 250×. All specimens were abraded on the fixed-abrasive lap before being polished. Compare Fig. 3.28.

individual grits are somewhat irregular in shape and are bounded by comparatively smooth conchoidal facets which intersect at a small number of sharp points (cf. Fig. 5.16a and b). Each grit is a single crystal.

2. *Monocrystal synthetic:* A crushed product that is morphologically similar to the monocrystal natural diamond, but which is produced from diamonds made by the controlled transformation of graphite at high temperature and pressure.

3. *Polycrystal synthetic:* Diamond particles that have been produced by the transformation of graphite by explosive shock.[26] The individual particles are composed of many small crystals, are graded without further comminution, and are roughly spherical in shape. They have an irregular outline with many sharp corners (Fig. 5.16c), and are bounded by surfaces that contain many small facets (Fig. 5.16d). A polycrystalline form of diamond is found naturally and is known as *carbonado,* but this product is not readily available.

Although blocky shapes of the type illustrated in Fig. 5.16(a) are sought in the comminuted monocrystals, other shapes may be present in commercial powders. Particles of rodlike shape for which the aspect ratio is greater than three are known as *slivers.* Particles of thin platelike shape are known as *shales* (Fig. 5.17). Clearly, powders should contain the minimum possible number of slivers and shales.

These powders are available in a range of graded sizes,[27] the particle size being defined as the diameter of a circle having the same area as the profile of the particle. They are generally prepared by some type of Stokesian settling separation, and the actual range of particle sizes obtained consequently is determined partly by the distribution of particle shapes. For example, shales will settle comparatively slowly so that, if present, they will tend to have lateral dimensions larger than nominal. Diamond powders from highly reliable suppliers can in fact contain shales whose dimensions are considerably larger than nominal (Fig. 5.17), in which event the performance of the abrasive in metallographic polishing is seriously degraded. Even apart from the presence of slivers and shales, the distribution of sizes in commercial monocrystal powders from different sources varies considerably.[28] The products of most suppliers are remarkably uniform in size and shape but, nevertheless, supplies for metallographic use need to be chosen with care. The various grades of diamond will be referred to here by median size when they are graded to an upper and a lower size limit, and by maximum size when they are graded only to an upper limit.*

A diamond abrasive may be applied to a polishing pad as a suspension in an oily liquid, by means of an aerosol spray, or in a carrier paste. Many proprietary pastes are available commercially but a paste suitable for

---

*For example, a 4-8-μm grade (see table in Appendix 5-B) will be called a 6-μm grade; a 0-1-μm grade will be called a 1-μm grade.

TABLE 5.1. Typical Abrasives Used in Metallographic Polishing and Their Polishing Rates for 70:30 Brass Relative to Diamond Abrasive (6-μm Grade)

Type	Abrasive Size Range, μm	Relative Polishing Rate
Silicon carbide	10-20	1.5
Aluminum oxide, α type	10-20	1.5
	0-1	0.3
	0-0.3	0.15
Aluminum oxide, γ type	0-0.1	0.1
Diamond	4-8	1.0
	0-1	0.3

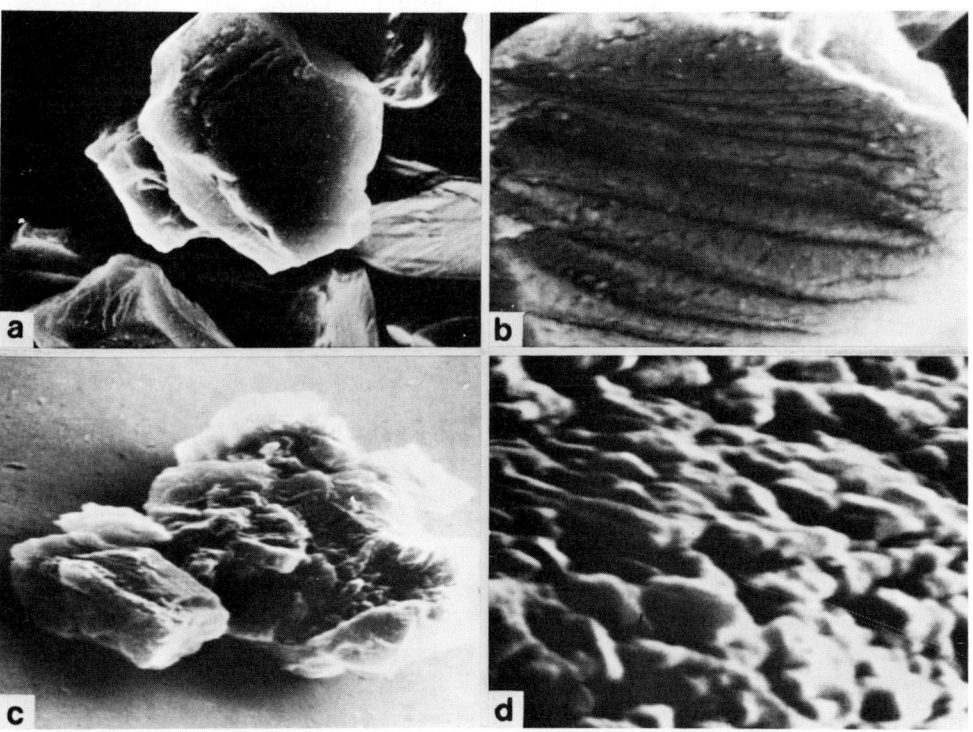

FIG. 5.16. Scanning electron micrographs of particles of 6-μm-grade diamond abrasive. **(a) and (b)** Monocrystalline. Magnifications: **(a)**, 2000×; **(b)**, 20,000×. **(c) and (d)**, Polycrystalline. Magnifications: **(c)**, 2000×; **(d)** 20,000×. The general shapes of the particles are shown in **(a) and (c)**. The topographies of the bounding surfaces of the particles are shown in **(b) and (d)**.

metallographic purposes can be made easily in the laboratory,[29] as detailed in Appendix 5-B. This formula contains about 1 wt % diamond, but pastes of higher diamond concentration could be made if desired. Commercial pastes contain up to 24 wt % abrasive,[28] a factor which needs to be taken into account when relative costs are considered. The polishing rates reported here were obtained with a carrier paste made to the formula and diamond concentration given in Appendix 5-B unless otherwise stated. Commercial pastes may be color coded for ready identification of abrasive particle size (Appendix 5-B).

It is not clear why diamond is so superior to conventional abrasives for metallographic polishing. Diamond has several unusual properties, the most obvious of which is high hardness. High hardness can indeed be expected to be an advantage with hard specimen materials because, assuming that the analogy with abrasion holds (p. 52), it is desirable that the hardness of an abrasive be several times that of the workpiece. There is no reason to suppose, however, that high hardness is an advantage with soft specimen materials. Another special characteristic of diamond is that it has an unusually low coefficient of friction against metals, and this could be favorable if it resulted in a lower critical rake angle for cutting (see p. 50). The details of the geometry of the points that contact the specimen, and the manner in which these points resist fracture in use, are also likely to be of importance. Only a minute portion of a point or an edge, on the order of 0.1 μm wide (about 0.5 mm in Fig. 5.16a and c), is involved in

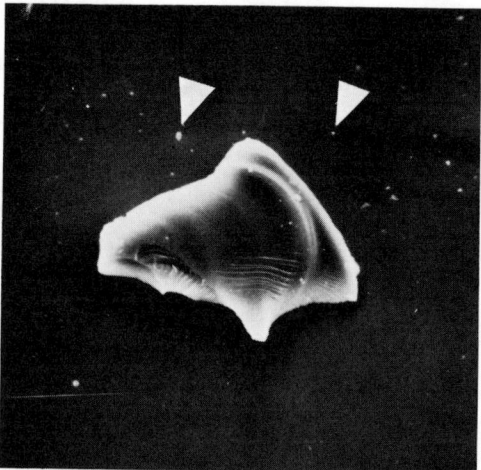

FIG. 5.17. A large shale (approximately 45 by 35 μm) in a diamond abrasive supplied as having a maximum particle size of 0.1 μm.

Agglomerations of particles of the correct particle size are indicated by arrows. Scanning electron micrograph. Magnification, 750×.

cutting a chip during polishing, so that it is the fine details of the point geometry and the manner in which these details are maintained in use that matter. The ease with which particles embed in, or can be held by, the fibers of the polishing cloth might also be significant.

Type of Polishing Cloth

The polishing abrasive is applied to a cloth (sometimes more correctly described as a paper) held against a flat backing plate. A fluid usually is then added, and the specimen surface is pressed against and moved relative to the cloth. The mechanism of polishing proposed earlier suggests that the removal of material will be determined by the manner in which the particles of diamond are embedded in, or held to, those portions of the cloth fibers that contact the specimen surface. This model of polishing also suggests that the smaller the elastic movement of the fibers the less the relief that will develop between different constituents in the surface. Hard, rigidly held fibers, on the other hand, are more likely of themselves to scratch the specimen surface without removing material. Moreover, space needs to be available between the fibers so that the chips of metal cut from the specimen surface can accumulate away from the active regions of the pad where they might interfere with the polishing process or damage the surface being polished. This is more likely with open cloths. Some of these requirements thus are incompatible with one another, and so a range of cloths is needed from which a choice can be made for optimum performance in a particular application. The cloths that are used fall into four categories:

1. *Papers.* Irregular mattes of short fibers, usually containing an extraneous filler material. Natural papers are composed of cellulose fibers and are sometimes used in metallography. The paper that will be used as an example here is a proprietary synthetic paper composed mainly of cotton and rayon fibers coated with an acrylonitrile/butadiene/styrene copolymer (Fig. 5.18a and b).* The fibers are mostly aligned parallel to the working surface, and large cavities are present close to this surface. This paper is resistant to many corrosive fluids that might be used in chemical-mechanical polishing techniques. It will be referred to as an *impregnated paper*.
2. *Felts.* Thick compacted masses of loose intertwined fibers, usually of wool (Fig. 5.18c and d). Billiard cloths are wool felts. Felts give poor results when used by most techniques of metallographic polishing, even with diamond abrasives. They will not be considered further.
3. *Woven Cloths.* Cloths woven from yarns of fibers that have been twisted into threads; the fibers in metallographic polishing cloths are usually of nylon (Fig. 5.18e, f and g), of silk or of cotton (Fig. 5.19a and b). The diameter and spacing of the fibers and threads may vary considerably, and a wide range of types of weave is pos-

---

*Trade name: Pellon.

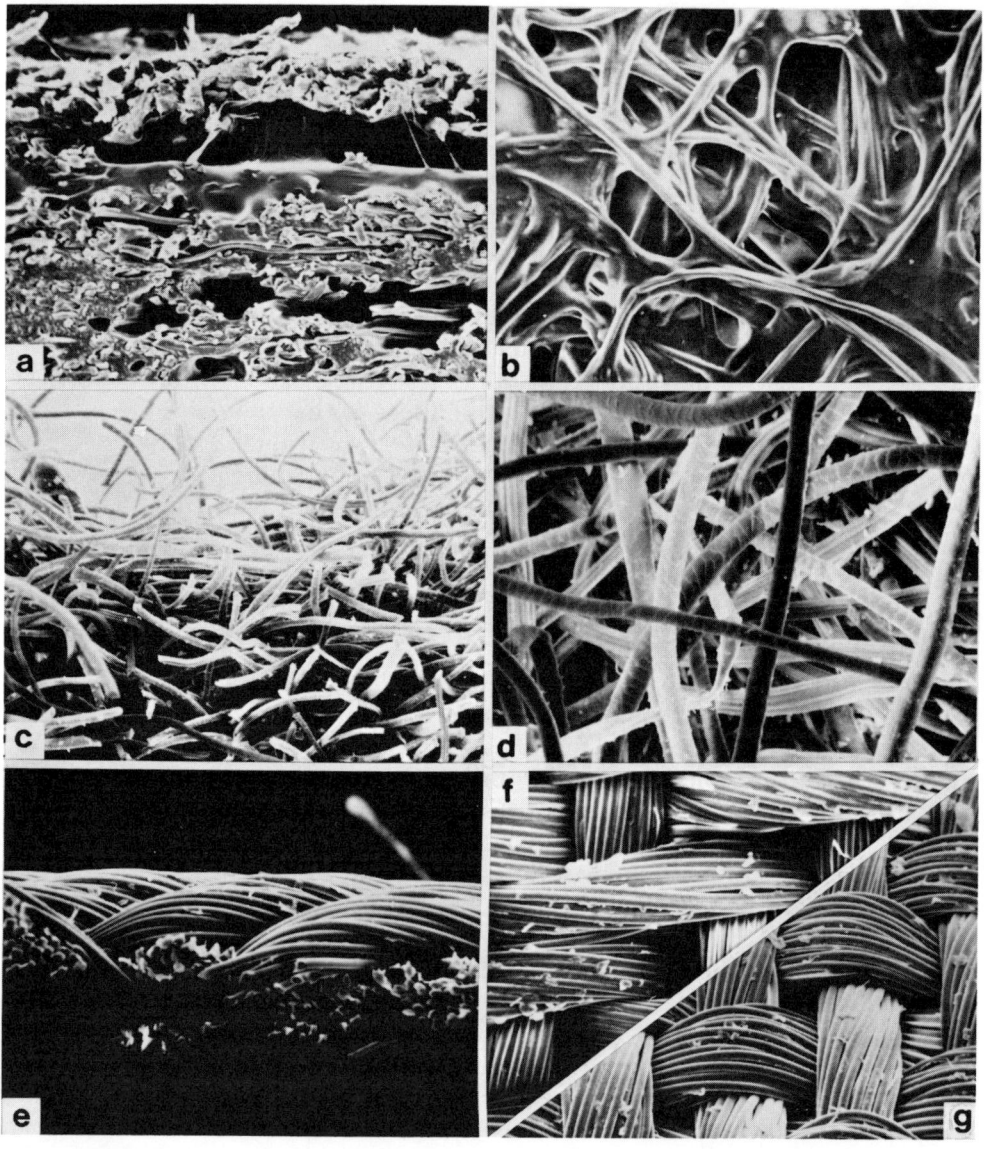

FIG. 5.18. Scanning electron micrographs of representative polishing cloths, showing sections that include the working surfaces (left column) and plan views of the working surfaces (right column).

(a) and (b) A plastic-impregnated synthetic paper formed from cotton and rayon fibers. Note the many voids. Magnifications: (a), 70×; (b), 80×. (c) and (d) A wool felt. Note the open texture and the random arrangement of the wool fibers. Magnifications: (c), 45×; (d), 150×. (e), (f) and (g) A nylon cloth. Note the regular section of the fibers and the small degree of twisting of the fibers in each yarn. This cloth is woven with two wefts and one warp yarn, hence the difference in the arrangement of the yarns on the two surfaces of the cloth (cf. f and g). Magnifications: (e), 80×; (f) and (g), 75×.

sible. The two types of weave that are used most in metallography are a plain weave, in which each transverse weft (filling) yarn passes over and under the longitudinal warp yarn, and a twill weave, in which there are several weft yarns to each warp yarn. The two sides of the cloth have different arrangements of the yarn when multiple wefts are used (cf. Fig. 5.18f and g), more of the thread being approximately parallel to the surface on one side than on the other; this is the side that usually is recommended for polishing (Fig. 5.18g and 5.19b). A twill may have an obvious diagonal rib pattern on the multiple-weft side. A cotton twill is known as a *drill*. The individual fibers of any woven cloth are mostly aligned parallel to the working surface, although more irregularly so in some than in others (cf. Fig. 5.18f and 5.19b). These cloths can be regarded as being napless.

4. *Napped or Piled Cloths.* A woven cloth backing on one surface of which a layer of pile fibers, of a similar or different sort from that used in the backing, is present. There are two general methods by which the nap may be formed. In one, a layer of fibers is deposited electrostatically onto the backing and attached to it by means of an adhesive (Fig. 5.19c); the surface of the pile is then clipped flat. Almost any combination of textile fiber can be arranged in this way, and the length and density of the pile fibers can be varied. The pile fibers are aligned roughly normal to the working surface in unused cloth (Fig. 5.19c). Modern velvets are usually made in this way, and such a cloth in which a rayon pile is attached to a cotton or polyester backing cloth is widely used in metallography for rough polishing with diamond abrasives. The latter type of cloth will be referred to as a *synthetic suede* (Fig. 5.19c and d); a number of suedes with slightly different characteristics are available, but the following discussion is confined to the product of one supplier.* In the second method of making a piled cloth, the pile fibers are inserted into the base cloth during weaving, and are cut after weaving. The resultant pile fibers are more irregularly aligned than for electrostatically deposited piles. The classic velvets are made in this way using fine cotton for both backing and pile, but are now rare and expensive. A cloth of this general type, commonly known by its trade name *Selvyt*, has a base of mercerized cotton and a pile of an unmercerized cotton. The pile fibers are tangled (Fig. 5.19e and f).

POLISHING RATES ACHIEVED IN PRACTICE

Determination of Polishing Rates

The polishing rates for which quantitative values are quoted below were obtained using the apparatus described in Appendix 3-B (p. 80), the loss-of-weight method being used to determine thickness removed. Unless otherwise stated, an unmounted specimen 1.25 cm in diameter, pre-

---

*"Microcloth", Buehler Ltd.

FIG. 5.19. Scanning electron micrographs of representative polishing cloths, showing sections that include the working surfaces (left column) and plan views of the working surfaces (right column).

(a) and (b) A cotton drill cloth. Note the flattened irregular section of the fibers. This cloth was woven with two wefts and one warp yarn, and has a ribbed surface. Magnifications: (a), 45×; (b), 70×. (c) and (d) A synthetic suede cloth. Note that the nap (rayon) fibers stand approximately perpendicular to the backing, which is a woven cotton. The pile has been cemented to the backing. Magnifications: (c), 45×; (d), 150×. (e) and (f) A cotton velvet (Selvyt) cloth. The pile has been woven into the backing. Magnifications: (e), 45×; (f), 150×.

polished to the polishing stage under investigation, an applied load of 500 g (395 g/cm^2), and a rotational speed of 200 rpm on a track 16 cm in diameter (length, 50 cm; area, 64 cm^2) were used. The results are reported as the depth, in micrometres, removed per 100 m of traverse* on the polishing track, which corresponds under the particular circumstances to the depth removed in micrometres per minute of polishing time.

The apparatus used allows control of a number of variables while reasonably reproducing practical polishing conditions, with the exception that the relative motion is rectilinear instead of approximately random as in practical polishing. The loss-of-weight method is used because the method based on the change in the dimensions of a hardness indentation cannot be used universally in polishing; specifically, rounding of the edges and corners of the indentation may occur with napped cloths and soft specimen materials, in which event the results become meaningless. Unmounted specimens must be used to avoid problems with variable and uncertain absorption of polishing and cleaning liquids, particularly if fissures are present between the mount and the specimen. The load that was applied is at the high end of the range normally used in practice. Prepolishing is desirable to eliminate uncertainties due to changes in the surface topography during the test. The results are reproducible to within ±10%.

We shall see, as the discussion progresses, that many factors affect the polishing rate. The discussion consequently will be confined to one abrasive (diamond) and will be oriented towards establishing how the maximum polishing rate can be achieved, most particularly with soft specimen materials. It is with soft specimens that high polishing rates are needed because the abrasion-deformed layer that has to be removed is comparatively deep. Nevertheless, so far as has been ascertained, the principles that emerge are generally applicable.

Method of Adding the Abrasive

For equal weights of abrasive (10 mg)† spread as uniformly as possible over the polishing track and using kerosine as the polishing fluid, much higher polishing rates are obtained when the abrasive is applied in the carrier paste described in Appendix 5-B (p. 192) than when it is applied as a dilute suspension in a light oil (Fig. 5.20); in fact, very small polishing rates indeed are obtained in the latter case. The effectiveness of application by aerosol spray cannot be compared quantitatively with that of other methods because the weight of abrasive applied to the polishing track by spraying is not known and cannot be controlled. However, when

---

*The abrasion rates discussed in Chapter 3 were determined with the same parameters but are reported as the depth removed per metre of traverse. This reflects the difference in magnitude of the two removal rates, and should be kept in mind when comparing the absolute values of the two.

†The unit of weight used in the diamond industry is the *carat*. 1 carat = 200 mg.

spraying is done generously according to the manufacturer's recommendations, the abrasion rates obtained are still significantly smaller than those achieved by application in a carrier paste in the experiment summarized in Fig. 5.20. Nevertheless, polishing rates higher than for application in a suspension are obtained.

All the subsequent discussion refers to application by a carrier paste, which will be referred to as the standard carrier paste. It is the paste described in Appendix 5-B (p. 192). The conclusions reached consequently apply directly only to this carrier paste, or to pastes of similar basic composition.

The superior performance of the carrier paste could be due to a number of factors. First, the paste, and hence the abrasive, would after a little use distribute itself more uniformly around the polishing track than would the abrasive applied by other methods. Secondly, the paste may help to hold the abrasive in the upper regions of the polishing cloth where it is actually available to contact the specimen surface. Finally, the paste may be the most effective means of holding the individual abrasive particles to the cloth fibers so that they can operate as cutting tools. This last reason is probably the most important one of all, as later discussion will demonstrate.

It will be apparent from the upper curves in Fig. 5.20 that there are several criteria by which the characteristics of a polishing system could be assessed. The important ones are the maximum polishing rate that is achieved and the total thickness of material that can be removed with one charging of abrasive; the latter may for most practical purposes be taken to be the thickness removed after 2×10^4 traverses. The first of these criteria will mostly be used, although generally the two would give the same ratings.

Grade of Abrasive

The variation of polishing rate with particle size of abrasive for one material (70:30 brass) is illustrated in Fig. 5.21. The polishing rates plotted are the maximum polishing rates found in experiments of the type illustrated in Fig. 5.20, and the polishing conditions are the optimum conditions established elsewhere in this chapter. There is in this case a sharp maximum in polishing rate at about a 5-μm mean particle diameter. A similar but less sharp maximum at the same particle size was found for copper and an aluminum alloy in earlier, but less quantitative, experiments using the same polishing conditions.[30] However, the maximum found then for steel was less marked and occurred at about a 1-μm mean particle diameter. Exner and Kuhn[30] found that the polishing rate was also a maximum at about a 7-μm-grade abrasive for a steel and a sintered tungsten carbide polished on a "soft" polishing cloth, although they found that the polishing rate increased continuously, if only slightly, with mean particle size when a "hard" cloth was used. The nature of the cloths used in their experiments was not specified in detail.

Thus, although the experimental data available are limited, it is clear that there is no advantage in using abrasive grades coarser than about 6-μm mean particle diameter. Coarser grades produce a poorer polish

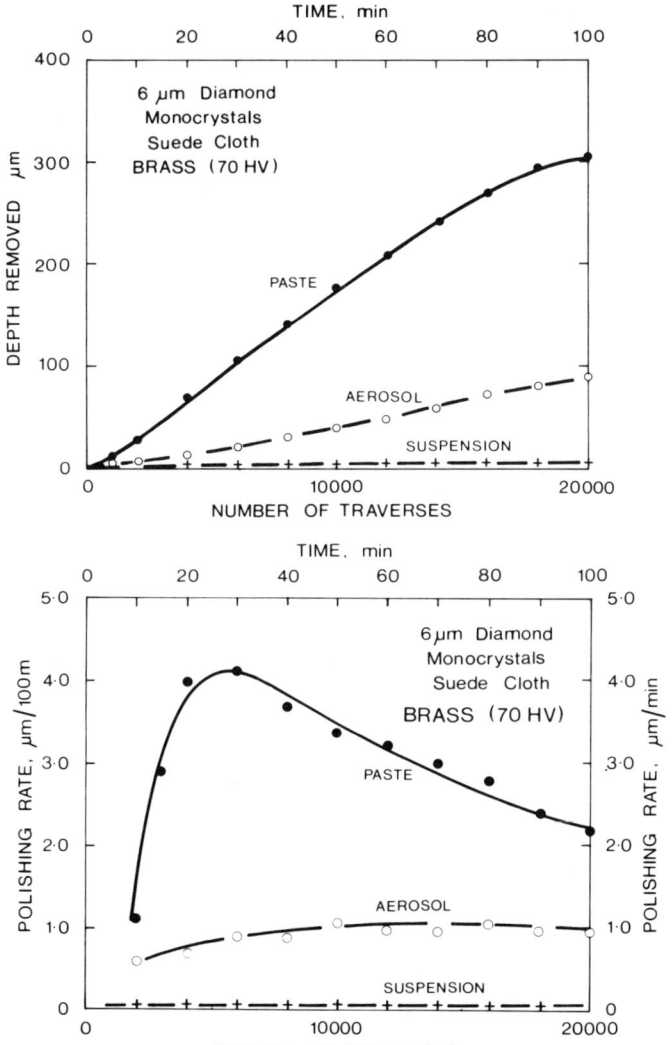

FIG. 5.20. Comparison of the variation in material removal with number of traverses of the specimen for different methods of applying the abrasive to the polishing cloth.

In the upper figure, the primary data of the thickness removed progressively from the surface are plotted. From these data, polishing rates after a particular number of traverses can be derived, and this is the information plotted in the lower figure and in subsequent figures characterizing the performance of polishing systems.

For the curve marked "paste", 1 g of the standard carrier paste, which provides 10 mg of abrasive, was applied to the polishing track. For the curve marked "suspension", the same quantity of abrasive was applied as a suspension in a light oil. For the curve marked "aerosol", the track was sprayed from a diamond-containing aerosol pack.

without any compensating advantage in the form of a significant improvement in polishing rate; in many cases there is likely to be a positive disadvantage in the form of a significant decrease in polishing rate. An upper limit to effective abrasive particle size is reasonably to be expected from the mechanisms of polishing developed elsewhere in this chapter. The diameter of the fibers of the polishing cloth used in experiments summarized in Fig. 5.21 is about 20 μm (Fig. 5.14a and 5.19d). It can scarcely be expected that abrasive particles larger than 20 μm in diameter would be held to the fibers of this cloth in the manner that seems to be necessary to achieve high polishing rates. It may be difficult for particles somewhat smaller than 20 μm in diameter to be held onto the fibers. This explanation is supported by the observation that globules of paste were not found adhering to the fibers of the polishing cloth in the manner illustrated in Fig. 5.14(a) when the coarser grades of abrasive were used for polishing. An implication of this hypothesis is that the maximum effective abrasive size might vary with the polishing cloth in a manner related to the diameter of the fibers that contact the specimen surface. Although intuitively it would be expected that the polishing rate would decrease with decrease in abrasive particle size below the maximum effective size, the reason for this is obscure at present.

The practical implications are that laboratories which handle a limited range of specimen materials and use a limited range of polishing conditions should establish by quantitative trial the most appropriate abrasive

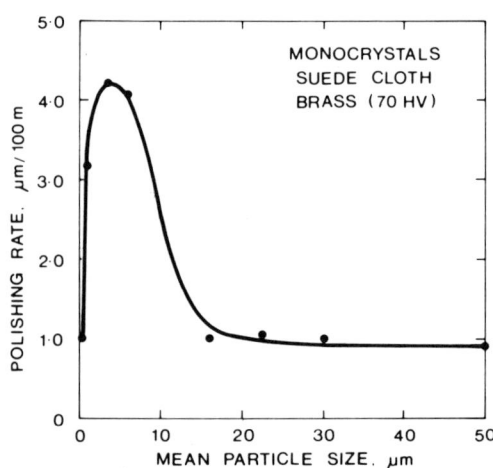

FIG. 5.21. Variation of maximum polishing rate with particle size of abrasive for 70:30 brass.

The specimens were polished under the optimum conditions described elsewhere in this chapter. Natural monocrystalline diamond was applied in the standard carrier paste; a synthetic suede polishing cloth and a kerosine polishing fluid were used.

grade for the first rough-polishing stage. Failing this, and for general applications, it is probably best to standardize on the 6-μm grade of abrasive for the first rough-polishing stage. In this event, it is necessary to use a finer supplementary stage employing either a 1-μm or a 0.5-μm grade of abrasive. It is pointless, however, to use more than two stages of rough polishing. If simplicity is of paramount importance, a reasonable compromise would be to use a single 1-μm stage for rough polishing.

Type of Diamond

The differences in performance between natural and synthetic monocrystalline diamond are at best small when the two are of comparable quality with respect to particle size and shape.[30] However, considerable differences in performance have been found for both natural and synthetic monocrystalline diamond purchased from different suppliers.[28]

On the other hand, the polishing rates obtained with polycrystalline diamond are consistently higher than those obtained with monocrystals* of the same nominal grade when applied in the standard carrier paste (Fig. 5.22 and Table 5.2);† not only are the polishing rates higher, but the total thickness that can be removed with one charging of a given mass of abrasive is also greater. The improvement factor varies with the specimen material, the polishing cloth, and the time of use (Fig. 5.22), but usually is over two and may be as high as five.

There are several possible explanations for the superior performance of polycrystalline diamond. The first, and most likely, is that the individual polycrystalline particles contain many more angular points of the size needed to provide cutting points than do the monocrystalline particles (Fig. 5.16). The second is that more points suitable for cutting might be regenerated if the initial contacting point should fracture in use; however, there is no evidence that diamond particles wear significantly during metallographic polishing.[28] A third possibility is that the polycrystalline particles are more readily held to the fibers of the polishing cloth than are monocrystalline particles, but there is no evidence available to support this suggestion either. Finally, the polycrystalline abrasive may, because of the more regular shape of its particles, be more closely graded than the monocrystalline abrasive.

Quantity of Abrasive

The amount of abrasive applied to the working area of a polishing cloth can be varied in either of two ways — namely, by varying the quantity of carrier paste applied or by varying the concentration of the abrasive in the carrier paste.

---

*Premium-grade natural diamond was used as the monocrystalline diamond in the experiments discussed in this chapter.

†Exceptions in Fig. 5.22 are brass polished on napless cloths. These, as will be discussed soon, are conditions under which the cloth clogs with polishing debris. There are also exceptions with a few metals (Table 5.2, p. 117).

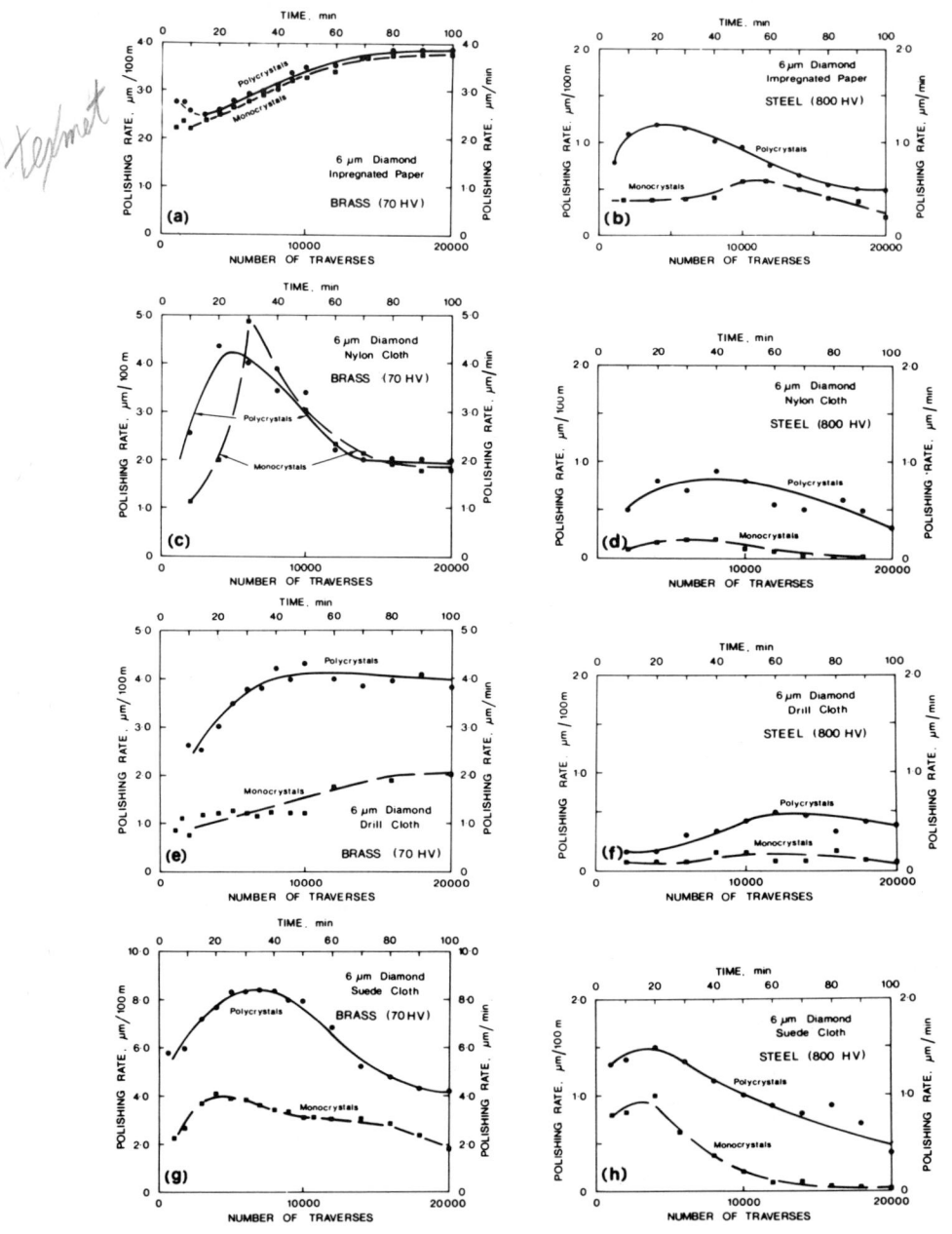

FIG. 5.22. Variation of polishing rate with number of traverses for polycrystalline and monocrystalline diamond abrasives.

The characteristics of a soft brass (left column) and a hard steel (right column) polished on a representative range of polishing cloths are compared. The abrasive was applied in the standard carrier paste, optimum amounts of kerosine were added as the polishing fluid, and the applied load was 395 g/cm^2.

TABLE 5.2. Maximum Polishing Rates Obtained in Polishing Various Metals and Alloys With Polycrystalline Diamond (6 μm) on a Synthetic Suede Cloth

Metal or alloy	Hardness, HV	Maximum polishing rate(a), μm/100 m	Thickness removed after 2×10^4 traverses, μm
Aluminum: high purity	24	9.9	780
alloy (4.5% Cu)	105	9.0	700
Chromium: high purity	200	0.42	12
Copper: commercial purity	50	5.6	350
brass (30% Zn)	70	8.5	630
brass (40% Zn), leaded	155	10	760
aluminum bronze (11% Al)	200	4.0	205
Gold: high purity	22	0.10	4.5
Lead: commercial purity	4	4.0	135
alloy (50% Sn)	13	5.6	400
Nickel: commercial purity	130	2.8	165
alloy(b)	260	1.8	80
Platinum: high purity	40	1.9	90
Steel: 0.15% C, annealed	130	2.2	140
0.65% C, heat treated	340	1.8	100
0.75% C, heat treated	800	1.4	100
1.4% C, heat treated	840	2.1	170
high speed steel (M2)	800	0.65	30
austenitic, type 304	305	2.2	127
Silver: high purity	35	3.0	240
alloy (7.5% Cu)	115	7.2	440
Tin: high purity	9	2.5	180
Titanium: commercial purity	200	1.8	130
alloy (6% Al, 4% V)	295	1.6	195
Tungsten: high purity	300	0.16	7
Tungsten carbide (12% Co)	1550	1.3	100
Zinc: commercial purity	35	8.3	575
alloy (4% Al, 1% Cu)	90	7.3	425

(a) Polishing was carried out under optimized conditions, as described elsewhere in this chapter. The applied load was 395 g/cm². (b) A complex creep-resistant alloy.

The effect of varying the quantity of the carrier paste, keeping the concentration of abrasive in the paste constant at 1 wt %, is illustrated in Fig. 5.23(a) and (b) for brass polished on a synthetic suede cloth. The maximum polishing rate that can be achieved increases, in a manner that is approximately directly related to the mass of abrasive added, as the mass of abrasive is increased to a value of about 20 mg per 100 cm², but does not change significantly thereafter (Fig. 5.23b). Note also that the number of traverses required for the maximum polishing rate to be reached increases with decreasing amount of paste; presumably, more traverses are required to spread the smaller quantities of paste uniformly over the polishing track.

When the concentration of abrasive in the carrier paste is varied, but the amount of paste added is kept constant at 1.0 g for the track area of 65 cm^2 (say ~2 g per 100 cm^2), the maximum polishing rate increases in a manner that is approximately directly related to the mass of abrasive added as that value is increased to about 20 mg per 100 cm^2, which in this case corresponds to a diamond concentration in the paste of 1 wt %; the polishing rate then continues to increase at a diminishing rate as the mass of abrasive added is increased to about 160 mg per 100 cm^2 (concentration of 8 wt % in the carrier paste), and thereafter is constant or even decreases slightly (Fig. 5.23c and d). Buchheit and McCall[28] also reported that the maximum polishing rate obtained with a variety of proprietary diamond

FIG. 5.23. Effects of the quantity of abrasive on polishing rate and on total thickness removed after 2 × 10^4 traverses.

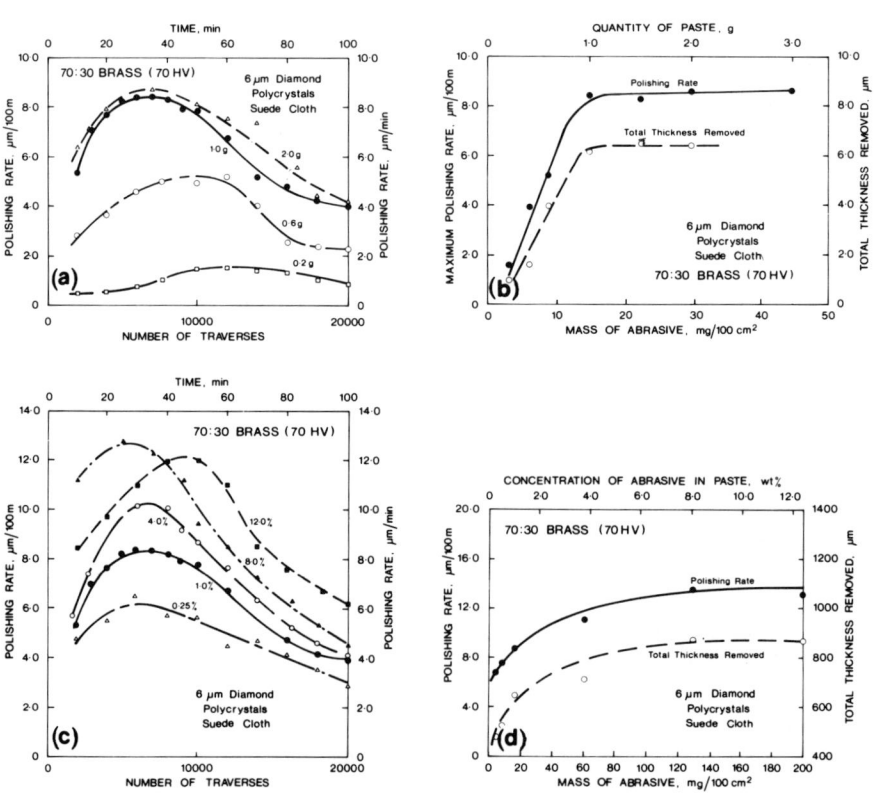

In **(a)** and **(b)**, the quantity of abrasive was varied by changing the mass of carrier paste added to the polishing track. In **(c)** and **(d)**, it was varied by changing the concentration of abrasive in the carrier paste. These data are for brass polished on a synthetic suede cloth using 6-μm-grade polycrystalline diamond abrasive.

abrasive compounds was achieved at a particular concentration of abrasive, but they did not identify that concentration.

It is instructive also to examine the influence of the two concentration parameters on the total thickness of material that can be removed with one charging of abrasive, because this has a bearing on cost effectiveness. There is no further increase in the total thickness removed when the concentration of diamond is increased beyond about 20 mg per 100 cm^2 by increasing the amount of carrier paste (Fig. 5.23b), and there is only a modest increase when this is done by increasing the concentration of abrasive in the carrier paste (Fig. 5.23d). The cost effectiveness of the more concentrated paste is not restored by using proportionately smaller quantities of carrier paste.

The two parameters have been investigated in detail only for annealed 70:30 brass polished on synthetic suede cloth under otherwise optimum conditions. Exploratory trials indicate, however, that the above conclusions apply to a wide range of metals and to a range of cloths as well. In fact, the indications are that the use of pastes with high concentrations of diamond is even less cost-effective with napless than with napped cloths.

To summarize, although there may be occasions requiring the maximum possible polishing rate in which it would be cost-effective to use carrier pastes with high concentrations of abrasive, generally it is not effective to increase the abrasive concentration in the carrier paste above 1 wt %. There is also an optimum effective quantity of carrier paste that should be applied to the polishing cloth irrespective of the abrasive concentration, and this quantity is about 2 g per 100 cm^2 of the polishing track being used. Operators develop a feel for the polishing condition when sufficient carrier paste has been added; it is associated with maximum slipperiness for minimum amount of paste. In any event, it is likely to be less wasteful to use excessive quantities of paste than to use excessive concentrations of abrasive in the paste.

Type of Polishing Cloth

The characteristics exhibited by four representative polishing cloths when used to polish two different specimen materials, one soft and one hard, are illustrated in detail in Fig. 5.22 and summarized in Fig. 5.24. Characteristically, the polishing rate increases for some time after the cloth has been charged with abrasive, reaches a maximum value which is maintained for a time, and then declines to a comparatively low value. The rates at which the rise and the decline occur, and the time for which the maximum rate is maintained, vary considerably with the cloth and the specimen material. In general, higher polishing rates are achieved with napped than with napless cloths and, of all cloths, the suede cloth demonstrates outstanding performance. As another generalization, the material-removal rates for polishing are two orders of magnitude lower than those for abrasion, but are still significant.

The initial rise in polishing rate could result from any or all of the following factors: (*a*) an initial increase in the uniformity with which the carrier paste (and hence the abrasive) has been spread around the pol-

> **FIG. 5.24. Polishing rates obtained in polishing a soft brass (left) and a hard steel (right) on a range of cloths.**
>
>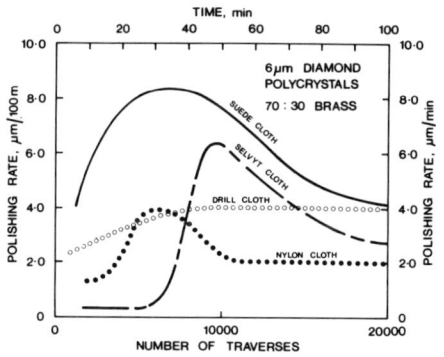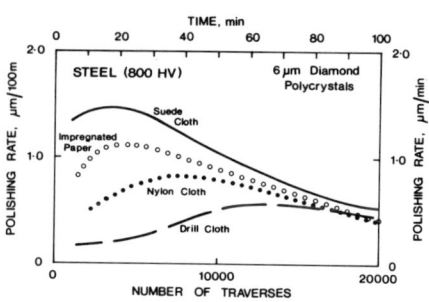
>
> These curves have been extracted from Fig. 5.22. The curve for brass polished on the impregnated paper would, at this scale, be coincident with that for the drill cloth.

ishing track by the specimen (the rate of this increase might be expected to vary with the type of cloth); (b) an initial increase in the number of abrasive particles that have become attached to the cloth fibers in positions where they can contact the specimen surface; and (c) conditioning of the surface of the cloth, causing an initial increase in the real area of contact. This last factor will now be discussed.

The pile fibers of napped cloths soon bend over with use until most become aligned parallel to the contacting surface (Fig. 5.25a; cf. Fig. 5.19d). The real area of contact thereby increases, and this takes longer with some types of cloth than with others (e.g., it takes longer with Selvyt cloth than with synthetic suede). The real area of contact with these cloths thus becomes comparatively large — much larger, as we shall soon see, than for napless cloths. This may account for the comparatively high polishing rates obtained with napped cloths.

The real area of contact for a napless cloth initially is quite small; it increases, although not rapidly, as flats are worn at the contacting areas. For example, in the portion of an impregnated paper shown in Fig. 5.25(b), only two areas (indicated by arrows) have been in contact with the specimen surface even though the paper has been used for a considerable number of specimen traverses. Similarly, only the high spots on some of the kinks in the threads of the woven nylon cloth illustrated in Fig. 5.25(c) have been in contact. The real area of contact for a drill cloth increases by flattening of the threads (Fig. 5.25d; cf. Fig. 5.19b) and becomes larger than for the other napless cloths considered. Thus the real area of contact is always likely to be smallest with the "hardest" cloths, a factor which is related to the structure and weave of the cloth and to the wear resistance of its fibers.

The decline in polishing rate seems primarily to be due to loss of abrasive from the areas of contact. When all the paste is gone, a residual polishing rate probably is maintained by the abrasive particles that remain embedded in the fibers of the polishing cloth. The carrier paste is scraped off the surface of the polishing cloth by the leading corner of the specimen or specimen mount, although this can be reduced considerably by chamfering the corner. Paste also is forced into the base of the polishing cloth. Drag-off losses occur each time a specimen is removed and cleaned. Moreover, fibers are lost from the polishing cloth by fracture and wear, so eventually reducing the real area of contact, but this does not seem to become of major importance until the cloth has been used for a very long time.

The experiment described in Fig. 5.26 illustrates the relative importance of some of these factors. The polishing rate fell dramatically, but

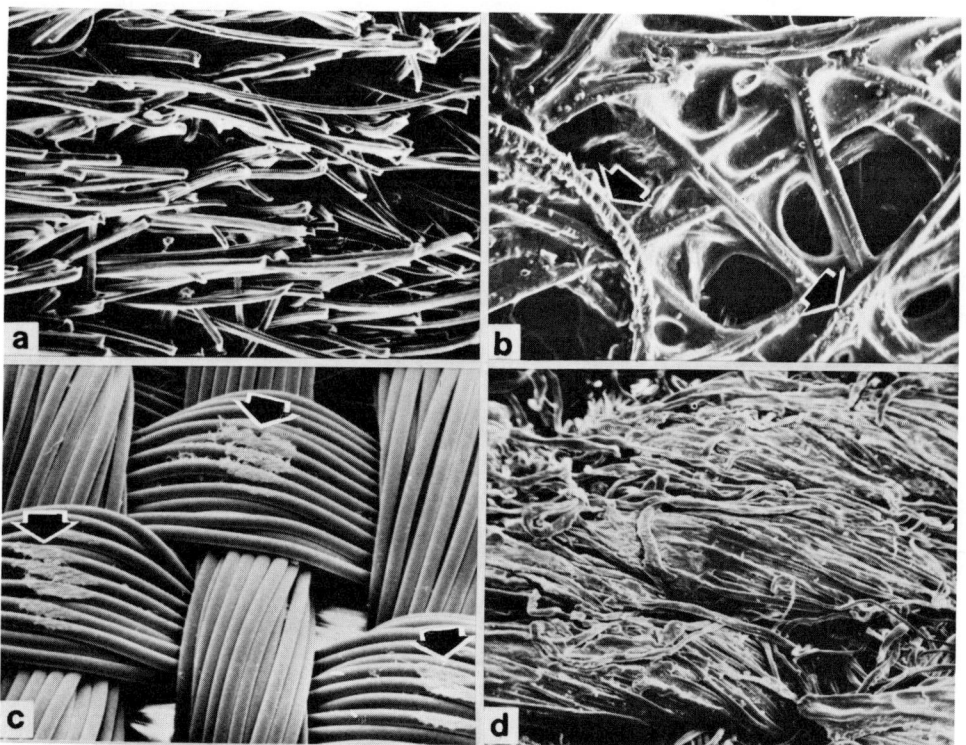

FIG. 5.25. Surfaces of typical cloths that have been used for polishing a hard steel for extensive periods of time.

(a) Synthetic suede cloth. The nap fibers are now mostly aligned parallel to the polishing plane (cf. Fig. 5.19d). Magnification, 70×. (b) Impregnated paper. Only the fibers indicated by arrows have contacted the specimen surface, as evidenced by the wear flats (cf. Fig. 5.18b). Magnification, 120×. (c) Nylon cloth. Only the wear flats indicated by arrows have contacted the specimen surface (cf. Fig. 5.18g). Magnification, 125×. (d) Cotton drill cloth. The threads have flattened, so increasing the real area of contact (cf. Fig. 5.19b). Magnification, 60×. All scanning electron micrographs. The cloths have been washed to remove the diamond carrier paste and any polishing debris.

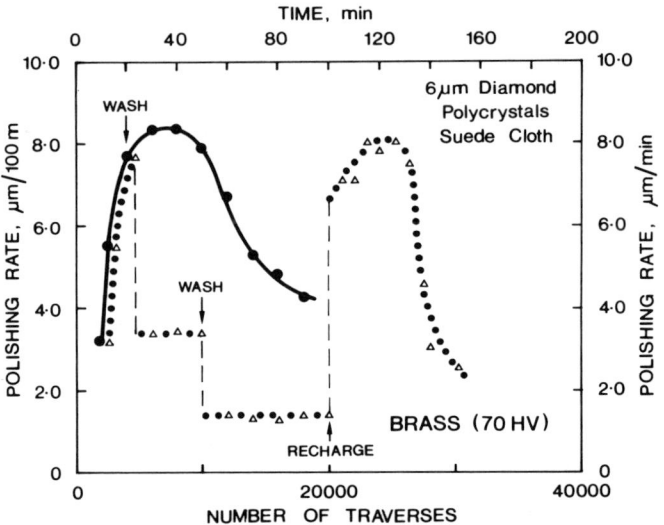

FIG. 5.26. Effects of washing and recharging a polishing cloth.

The cloth was washed gently, where indicated, by spraying with a solvent for kerosine and then by spraying with detergent and water. The solid curve is for a cloth that was charged with abrasive only once. The curve of dots is for a cloth that was charged, washed twice, and then recharged.

not to zero, when this polishing cloth was washed gently in a manner that removed the carrier paste but not the abrasive particles embedded in the fibers of the cloth. The polishing rate returned to the before-washing value when the cloth was recharged. Moreover, a cloth which has been through a full cycle with one charging of abrasive (Fig. 5.22 and 5.24) returns rapidly to a high polishing rate when it is recharged; a cloth may be used for several such cycles. Thus the primary factor determining polishing rate seems to be the amount of carrier paste that is attached to the contacting fibers (see p. 159). The condition of the polishing cloth, including the number of abrasive particles embedded in its fibers, plays a secondary role.

Another factor which can play an important role in the decline in polishing rate is the accumulation of agglomerates of polishing debris in the cloth. With soft materials, such as brass, the micromachining chips produced during polishing tend to agglomerate into packed masses. In the worst cases, and nylon cloth is an example, the agglomerates cap over most of the real areas of contact (Fig. 5.27a). This reduces the real area of contact and therefore probably reduces the polishing rate. The formation of agglomerates in fact correlates with the marked decrease in polishing rate illustrated for nylon cloths after about 5000 traverses in Fig. 5.24 (left). Similar agglomerates form in other napless cloths but tend to accumulate more away from the contacting regions. In drill cloths, for example, they form mostly in regions between threads and only partly obscure the contacting regions (Fig. 5.27c). They tend mostly to fill the many cavities and interstices between the contacting fibers in the particular impregnated paper being considered; the phenomenon might however be more serious for papers in which the fibers are more closely packed. On the other hand, the agglomerates accumulate well down among the nap fibers in napped cloths, where they have little or no effect (Fig. 5.27d). This may also contribute to the higher polishing rates and superior finishes obtained with napped cloths (except with respect, of course, to the production of flat surfaces). The agglomerates can be removed from all cloths by a simple washing process, but the cloth then has to be recharged with abrasive, as already discussed.

FIG. 5.27. Surfaces of polishing cloths that have been used for 2×10^4 specimen traverses.

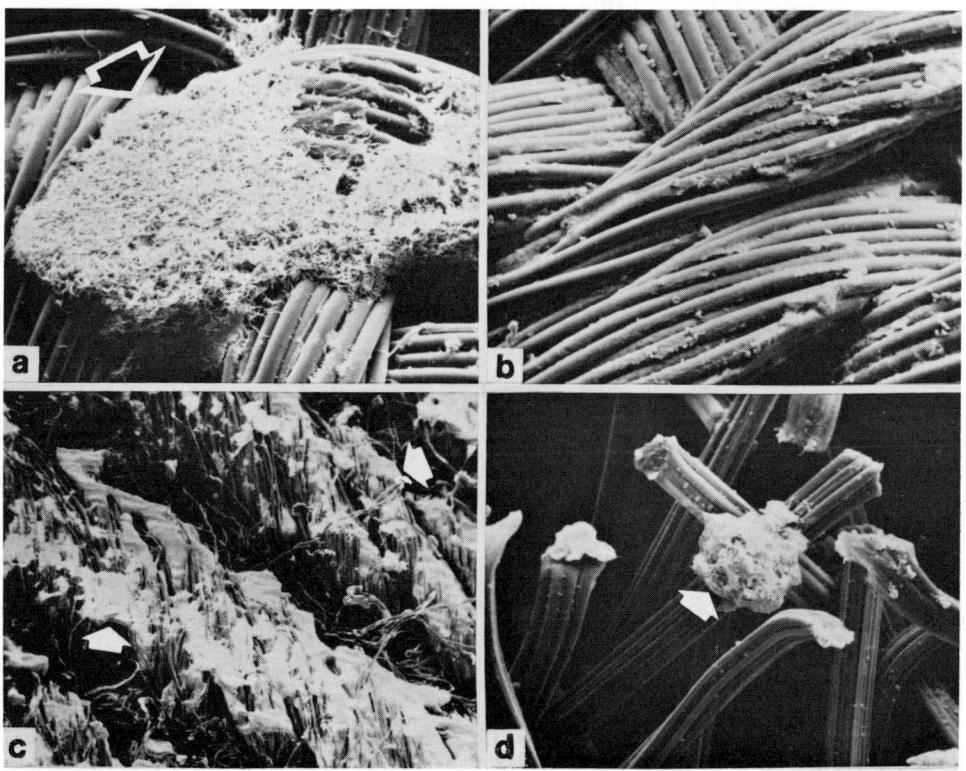

(a) Nylon cloth used in polishing brass. Magnification, 150×. (b) Nylon cloth used in polishing a hard steel. Magnification, 150×. (c) Drill cloth used in polishing brass. Magnification, 30×. (d) Synthetic suede cloth used in polishing brass. Magnification, 250×. All scanning electron micrographs. The cloths were not washed after use. Arrows indicate agglomerates of polishing debris.

The agglomerates of polishing chips do not form during polishing of all specimen materials. For example, they do not form with steels (Fig. 5.27b), and the early decline in polishing rate found when brass is polished on nylon cloth does not occur (Fig. 5.24b). In fact, polishing chips cannot be seen by high-resolution scanning electron microscopy on any type of polishing cloth after polishing of steels, although electron probe microanalysis indicates that much iron is present. Either the chips are too small to be detected; or they have oxidized, fragmented and dispersed; or they have been assimilated into the carrier paste.

Insufficient information is available at present to classify specimen materials into categories of those for which clogging is likely to be severe on a particular polishing cloth and those for which it is not. The development of patches of debris can be seen readily enough on a used polishing cloth by examining it at low magnifications, or even with the naked eye; but it is not possible by these simple means to tell whether the agglomerates have formed at positions where they seriously affect the polishing process. Examination at high magnifications, preferably by scanning electron microscopy, is necessary for this. A close watch needs to be kept for the phenomenon.

Polishing Fluid

The fluid added to the polishing track has an important influence on polishing rate in several respects. Both the type of fluid used and the quantity in which it is maintained on the polishing track are important. For example, the polishing rate obtained with the carrier paste being discussed (see Appendix 5-B) is considerably lower when water is used as the polishing fluid than when kerosine is used (Fig. 5.28); this applies to both napped and napless cloths and to a wide range of specimen materials. Moreover, the polishing rate changes if the fluid is changed. For example, the polishing rate follows the lower curve in Fig. 5.28 if polishing is started with water, but rises immediately to the upper curve if the supply of water is stopped and replaced by kerosine (Fig. 5.29). On the other hand, the characteristics of the lower curve are resumed if the supply of kerosine is then stopped and replaced by water (Fig. 5.29). The cycle can be repeated several times.

The reason that polishing rate varies with polishing fluid is that the polishing fluid affects the manner in which the carrier paste is dispersed among the fibers of the polishing cloth. The paste disperses into the nap or the lower threads of the cloth when water is present (Fig. 5.14b), but accumulates in lumps attached to the nap fibers when kerosine is present (Fig. 5.14a). Moreover, the two conditions of dispersion illustrated in Fig. 5.14(a) and (b) can be alternated by alternating the polishing fluid; that is, a cloth in the condition illustrated in Fig. 5.14(b) will resume the condition illustrated in Fig. 5.14(a) when water is replaced by kerosine. This is a further indication of the importance of the role of the carrier paste in holding abrasive particles to the fibers of the polishing cloth. This it does only in the presence of an appropriate fluid.

The carrier paste under consideration is a water-base emulsion of a triethanolamine – stearic acid soap. It is plausible to assume that water

Polishing With Abrasives: Principles / 185

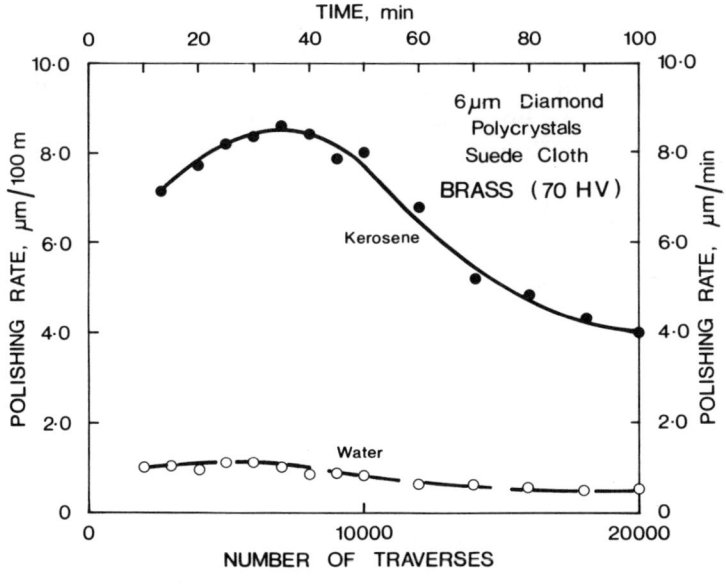

FIG. 5.28. Comparison of the polishing rates obtained when kerosine and water are used as polishing fluids with the standard carrier paste.

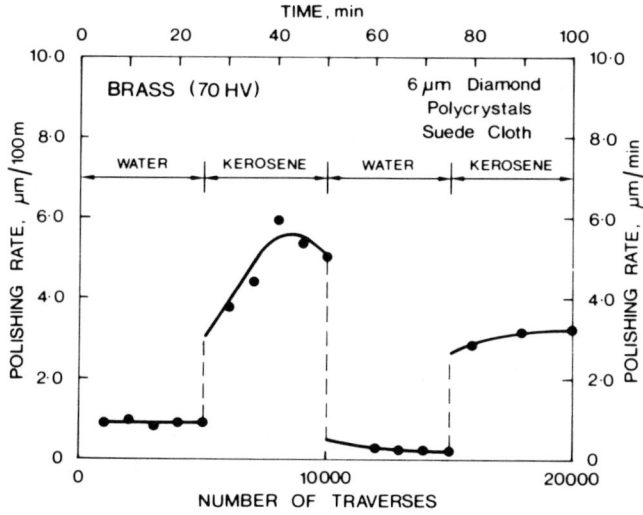

FIG. 5.29. Effect of alternating water and kerosine as the polishing fluid with the standard carrier paste.

would wash such an emulsion into the base of the polishing cloth. Kerosine, on the other hand, would probably float globules of the paste up to the working regions of the cloth.

Even when kerosine is used as the polishing fluid, it still has to be added in a sufficient quantity to keep the polishing track moist if the maximum polishing rate is to be obtained. Moreover, this fluid must be replenished regularly. The polishing rate falls off markedly if the addition of fluid is discontinued, but is restored when addition is resumed (Fig. 5.30). This applies to both napped and napless cloths. The addition of excessive amounts of fluid, on the other hand, reduces the polishing rate with napped cloths (Fig. 5.31), but not with most napless cloths. Napless cloths, such as drill cloth, absorb and distribute the excess fluid. In any event, the optimum amount of fluid that should be maintained on the polishing cloth can usually be judged qualitatively: the pad must be quite moist, but not sloppy. There are therefore advantages in adding the fluid regularly at a controlled rate. On the other hand, volatile solvents would have disadvantages.

Kerosine was added continuously at the rate of 0.5 ml per 1000 traverses in the other polishing-rate trials discussed in this chapter.

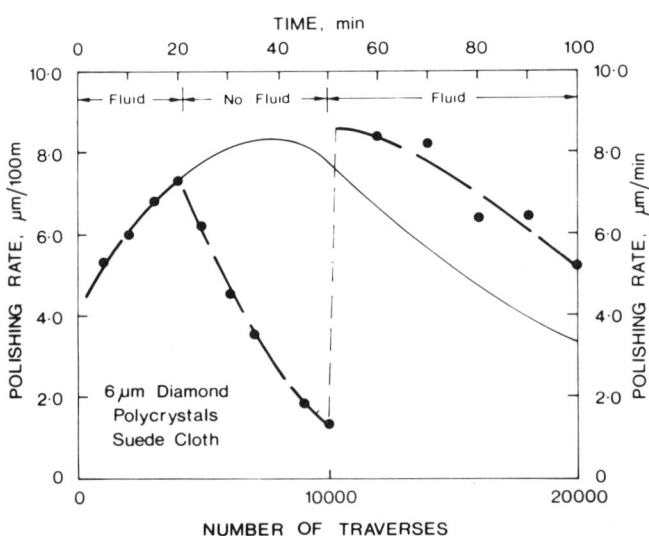

FIG. 5.30. Experiment illustrating the effect of interrupting the addition of polishing fluid to the polishing cloth.

The continuous curve indicates the variation of polishing rate with number of traverses when kerosine was added continuously at a rate of 0.5 ml per 1000 traverses. The broken curve indicates the result of adding kerosine at the same rate during the periods labeled "Fluid" but discontinuing its addition during the period marked "No Fluid".

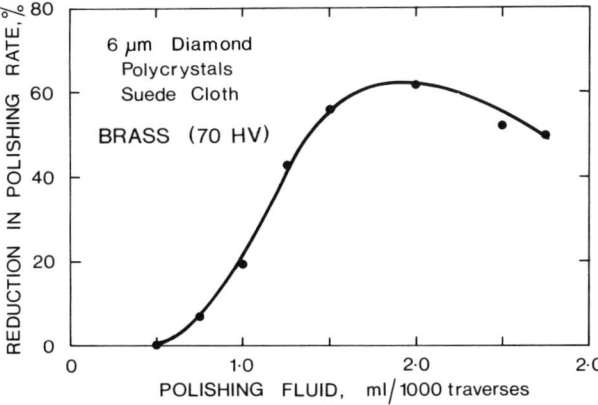

FIG. 5.31. Effect of adding excessive amounts of polishing fluid (kerosine) to the polishing pad.

In this experiment, the supply of kerosine was gradually increased during one charging of abrasive to the polishing track.

In addition to reducing the polishing rate, very dry polishing cloths produce blemishes, of the type illustrated in Fig. 5.32, on the polished surface.

Load Applied to the Specimen

The polishing rate and the total thickness of material that can be removed by a specific number of traverses increase linearly with increases in the pressure applied to the specimen once it has exceeded a certain value. The value for synthetic suede cloth is about 75 g/cm^2.

Specimen Material

Material-removal rates obtained for a number of representative metals and alloys under a particular set of polishing conditions are listed in Table 5.2. The total thickness removed after 2×10^4 traverses with one charging of carrier paste is also listed. The polishing conditions were optimized, as discussed earlier in this chapter, to give the best conditions for material removal. The materials listed are mostly the same as those for which abrasion rates are given in Table 3.3 (p. 57).

For polishing even more so than for abrasion, there is no apparent correlation between the material-removal rate and any simple physical or mechanical property of the specimen material. Certainly there is no direct relationship with hardness (Fig. 5.33a). Some soft metals (e.g., gold) have very low polishing rates, and some (e.g., aluminum and leaded brass) have anomalously high polishing rates. These same metals have anomalous abrasion rates (cf. Table 3.3). Some quite hard metals (e.g., the quench-hardened 1.4%-C steel) have comparatively high polishing rates. Often, pure metals have polishing rates lower than those of

188 / **Metallographic Polishing by Mechanical Methods**

FIG. 5.32. Blemishes produced on a brass surface that was polished on an excessively dry suede cloth charged with 6-μm-grade diamond abrasive. Magnification, 250×.

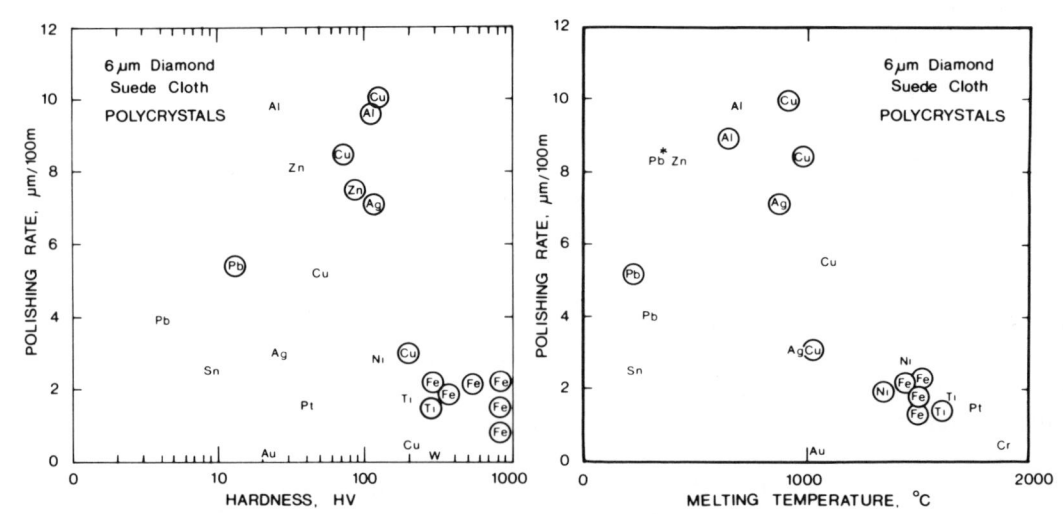

FIG. 5.33. Variation of maximum polishing rate with hardness and with melting temperature.

Circled symbols indicate alloys of the base metal. Other symbols indicate commercially pure metals. The metals and alloys are those listed in Table 5.2.

their harder alloys. All that can be said is that many materials with hardnesses below about 125 HV have polishing rates higher than those of harder materials. But it can also be said that many materials with melting temperatures below about 1200°C have polishing rates higher than those of materials with higher melting temperatures (Fig. 5.33b).

The values listed in Table 5.2 are for polycrystalline diamond, the ratio of the polishing rates of polycrystalline and monocrystalline diamonds being shown graphically in Fig. 5.34. With a few exceptions, a higher polishing rate is obtained with polycrystalline diamond (by a factor of between 1.5 and 2.5) than with monocrystalline diamond when specimen hardness is greater than about 20 HV. The polishing rates for the two types of diamond are about the same when specimen hardness is from 10 to 20 HV, and the monocrystals polish faster than the polycrystals when specimen hardness is less than about 10 HV. The same general pattern is observed when the comparison is based on the total thickness removed after 2×10^4 traverses. A possible explanation is that the advantage of the multitudinous small facets on the polycrystalline particles is lost when the general depth of indentation of the particle during polishing exceeds a certain value. The general shape of the particles then becomes of dominating importance, in which event the more rounded shape of the polycrystalline particles eventually becomes a comparative disadvantage. However, metals which polish faster with monocrystalline

FIG. 5.34. Variation with specimen hardness of the ratio of maximum polishing rates obtained with polycrystalline and monocrystalline diamond abrasives of similar grades.

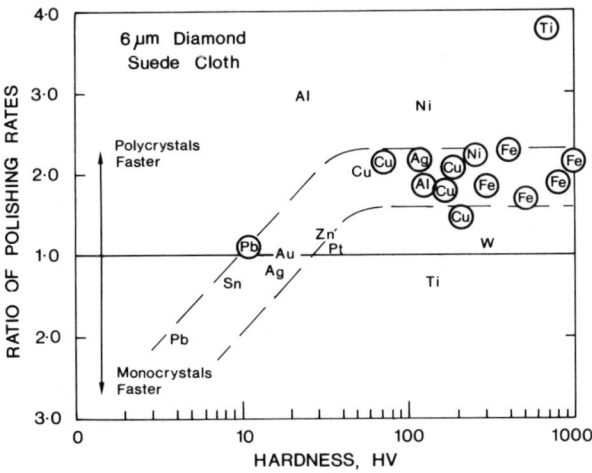

Circled symbols indicate alloys of the base metal. Other symbols indicate commercially pure metals. The metals and alloys are those listed in Table 5.2.

than with polycrystalline diamond are very much the exception. Polycrystalline diamond is the more effective abrasive of the two for general use.

Summary of Optimum Polishing Conditions

As was the case for abrasion, many parameters have significant effects on the rate of material removal achieved during polishing with diamond abrasives. It is clear that use of diamond abrasives does not ensure that a high polishing rate will be obtained. The rate obtained can vary over at least an order of magnitude if attention is not paid to details of the conditions of use.

The first factor to be considered is the manner in which the diamond abrasive is added to the polishing cloth. It is vital that it be added in a carrier paste or compound which holds the abrasive particles to those fibers of the polishing cloth that actually contact the specimen surface. The characteristics of the paste described in Appendix 5-B that does this effectively have been explored and, for this paste, the best conditions of use are:

1. Do not use abrasive coarser than 6-μm mean particle-size diameter.
2. Use polycrystalline diamond.
3. Use a concentration of 0.5 to 1.0 wt % of abrasive in the carrier paste.
4. Add about 2.0 g of paste for every 100 cm² of surface of the polishing cloth that is to be used, and spread the paste as uniformly as possible.
5. Use a wear-resistant napped cloth. Napless cloths should be used only when flatness of finish is of overriding importance.
6. Add kerosine to the polishing track, preferably in a regular, controlled manner, at a rate which keeps the polishing track moist but not sloppy.
7. Apply as high a load to the specimen as is practicable.
8. Determine by prior experiment the stage at which the polishing rate falls to a comparatively low level; wash and/or recharge the polishing cloth when this stage has been reached.
9. Wash and recharge the cloth at even shorter intervals if there are visible signs of the polishing cloth clogging with polishing debris; take particular care in this respect with napless cloths.
10. Replace the polishing cloth when it shows obvious signs of wear.

Diamond-containing pastes and compounds in a variety of base compositions are available commercially, and it does not follow that the above recommendations apply to them all.* Nevertheless, it does seem reasonable to conclude that, in metallographic polishing, a compound must be used which attaches itself to the fibers of the contacting regions of the

---

*Many of them are, however, emulsions of soaps or waxes not too different in principle from the soap emulsion considered here.

polishing cloth. Moreover, it is probable that a polishing fluid must be used which is matched specifically to the compound so that it causes the compound to remain at these positions during use. Several other recommendations in the list above are probably also generally applicable. Examples are the influence of the type of polishing cloth used and the effects of wear of the polishing cloth; the effect of clogging of the polishing cloth; and the effect of the pressure applied to the specimen. On the other hand, some of the recommendations may not be so generally applicable, such as the optimum concentration of abrasive in the compound, the optimum quantity of compound and the optimum amount of polishing fluid.

These matters can be resolved only through experimental investigation by methods which are as quantitative as possible. Many of the parameters have such large effects that even the most semiquantitative estimate of polishing rate should indicate whether or not they are significant under a particular set of circumstances. If these investigations are not carried out, performance far from optimum may unknowingly be obtained. This is a situation where subjective impressions cannot be relied upon, but unfortunately this is just what frequently is done.

It is unlikely, of course, that all factors will be optimized precisely in practical polishing. However, a knowledge of how and why the maximum polishing rate can be achieved without degrading the quality of polishing will help greatly in ensuring that nearly optimum conditions are achieved.

APPENDIX 5-A

Method of Grading Aluminum Oxide Polishing Abrasive by Levigation

General Procedure*

1. To each 100 g of raw abrasive add 1 ml of 40% sodium silicate and 500 ml of clear water. Either run through a colloid mill three times or grind for 2 h in a pebble mill. The purpose of the silicate addition and the grinding is to obtain thorough wetting and dispersion of individual abrasive particles.

2. Transfer to a suitable settling vessel and stir thoroughly.

3. Allow to settle for a known period and then siphon off the top layer, removing 2.5 cm (1 in.) for each 4 h of settling time. Retain this material, which can be called the "4-hour" grade.

4. Make up to original volume with water, stir, and repeat step 3. Keep repeating until the liquid drawn off contains insufficient abrasive to warrant further collection.

5. Repeat the procedure to collect the next grade, with a settling rate of 2.5 cm (1 in.) in 30 min.

---

*Based on a procedure developed by Rodda.[24]

6. Repeat the procedure to collect a grade with a settling rate of 2.5 cm (1 in.) in 3 min.

7. Repeat the procedure to collect a grade with a settling rate of 2.5 cm (1 in.) in 20 s.

8. For each grade, collection should be followed by a settling period, during which much of the water may be removed by decantation. Settling, particularly of the finer grades, may be greatly accelerated by adding about 1.5 g of a detergent for each 3 l of suspension, which is then acidified with a few drops of concentrated hydrochloric acid. This also reduces the tendency of the abrasive to cake when dried further. If caking still occurs, repeated washing and decantation may be necessary. In any event, the "4-hour" grade should be kept in suspension. The remainder should be dried out and stored in stoppered bottles.

Levigation can be commenced at any stage of the process. For example, if only the "20-second" grade is desired, the first stage is removal of all material with a settling rate of 2.5 cm in 3 min or less, and the second is collection of the 2.5-cm-in-20-s grade.

Starting Materials

A 500- or 600-mesh bulk alumina is a suitable starting material when only the two coarsest grades are to be collected. The yield of the "4-hour" grade from such material is, however, vanishingly small, and that of the "30-minute" grade is rather poor. Reasonable yields of these grades can be obtained, however, from aluminas sold as commercial buffing compounds.

Particle Sizes

The approximate diameter range (in microns) of the particles in each grade, as calculated from Stokes' Law, is as follows:

20-second grade . . . . . 10-30 μm
3-minute grade 3.5-10
30-minute grade 1.5-3.5
4-hour grade 1.5

APPENDIX 5-B

Method of Preparing a Carrier Paste for Diamond Abrasives *(Samuels[29])*

Constituents

Stearic acid   25 g
Triethanolamine*   12 ml
Water   50 ml
Diamond abrasive   1 g (5 carats)

---

*It is a wise precaution to wear protective gloves when handling triethanolamine.

Method of Manufacture

The stearic acid is melted and heated to 80 to 90°C (higher temperatures will cause frothing when the water is added later). The triethanolamine and most of the water are mixed and heated to a temperature in the same range, a small amount of a wetting agent and the diamond dust are added, and the abrasive is shaken into a uniform suspension. The molten stearic acid is stirred vigorously with a mechanical stirrer and the abrasive suspension is introduced rapidly; the water not used in the original suspension can then be used to wash in any abrasive remaining in the container. The mixture emulsifies immediately, but stirring should be continued until the emulsion cools and thickens.

Alternatively, and more satisfactorily if the equipment is available, the diamond abrasive is added to all of the water in a glass container and the wetting agent added. The container is then placed at the focus point in the bath of an ultrasonic cleaner and irradiated until all of the abrasive has been found into suspension. The triethanolamine is added to this suspension and the same procedure as above then followed.

The paste can be conveniently stored in and dispensed from collapsible tubes, although more elaborate syringe-type dispensers are also available.

The paste may be colored for grade identification by adding an alkali-stable dye to the water during manufacture of the emulsion.

Identification colors that have been standardized are set out in the following table:

Standard Color System for Diamond Pastes
(From British Standard 2886, 1957)

Nominal size range, μm	Color
8-12	Green
4-8	Orange
2-4	Yellow
1-2	Ivory
0-1	White

REFERENCES

1. L. H. Tanner and M. Fahoum, *Wear*, 1976, *36*, 299.
2. Sir George Beilby, "Aggregation and Flow of Solids", MacMillan, London, 1921.
3. G. P. Thompson, "Structure and Properties of Solid Surfaces", edited by R. Gomer and C. S. Smith, Univ. Chicago Press, Chicago, 1953, p. 185.
4. F. P. Bowden and T. P. Hughes, *Proc. Roy. Soc.*, 1937, *160A*, 575.
5. F. P. Bowden and D. Tabor, "The Friction and Lubrication of Solids", Clarendon Press, Oxford, 1950.
6. L. E. Samuels, *J. Inst. Metals*, 1956-57, *85*, 51.
7. F. P. Bowden and P. H. Thomas, *Proc. Roy. Soc.*, 1954, *223A*, 29.
8. R. J. Heighway and D. S. Taylor, *Wear*, 1966, *9*, 310.
9. E. C. Rabinowicz, *Scientific American*, 1968, *218* (6), 91.

10. W. Blum and H. S. Rawdon, *Trans. Amer. Electrochem. Soc.*, 1923, *44*, 397.
11. A. M. Portevin and M. Cymboliste, *Trans. Faraday Soc.*, 1935, *31*, 1211.
12. P. A. Jacquet, *J. Chemie. Phys.*, 1936, 33, 226.
13. L. E. Samuels, *J. Inst. Metals*, 1956-57, *85*, 177.
14. D. M. Turley and L. E. Samuels, *Metallography*, 1981, *14*, 283
15. G. P. Thompson, *Proc. Roy Soc.*, 1930, *128A*, 649.
16. R. C. French, *Nature*, 1932, *129*, 169.
17. H. Raether, *Z. Physik*, 1933, *86*, 82.
18. W. Kranert and H. Raether, *Ann. d. Physik*, 1943, *43*, 520.
19. P. E. Axon, *Proc. Phys. Soc.*, 1940, *52*, 312.
20. L. E. Samuels and J. V. Sanders, *J. Inst. Metals*, 1958-59, *87*, 129.
21. N. Takahashi, *Métaux et Corrosion*, 1950, *25*, 37.
22. R. L. Aghan and L. E. Samuels, *Wear*, 1970, *16*, 293.
23. W. J. McG. Tegart, "The Electrolytic and Chemical Polishing of Metals", Pergamon, London, 1959.
24. J. L. Rodda, *Trans. Amer. Inst. Min. Met. Eng.*, 1932, *99*, 149.
25. E. J. Schneider, "Science and Technology of Industrial Diamonds", edited by J. Burls, Industrial Diamond Information Bureau, London, 1967, Vol. 1, p. 161.
26. N. F. Bailey, "Proceedings: Diamonds in the 80s", Industrial Diamond Association of America, Chicago, 1980.
27. "Grading of Diamond Powders in Sub-Sieve Sizes", Commercial Standard CS261-63, U.S. Government Printing Office, Washington, D.C.
28. R. D. Buchheit and J. L. McCall, "Metallographic Specimen Preparation", edited by J. L. McCall and W. M. Mueller, Plenum, New York, 1974, p. 77.
29. L. E. Samuels, *J. Inst. Metals*, 1952-53, *81*, 471.
30. H. E. Exner and K. Kuhn, *Practical Metallography*, 1971, *8*, 453.

CHAPTER 6

Polishing With Abrasives: Surface Deformation

PLASTICALLY DEFORMED LAYERS similar to, but shallower than, those present on abraded surfaces (see Chapter 5) can be expected on mechanically polished surfaces if and when polishing occurs by micromachining. In the present chapter we shall describe current knowledge on the structure of the deformed layer produced by polishing, using the same approach as that used in Chapter 4.

GENERAL CHARACTERISTICS OF THE PLASTICALLY DEFORMED LAYER

Structure of the Layer

The general pattern of the plastically deformed layer introduced during polishing is most conveniently revealed by examination by optical microscopy of taper sections of suitably prepared specimens of 70:30 brass.[1] Sections of a moderately finely polished surface etched by three selected methods are shown in Fig. 6.1 (cf. Fig. 4.10 for a corresponding series for an abraded surface). Etching in the ferric chloride reagent has developed an unresolved dark-etching band at the surface similar in appearance to the fragmented layer in abraded surfaces (cf. Fig. 6.1a and 4.11), but twin or deformation-band strain markings are not developed immediately beneath the fragmented layer. Slip strain markings, however, can be developed by the use of appropriate etchants (Fig. 6.1b and c). These more lightly strained regions extend in bands beneath individual surface scratches in the same way as for finely abraded surfaces (cf. Fig. 6.1c and 4.12). Thus, the strains immediately beneath the surface fragmented layer are comparatively small in magnitude, and there is a shallow strain gradient to the elastic-plastic boundary. The strains are, however, inhomogeneously distributed laterally in a manner that is closely related to the surface scratches produced during polishing.

The fine structure of the fragmented layer has been elucidated by examination by transmission electron microscopy of sections cut through the layer formed on polycrystalline copper.[2] The structure for a relatively coarsely polished surface is illustrated diagrammatically in Fig. 6.2 (cf. Fig. 4.15, p. 106, for the corresponding structure of an abraded surface).

196 / Metallographic Polishing by Mechanical Methods

FIG. 6.1. Taper section of the surface of annealed polycrystalline 70:30 brass that has been polished on a synthetic suede cloth charged with 1-μm-grade diamond abrasive.

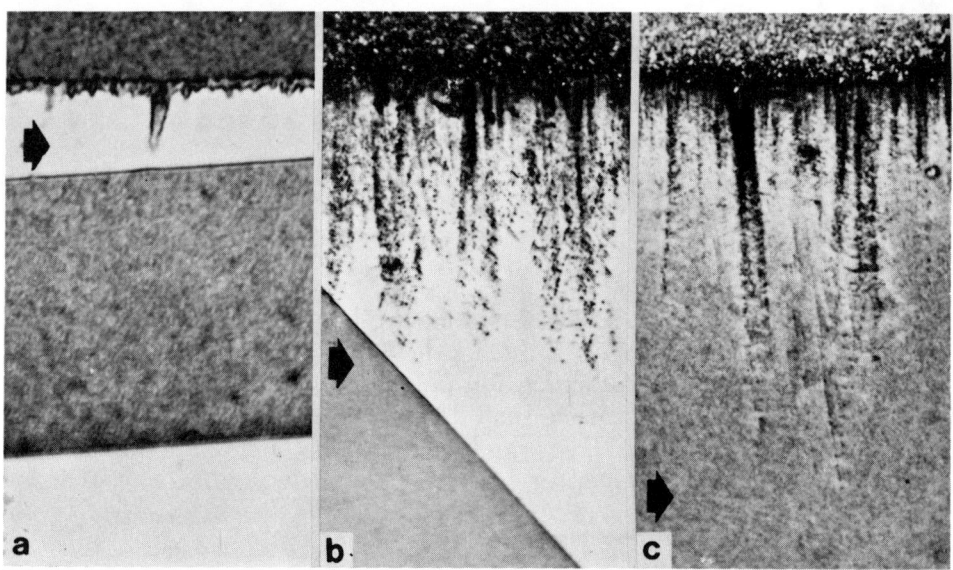

The section has been etched by three methods that have different threshold strains for revealing deformation, as follows. **(a)** Ferric chloride reagent (threshold strain: 5% compression). **(b)** Low-sensitivity sodium thiosulfate etch (threshold strain: 0.1% compression). **(c)** High-sensitivity sodium thiosulfate etch (threshold strain: elastic limit). In each case, the base of the layer in which the manifestations of deformation have been developed is indicated by an arrow. Taper ratio, 10.9. Magnification, 2000×. (See Fig. 4.10 for a corresponding section of an abraded surface.)

FIG. 6.2. Diagrammatic illustration of the structures observed by transmission electron microscopy in the fragmented layer in a copper surface polished on 6-μm-grade diamond abrasive. *(Turley and Samuels, Ref 2)*

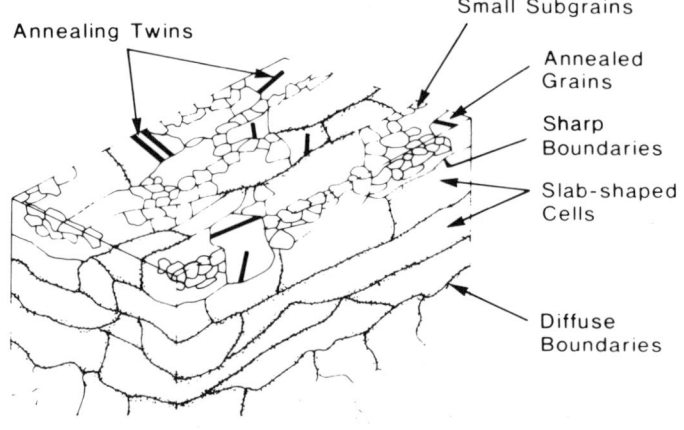

At some positions, a layer of small equiaxed grains (approximately 0.1 μm in diameter), or of larger equiaxed grains (approximately 0.5 μm in diameter) containing annealing twins, is present at the surface (a layer of the smaller equiaxed grains is visible in Fig. 5.7, p. 151). The equiaxed grains at these positions overlay slab-shape cells 1 to 2 μm long and 0.1 to 0.2 μm thick with sharp boundaries (a subsurface layer of these cells is also visible in Fig. 5.7). These cells in turn overlay slab-shape cells of similar size but with diffuse boundaries. This is exactly the same sequence of structures as that found in abraded surfaces (Fig. 4.15). The slab-shape cells can be identified as having resulted from the development of deformation microbands and shear bands during micromachining,[3] and the equiaxed grains can be attributed to relaxation of these deformation structures, as described earlier for abraded surfaces (p. 106). The presence of the equiaxed grains again probably indicates that some general heating of the surface occurred during polishing, probably to a higher temperature than for abrasion but still within the range 100 to 150°C.

At other positions on the polished surfaces, however, the slab-shape cells with sharp boundaries emerge at the surface, but still overlay cells with diffuse boundaries. At still other positions, only the cells with diffuse boundaries emerge at the surface. Moreover, the proportion of positions at which slab-shape cells emerge at the surface increases with increasing fineness of polish. The implication is that the maximum strain is smaller beneath some polishing scratches than beneath others, and that the proportion of low-strain scratches increases with increasing fineness of polish. This confirms the interpretations of overgrowth experiments and electron diffraction experiments discussed on pp. 149 and 150, respectively, but it is not clear why it should be so. The maximum strain might be expected to be the same for all scratches, as it is for abrasion scratches. One possibility is that the maximum strains initially were the same but that chemical solution or oxidation has removed the higher-strained surface of the deformed layer. It might be expected that a proportionately larger fraction of the high-strain regions would be removed from finely polished surfaces, which have shallower deformed layers. Even if only a very thin layer was removed by such means, a layer containing comparatively small strains would be exposed (see below), and these strains would then be smaller the finer the polish.

Irrespective of these considerations, the maximum strain at a polished surface is still large (natural strain of 4 or more; engineering strain of 98% compression or more). Nevertheless, the highly strained regions are so thin that removal of a few tenths of a micrometre, as would occur in metallographic etching, exposes material that has been much less severely strained. For example, the reflection electron diffraction pattern shown in Fig. 5.9(b) was obtained from a surface of silver which has been polished on magnesium oxide and then etched in an ammonium hydroxide – hydrogen peroxide metallographic etchant. Kikuchi lines are visible in the pattern and there is very little arcing of the diffraction spots, both of which are indicative of a crystal of considerable perfection that is much more nearly perfect than before etching (Fig. 5.8d).

Note also that the grains and subgrains exposed at the polished surfaces, although small, are much too large to be described as amorphous or even amorphous-like. Diffraction patterns characteristic of fully crystalline material are obtained when areas containing even the smallest grains are examined in these thin sections by transmission electron diffraction.[2] Thus they cannot be regarded as having a structure anything like that which the Beilby layer was supposed to have (see p. 142). Even so, it is sometimes suggested that the layer whose structure was altered during polishing should still be referred to as the Beilby layer in deference to Beilby's contribution to this field. Such a course of action would be most confusing and misleading because of the deeply seated association between the name Beilby and amorphous structures, and so it will not be followed here.

Translations Parallel to the Surface

We saw in Chapter 4 (p. 107) that translations parallel to the surface occur during abrasion, but to an insignificant depth. Consequently, the question that now has to be addressed is whether or not similar translations occur during polishing and, if so, to what extent.

We have already noted in the discussion of the experiment illustrated in Fig. 5.3 (p. 146) that translations can occur during polishing in the form of development of burrs at free edges, but we have seen that the burrs formed by even comparatively coarse polishing in soft materials are too small to be detected by optical microscopy. The possible extent of the same type of phenomenon at the edge of a second phase is illustrated in Fig. 4.17(b) (p. 108). Burrs have formed at the exit edges of the plates of silver in this alloy but the burrs extend for only 0.1 to 0.2 μm; so again they are too small to be detected by optical microscopy.

It is sometimes claimed that small structural features visible after electrolytic polishing but not after mechanical polishing have been obscured by surface smearing during mechanical polishing. A critical re-examination of a number of such cases has indicated that they fall into one of two categories. In the first category, fine precipitates are really present and these precipitates can be detected after polishing by those mechanical methods which produce adequately scratch-free finishes;[4-8] an example is given in Fig. 8.21 (p. 241). In the second, the features detected after electrolytic polishing are not precipitates but rather etch pits.[6,9] It has also been reported that fine cavities in brass are obscured during polishing on diamond abrasives.[10] A re-examination of this case showed that the cavities could not be seen because they had been filled by loose polishing debris (see p. 303). It was found that the debris could be cleaned out of the cavities by an ultrasonic cleaning treatment, when the cavities became visible. The debris could also be removed by any of the final polishing treatments that should be used for brass. Some of the problems just referred to seem to arise because investigators assume that the mere use of diamond abrasives is a cure for all the problems of mechanical surface preparation. Effective though they are, the finest available diamond abrasives usually do not produce adequate final polishing of soft specimen materials, as will be discussed in Chapter 8.

TABLE 6.1. Depth of the Plastically Deformed Layer Produced on Surfaces of Annealed Polycrystalline 70:30 Brass by Metallographic Polishing

Polishing abrasive(a)		Depth of scratches, μm	Depth of fragmented layer, D_f(b), μm	Depth of deformed layer, D_d(b), μm
Type	Grade, μm			
Aluminum oxide, α type	10-20	...	0.2	1.7
	0-1	...	...	2.5
	0-0.5	...	0.08	1.1
Aluminum oxide, γ type	0-0.1	...	0.03	0.7
Diamond, monocrystalline	6	0.08	0.17	1.0
	1	0.05	0.1	0.7

(a) The data in this table are for hand polishing on a napped cloth using water as a polishing fluid with aluminum oxide abrasive, or kerosine with diamond abrasive. (b) Depth beneath the root of the polishing scratches.

The general conclusion is, therefore, that translations parallel to the surface, or *plastic smearing* as it may be called to distinguish it from *Beilby smearing* (p. 142), can be dismissed as a significant phenomenon in optical microscopy. We established earlier that Beilby smearing is not a real phenomenon.

Depths of the Deformed Layers

The depths of the deformed layers produced by typical metallographic polishing processes using aluminum oxide and diamond abrasives are listed in Table 6.1, the same definitions being used for the various layers as those listed on p. 110. In this instance, the depths of significant damage can be taken to be the same as the depths of the fragmented layer for all etchants that do not develop slip strain markings. The depths of the deformed layers can be taken to be the depths of significant damage for etchants that do.

Note that the depths of the deformed layers on polished surfaces are an order of magnitude smaller than those on abraded surfaces (cf. Tables 4.1 and 6.1), although the ratio of the depth of deformation to the depth of scratches is of the same order. Note also that the depth of deformation is somewhat less for diamond than for aluminum oxide abrasives of similar grade, although the difference is not a major one.

MODIFICATION OF THE STRUCTURE OF THE DAMAGED LAYER DUE TO FORMATION OF DEFORMATION TWINS

The plastic deformation caused by mechanical polishing can result in the formation of mechanical twins in noncubic metals, as it can in abrasion; but the disposition of these twins is quite different from the disposition of those in abraded surfaces.[11] Small discrete twins are formed beneath individual surface scratches (Fig. 6.3) but not beneath all scratches.

FIG. 6.3. Taper sections of polished surfaces of polycrystalline zinc, viewed in polarized light.

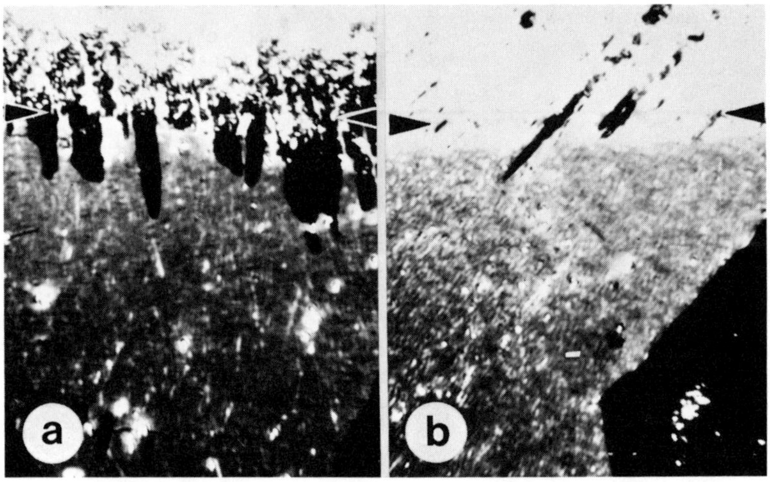

(a) Polished on 20-μm-grade aluminum oxide abrasive. Taper ratio, 7.8. **(b)** Polished on 1-μm-grade diamond abrasive. Taper ratio, 10.5. Magnification (both), 1500×. The arrows indicate the locations of the interfaces between the polished surfaces and the electrodeposits used to protect these surfaces during preparation of the taper sections. (See Fig. 4.29 for corresponding sections of abrasive-machined surfaces.)

Moreover, the proportion of scratches with which twins are associated decreases with increasing fineness of polish (cf. Fig. 6.3a and b).

The twin-containing layer can extend to quite appreciable depths. In zinc,[11] this layer is approximately as deep as the outer fragmented layer in brass (cf. Tables 6.1 and 6.2).

MODIFICATION OF THE STRUCTURE OF THE DAMAGED LAYER DUE TO RECRYSTALLIZATION AND STRAIN RELIEF

Recrystallization of the polished damaged layer that is detectable by optical microscopy has not been observed, even in those metals whose normal recrystallization temperatures are at or below room temperature. It is possible that, under these circumstances, the small volumes of strained material are absorbed by the base crystal instead of new crystals being nucleated. There is some evidence that something of this nature does occur. For example, subgrains do not form in electrodeposits of tin grown on quite coarsely polished surfaces of tin. This is most simply explained by assuming that the misoriented areas expected beneath the polishing scratches have been absorbed into the matrix without new grains with different orientations being formed. Likewise, those polishing scratches

TABLE 6.2. Depth of the Twin-Containing Layer Produced on Surfaces of Polycrystalline Zinc by Metallographic Polishing

Type	Polishing abrasive(a) Grade, μm	Depth of twin-containing layer(b), μm
Aluminum oxide, α type	10-20	3.8
	0-2	2.5
	0-0.3	0.7
Aluminum oxide, γ type	0-0.1	0.1
Diamond, monocrystalline	6	2.3
	1	0.6

(a) The data in this table are for hand polishing on a napped cloth using water as a polishing fluid with aluminum oxide abrasive, or kerosine with diamond abrasive. (b) Depth beneath the root of the polishing scratches.

in the surface illustrated in Fig. 6.3(b) which have not produced twins in the surface have not initiated subgrains in the covering electrodeposit. Moreover, surface twins of the type illustrated in Fig. 6.3 are absorbed into the matrix without trace when the polished specimen is heated to temperatures as low as 250 °C.

It is probable, therefore, that polished surfaces of metals such as zinc, tin and lead are virtually strain-free. It is possible, moreover, that heat treatment at temperatures at, or even below, the normal recrystallization temperature would produce strain-free surfaces in many metals without introducing new structural features.

EMBEDDED ABRASIVE

Embedding of abrasive can occur during polishing on diamond abrasive polishing pads; however, it is a problem only with very soft materials, such as annealed high-purity aluminum (20 HV). Several factors determine the extent to which embedding occurs.

First, the fluid used on the polishing pad is important. For example, marked embedding may occur when a diamond-impregnated paste of the type described on p. 192 is used with water as the polishing fluid but not when it is used with kerosine. Secondly, the effect is marked only with freshly charged polishing pads. Presumably, only loose abrasive particles are likely to embed in the specimen surface. Loose particles are likely to be less numerous when kerosine is used as the polishing fluid than when water is used (see p. 184). They can also be expected to be less numerous when the polishing pad has stabilized after a little use.

REFERENCES

1. L. E. Samuels, *J. Inst. Metals*, 1956-57, *85*, 51.
2. D. M. Turley and L. E. Samuels, *Metallography*, 1981, *14*, 283.
3. A. S. Malin and M. Hatherly, *Met. Sci.*, 1979, *13*, 463.
4. L. E. Samuels, *J. Inst. Metals*, 1953-54, *82*, 227.

5. L. E. Samuels and A. R. Bailey, *Metal Industry*, 1954, *85*, 143.
6. L. E. Samuels, *J. Inst. Metals*, 1955-56, *84*, 333.
7. L. E. Samuels, *J. Inst. Metals*, 1955-56, *84*, 509 (discussion).
8. A. R. Bailey, R. McDonald and L. E. Samuels, *J. Inst. Metals*, 1956-57, *85*, 25.
9. L. E. Samuels, *J. Inst. Metals*, 1949, *76*, 91.
10. G. J. Cocks and D. M. R. Taplin, *Metallurgia*, 1967, *75*, 229.
11. L. E. Samuels and G. R. Wallwork, *J. Inst. Metals*, 1957-58, *86*, 43.

CHAPTER 7

Brittle Materials: Principles

SOME ALLOYS CONTAIN PHASES which are brittle in bulk in the sense that they have poor resistance to the propagation of cracks and like defects; some of the intermetallic phases present in aluminum alloys are examples. Other specimens may contain extensive regions of a brittle material, an example being surface oxide scales. Some materials that a metallographer might be expected to examine are themselves brittle; examples include sintered carbides, semiconductors, ceramics and glasses. The behavior of brittle phases or brittle materials is different from that of metals, particularly in that abrasion occurs by a mechanism different from that by which polishing occurs.

It is somewhat ironic that the discussions earlier in this century on whether abrasion and polishing occurred by different mechanisms (see p. 142) centered mainly around brittle materials — namely, the glasses used in optical devices such as telescopes. The historic arguments that have arisen in this field have perhaps been due partly to the fact that observations made on one class of materials have been extrapolated to another without due consideration of the intrinsic differences in the manners in which the two behave during deformation.

MECHANISMS OF ABRASION AND POLISHING

Although in practice there are many variants, the general sequence of events followed when a sharp point is forced into the surface of a brittle material under an increasing load, and then unloaded, is as follows:[1]

1. A zone of irreversible deformation is produced beneath the indentation (Fig. 7.1a) and an impression remains in the surface after unloading. The mechanism by which this pseudoplastic deformation occurs is not known in most cases, but the fact that a permanent impression is left in the surface is a positive indication that it does occur.
2. At some critical indentation load, a crack is initiated below the point of contact where the stress concentration is greatest (Fig. 7.1b). This crack, commonly called a *median vent crack*, lies on a plane of symmetry in the applied stress field. The specific orientation of

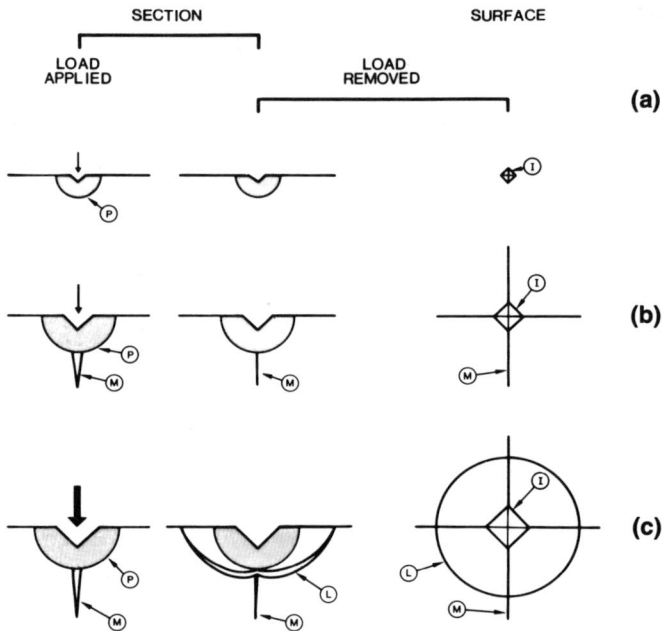

FIG. 7.1. Idealized sketch illustrating the sequence of events that is initiated when a sharp indenter is forced into the surface of a brittle material under an increasing load (top to bottom), and the events that occur during unloading.

P: pseudoplastic zone. I: permanent impression. M: median vent crack. L: lateral vent crack.

the plane depends on such factors as the geometrical configuration of the indenter and the structural anisotropy of the specimen material. The median vent crack closes but does not heal on unloading. Thus an impression and several cracks oriented perpendicular to the surface are left in the surface after unloading.

3. The deformed zone and the median vent crack grow in a stable manner as the load is increased further until, at a certain stage, sideways-extending cracks (termed *lateral vent cracks*) form during unloading. The lateral vent cracks begin to appear just before unloading of the indenter, because relaxation of the irreversibly deformed material superimposes intense residual stresses on the applied stress field. The lateral vent cracks continue to extend, eventually reaching the free surface, during further unloading (Fig. 7.1c).

Regardless of whether contact between the abrasive point and the specimen involves static indentation of the point into the specimen surface or translation of the point across the surface, material will be removed from the specimen whenever a lateral vent crack extends to the free surface; in fact, vent cracks develop at lower loads with sliding indenters.[2] A considerable volume of material will thus be removed — a much larger volume than that which is swept out of the surface by the indenting point. The material so removed can be called a *fracture chip*, as distinct from the machining chip referred to in Chapters 3 and 5, and the formation of such fracture chips is the basic material-removal mechanism for the abrasion of brittle materials.

The first implication of this mechanism is that rolling abrasive particles with sharp points (i.e., those operating in mode A, Fig. 3.3) may remove material each time that a point indents into the surface of the specimen. The second implication is that all contacting points, when translated under a high enough load in two-body abrasion, remove material irrespective of their rake angles. Because of this, and because a comparatively larger volume of material is removed by each contact point, the rate of material removal is potentially much larger for abrasion of brittle materials than for abrasion of metals. In general, it is an order of magnitude larger in practice.

The situation is a little different with blunt indenters.[3] A spherical point, for example, can maintain perfectly elastic contact up to a certain load. Beyond this load, fracture is initiated just outside the contact circle, where the Hertzian elastic tensile stresses are at a maximum, and the crack propagates by running around the line of contact to produce a *ring crack* (Fig. 7.2). This crack will be circular in an isotropic material, but

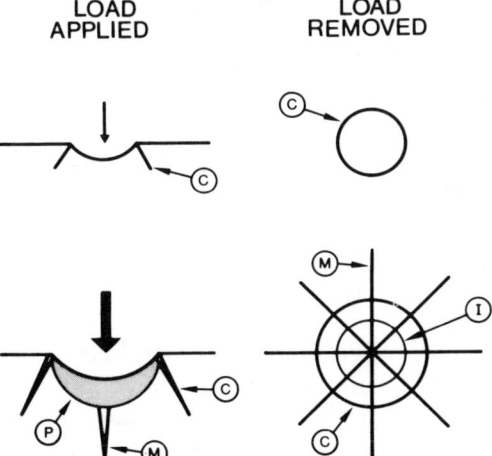

FIG. 7.2. Idealized sketch illustrating the sequence of events that is initiated when a spherical indenter is forced into the surface of a brittle material under an increasing load (top to bottom), and the events that occur during unloading.

P: pseudoplastic zone. C: cone crack, also known as a ring crack when viewed on the surface. M: median vent crack. I: permanent impression.

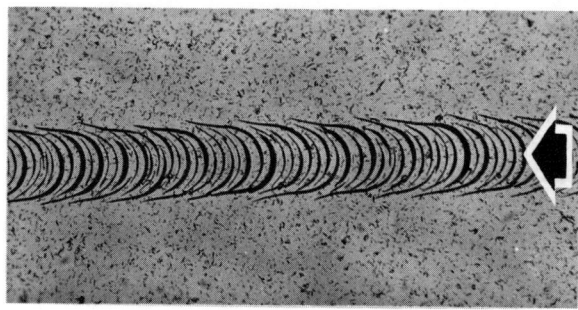

FIG. 7.3. Cracks produced in the surface of a soda-lime glass by a tungsten carbide sphere translated under load in the direction indicated by the arrow. *(Lawn and Wilshaw, Ref 3)*

The surface has been etched to make the cracks more visible. Magnification, 25×.

its path may be partly modified to reflect the crystal symmetry in crystalline materials. The ring crack propagates downward on the surface of an expanding cone of maximum Hertzian stress, so developing a *cone crack* (Fig. 7.2). The circular symmetry of the cone cracks is lost when the spherical point is slid across the surface, being replaced by a system of cracks which typically follow the trajectories illustrated in Fig. 7.3. Intersections of crack systems from adjoining grooves could result in the removal of fracture chips, but this at best would be a minor mechanism of material removal.

However, just as in abrasion by sharp indenters, increasing the applied load eventually introduces additional systems of median and lateral vent cracks (Fig. 7.2). Material removal can then occur by the same mechanism as for sharp points, although only after the application of a larger load. Thus, with blunt points, the morphology of the crack systems will be different and the efficiency of material removal will be smaller, but the major mechanism of material removal, when such removal occurs, will be much the same.

These principles are illustrated in Fig. 7.4. In two-body abrasion (such as on abrasive papers), sharp abrasive points have removed chips along their tracks (Fig. 7.4a) whereas blunt points have produced tracks of partial ring cracks (Fig. 7.4b). In three-body abrasion, on the other hand, sharp rolling abrasive particles have produced large, randomly arranged pits (Fig. 7.4c) whereas blunt particles have produced randomly arranged sets of ring cracks (Fig. 7.4d). The sharp abrasive has removed considerable amounts of material in both modes, whereas the blunt abrasive has removed virtually none under these particular circumstances.

A mathematical model of two-body abrasion by sharp abrasive points can be based on these concepts,[4] with the following result:

$$M = \eta\rho D/L(P\phi)$$

where M is the mass removed, η is a linear scaling factor relating the volume of material enclosed by lateral vent cracks to that of the plastic grooves, ρ is the density of the specimen material, D is the distance traversed, L is the load applied, P is the indentation hardness of the specimen material, and ϕ is a form factor for the indenter.

This equation is similar to that developed earlier for metals (Equation 5, p. 47) except that the factor f (which generally is less than one) in the earlier equation is replaced by η (which generally is greater than one). Some implications associated with this model are that the mass (or volume) of material removed is independent of the apparent area of contact between the workpiece and the abrasive device, is independent of the

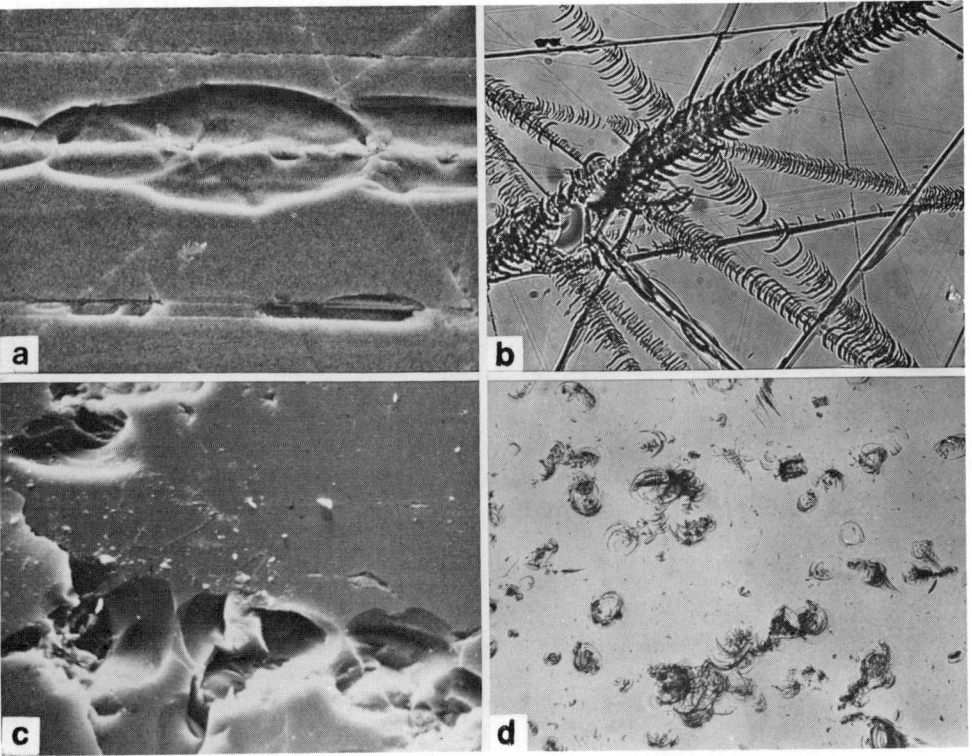

FIG. 7.4. Surface of fused silica abraded with 220-grade silicon carbide abrasive under constant load. *(Lawn and Wilshaw, Ref 3)*

(a) Two-body abrasion. Fresh, sharp abrasive particles. Scanning electron micrograph. Magnification, 300×. (b) Two-body abrasion. Worn, blunt abrasive particles. Optical micrograph after etching. Magnification, 50×. (c) Three-body abrasion. Fresh, sharp abrasive particles. Scanning electron micrograph. Magnification, 300×. (d) Three-body abrasion. Worn, blunt abrasive particles. Optical micrograph after etching. Magnification, 50×.

diameter and number of contacting particles, and is dependent on the total load applied rather than on the way in which this load is distributed. The equation also implies that abrasion rate is dependent on the hardness of the specimen material, but in brittle materials hardness can depend on both the rate of application of the load and on the environment. The abrasion rate possibly is also dependent on other material parameters that influence the development of the vent cracks, and consequently the value of η.

The above analysis applies only when the intensity of the residual stress field about the deformation track is sufficiently high to drive the lateral vent cracks to the surface. If not, material removal by a cutting mechanism is to be expected. Thus a brittle-to-ductile (i.e., a brittle-chipping to ductile-cutting) transition should occur when the load applied to a contacting point falls below a certain value. The transition does occur in practice, even in abrasion-type processes. Ductile machining chips similar to those cut from metals can be cut from even the most brittle materials (Fig. 7.5), remembering again that "brittle" here refers to behavior in bulk. When increasingly finer abrasion processes are applied, it is first found that a small proportion of the contacting points cut plastic grooves. Then the proportion of the plastic grooves increases with increasing fineness of the abrasion process. Eventually, all of the contacting points produce plastic grooves. For example, the large particles of primary silicon phase illustrated in Fig. 3.30(c) (p. 78) were chipped when abraded on 600-grade silicon carbide paper (Fig. 3.30c) whereas normal scratch grooves were produced during abrasion on the fine abrasive lap (Fig. 3.30d).

The stage at which the chipping-to-cutting transition occurs is determined by several factors. The first is the load applied to the contacting point, which in turn is determined by the total load applied to the specimen and the number of contacting points. The second factor is the shape of the contacting point; cutting will be obtained at lower loads with blunter points, so that the proportion of cutting points may differ with different abrasives of the same nominal size and may increase if the points become blunted during use. The third factor is the fracture toughness of the specimen material; the more brittle the material the smaller the proportion of cutting points.[6] Because cutting points remove much less material than chipping points, all of these factors affect the abrasion rate. Simple relationships between abrasion rate and mechanical properties consequently cannot be expected, and are not found.[6]

Irrespective of these complications, the transition from chipping to cutting invariably has occurred completely by the time that polishing processes of the type discussed in this book have been reached. That is, polishing occurs basically by micromachining, as it does for metals.

There are some exceptions to the last generalization, one of them being exemplified by glasses.[7] The surface of glass polished in the presence of water always contains a layer of hydrated material, and the formation and presence of this layer is generally thought to play a significant role in the polishing process. The polishing rate is certainly greatly influenced by the hydroxyl activity of the polishing fluid, the reactivity of the abrasive

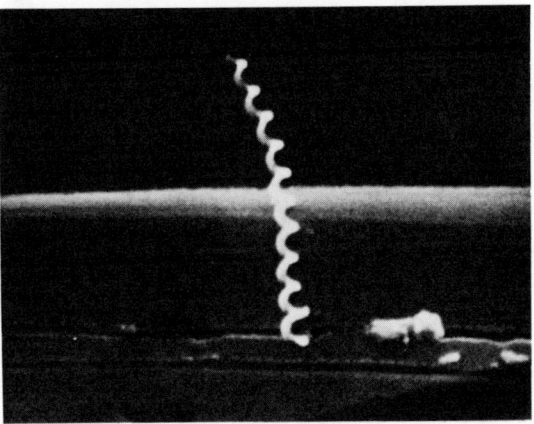

FIG. 7.5. Ductile machining chip cut in rutile (titanium dioxide) by abrasion on a 600-grade silicon carbide abrasive paper. *(Aghan and McPherson, Ref 5)*

Scanning electron micrograph. Magnification, 10,000×.

and the reactivity of the glass.[8] So, at the very least, polishing by micromachining must be modified in a major way by chemical processes. Similar modified mechanisms of polishing are possible for any material that hydrates or reacts with the polishing environment. For example, aluminum oxide can hydrate, and single crystals (sapphire) can be polished by rubbing them against a rotating disk of wood or graphite without abrasive if this is done in an atmosphere of steam at about 200°C.[9]

Polishing of diamond must also be an exception. Diamond has to be polished by diamond. Moreover, the polishing rate varies by several orders of magnitude depending on the orientation of the face being polished and the direction on that face in which the abrasive moves. Variations of only a few degrees in this direction can have a significant effect on the polishing rate.[10] Polishing of diamond therefore can scarcely occur by micromachining. The most likely mechanism of material removal is that of chipping on cleavage planes on an extremely fine (almost submicroscopic) scale, the highest polishing rates being obtained when the cleavage planes are most suitably oriented.[10]

SURFACE DAMAGE

The topography of the surface of a brittle material abraded under three-body conditions, when material removal occurs entirely by chipping fracture, consists of a random array of pits. These pits usually are bounded by conchoidal fracture surfaces. They may also be bounded partly by cleavage facets when the material has planes of easy cleavage (Fig. 7.6a). These are the pits out of which came the fracture chips.

For two-body abrasion, a mixture of grooves and randomly arrayed pits is possible (Fig. 7.6b). The grooves may be groupings of small fracture

210 / Metallographic Polishing by Mechanical Methods

pits or they may be plastically produced grooves. The proportion of plastic grooves tends, for the reasons discussed above, to be larger the finer the abrasion process and the higher the fracture toughness of the specimen material. Most of the grooves seen in Fig. 7.6(b) are bounded by small facets (Fig. 7.6d). Thus, in this particular case, they are mostly of the microfracturing type.

A layer containing fracture cracks is present beneath the fracture pits.[11,12,13] The cracks are randomly distributed after three-body abrasion (Fig. 7.6c). However, they tend to be grouped in arrays beneath the scratch grooves after two-body abrasion (Fig. 7.6d) — that is, a crack-

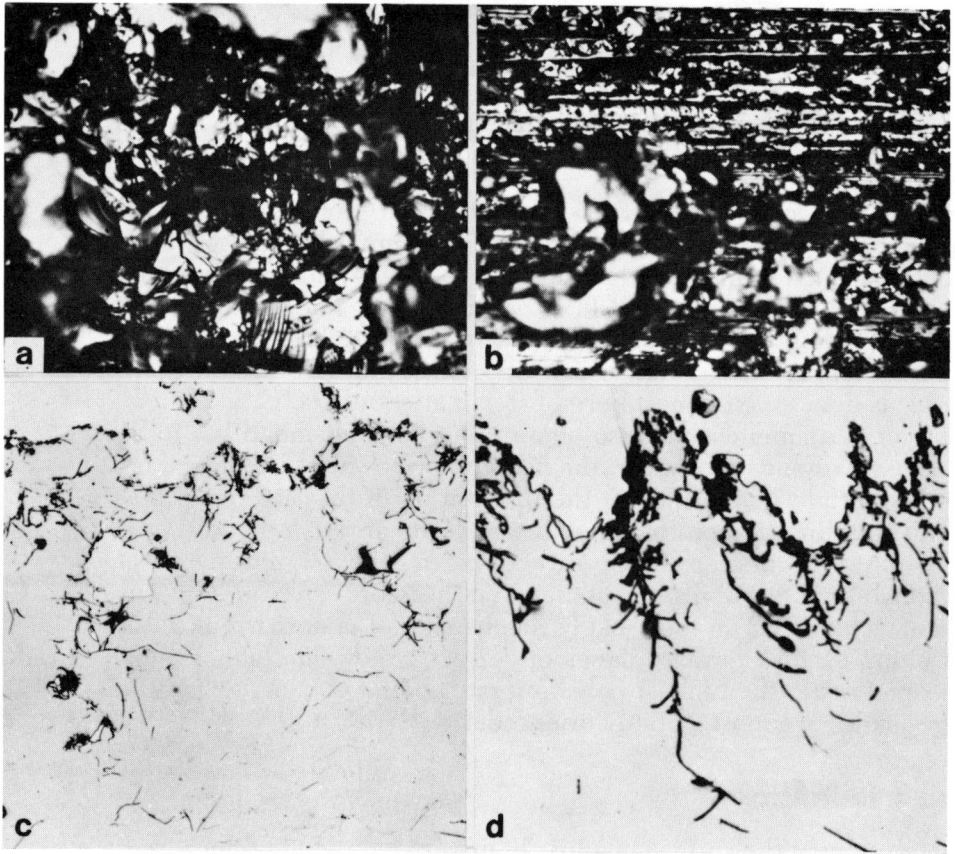

FIG. 7.6. Surfaces of a single crystal of germanium that have been abraded on 220-grade silicon carbide abrasive.

(a) and (c) Three-body abrasion. Magnification, 500×. **(b) and (d)** Two-body abrasion. Magnification, 500×. The appearance of the surfaces is shown in **(a) and (b)**. Taper sections of these surfaces are shown in **(c) and (d)**. The sections have been etched lightly in a ferricyanide reagent. Taper ratio, approx. 10.

TABLE 7.1. Maximum Depth of the Crack-Containing Surface Layers Produced by Abrasion of a Single Crystal of Germanium on a {111} Surface *(Pugh and Samuels, Ref 11)*

Abrasive			Depth of Surface Irregularities, μm	Depth of Cracked Layer(a), μm
Type	Grade	Method of use		
Silicon carbide	220	Two body	7	22
	400	Two body	2.5	10
	600	Two body	1.5	6.5
	220	Three body	23	85
	400	Three body	6	22
	600	Three body	3.5	11.5
Aluminum oxide	320	Three body	6	24
	600	Three body	2.5	9

(a) Depth beneath the root of the surface irregularities.

containing damaged layer is present. The crack-containing layers extend to considerable depths (Table 7.1), are deeper after three-body than after two-body abrasion, and are much the same depth for silicon carbide and aluminum oxide abrasives. The ratio of the depth of the cracked layer to the depth of the surface irregularities is, however, about the same for all conditions of abrasion. The absolute value of the depths varies with the nature of the specimen material. The damaged layer may also contain a small density of dislocations associated with the fracturing process, and even some glide dislocations for more ductile materials, but these features are not of importance in the present context.

The damage cracks themselves clearly are the residuals of lateral and median vent cracks — i.e., they are vent cracks that have not intersected with other cracks in such a way that a fracture chip has been generated.

The significance of these effects in the abrasion of alloys containing brittle constituents is illustrated in Fig. 7.7. Note, first, that the level of the silicon constituents is below that of the aluminum matrix. This is a result of the higher abrasion rate of the brittle compared to the ductile material. Secondly, the topography of the silicon constituents is more irregular than that of the aluminum matrix. Thirdly, cracks extend beneath the surface for both large and small silicon constituents.

Cracks will not be produced in the surface when the contacting points operate entirely in a ductile cutting mode, as occurs in polishing, but a layer of plastically deformed material will be produced. The deformed layer in crystalline materials usually is the result of dislocation glide. Dislocation arrays suggestive of this are observed in the surface layers of materials as diverse as silicon[12] and aluminum oxide (Fig. 7.8).[14] The mechanism involved in the deformation of noncrystalline materials, however, is not clear. Concentration flow on shear faults has been observed

FIG. 7.7. Perpendicular section of the surface of an aluminum – 13% silicon alloy abraded on 220-grade silicon carbide paper.

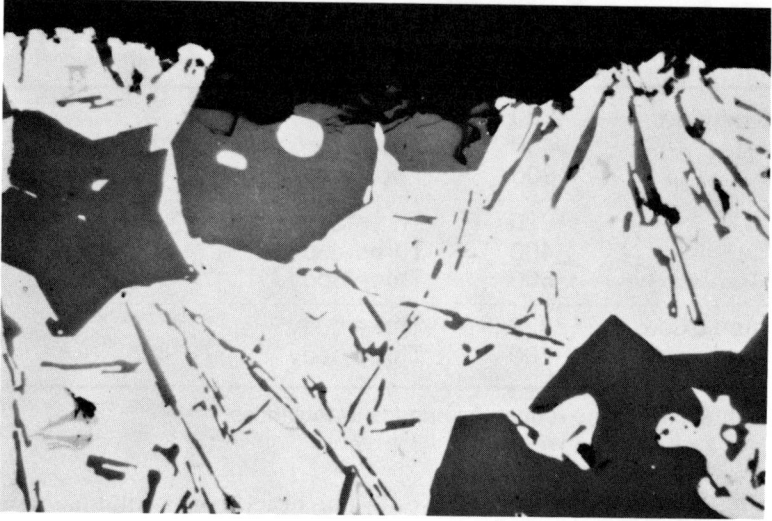

Note that both the large idiomorphic and the small lenticular particles of silicon have been abraded to a level below that of the aluminum matrix, and that both contain subsurface cracks. Magnification, 500×.

FIG. 7.8. Arrays of dislocations beneath scratches produced by polishing polycrystalline aluminum oxide on 0.25-μm diamond abrasive. *(Hockey, Ref 14)*

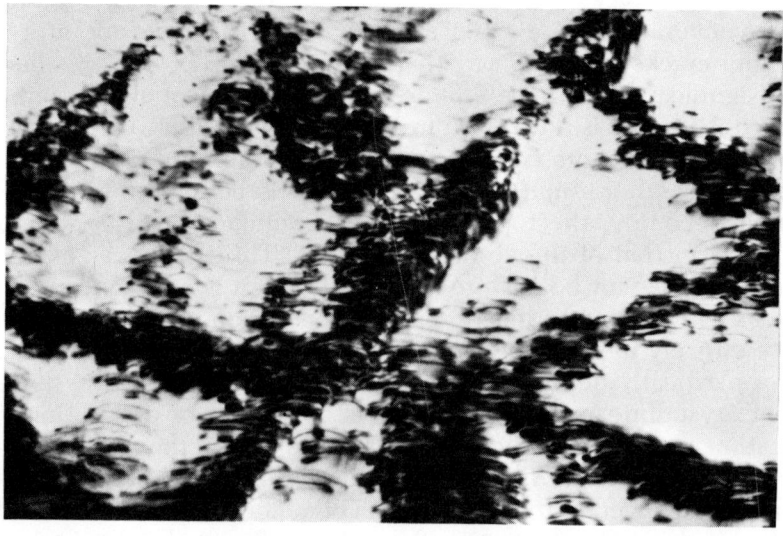

Transmission electron micrograph. Magnification, 20,000×.

beneath static indentations in glass,[15] but the nature of the actual process that occurs in these flow lines has not been established. Compaction of the molecular network may also contribute to the plastic deformation of these materials.

REFERENCES

1. B. R. Lawn and M. V. Swain, *J. Mat. Sci.*, 1975, *10*, 113.
2. B. Bethune, *J. Mat. Sci.*, 1976, *11*, 199.
3. B. R. Lawn and R. Wilshaw, *J. Mat. Sci.*, 1975, *10*, 1049.
4. B. R. Lawn, *Wear*, 1975, *33*, 369.
5. R. L. Aghan and R. McPherson, *J. Amer. Ceramics Soc.*, 1973, *56*, 46.
6. M. A. Moore and F. S. King, *Wear*, 1980, *60*, 123.
7. E. Brüche and H. Poppa, *Trans. Soc. Glass Tech.*, 1956, *40*, 513.
8. D. C. Cornish and I. M. Watt, "Proceedings of the Symposium on the Surface Chemistry of Glass", American Ceramics Society, 1966.
9. M. Okutomi and O. Imanaka, *Bull. Japan Soc. Prec. Eng.*, 1976, *10*, 156.
10. J. Wilks and E. M. Wilks, "Properties of Diamond", edited by J. E. Field, Academic Press, London, 1979, p. 351.
11. E. N. Pugh and L. E. Samuels, *J. Electrochem. Soc.*, 1961, *108*, 1043.
12. R. Stickler and G. R. Brooker, *Phil. Mag.*, 1963, *8*, 859.
13. E. N. Pugh and L. E. Samuels, *J. Electrochem. Soc.*, 1964, *111*, 1429.
14. B. J. Hockey, *J. Amer. Ceramics Soc.*, 1971, *54*, 223.
15. J. T. Hagan, *J. Mat. Sci.*, 1980, *15*, 1417.

CHAPTER 8

Principles of the Design of Preparation Systems

WE SAW IN CHAPTER 5 that a plastically deformed layer inevitably is formed on a metal surface when the surface is machined or abraded, and that microstructural changes caused by this deformation might be detected in a subsequent metallographic examination. The depth of the deformed layer in which these effects can be detected under a particular set of circumstances was defined as the depth of significant deformation (D_s). It will be apparent that a primary objective of each stage of a preparation sequence must be to remove the significant deformation that was produced by the preceding stage. A more relaxed criterion would require that each stage reduce the depth of the old significant deformation to the depth of significant damage being introduced by the new stage.

BASIC CONCEPTS

The sketches in Fig. 8.1 illustrate diagrammatically the deformed layers that are produced by a sequence of grinding, abrasion and polishing stages that might be used in a typical preparation system. The depths of the deformed layers are shown approximately to scale, and the surfaces are displaced vertically to indicate the maximum thickness that would have to be removed during each stage to meet the stricter of the two criteria discussed above. The vertically shaded bar to the left of each sketch indicates, approximately to scale, the relative material-removal rate that can be expected during each stage. The most significant feature illustrated qualitatively by these sketches is that there are two critical steps in the sequence — steps that involve large increases in the thickness of material that has to be removed and at the same time large decreases in the material-removal rate. The first critical step is from the last machining stage to the first abrasion stage, and the second is from the last abrasion stage to the first polishing stage. We shall first explore the consequences of the criterion not being met at the second of these critical steps.

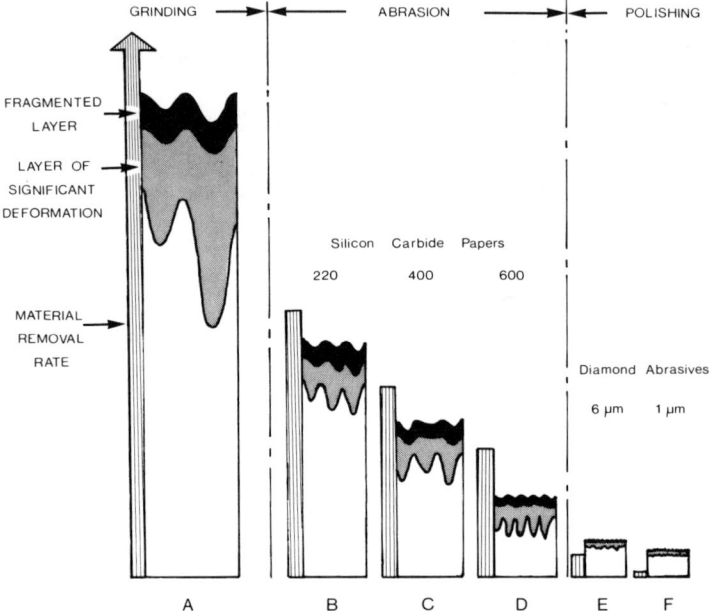

FIG. 8.1. Diagrammatic illustration of the surface deformed layers produced by the various stages of a typical preparation sequence.

The sketch for each successive stage has been displaced vertically to indicate the thickness of material that must be removed by that stage in order to remove the significant damage resulting from the immediately preceding stage. The vertically shaded bars indicate relative material-removal rates.

ABRASION ARTIFACTS

Artifacts Originating in the Fragmented Layer

The sketches in Fig. 8.2 illustrate a material in which only the fragmented layer is significant, an example of which might be a pearlitic steel (the structures developed in the fragmented layer in this material are illustrated in Fig. 4.24, p. 118). Suppose that an abraded surface of a pearlitic steel has been polished until the surface has been reduced to a level just below the roots of the abrasion scratches, as indicated in sketch B in Fig. 8.2. A typical structure that would then be observed after etching is illustrated in inset B_1; the structure is not clearly resolved but tends to have a vaguely spheroidized appearance. Suppose now that the abraded surface has been polished for a somewhat longer time to the level indicated in sketch C in Fig. 8.2. The more severely distorted regions of the fragmented layer would have been removed, but some of the regions in which the cementite lamellae have been bent during abrasion would remain. Bands of pearlite with a lamellar but distorted structure would then be seen when the surface was etched (inset C_1 in Fig. 8.2), each band being located beneath the site of a pre-existing abrasion scratch. It is only when the surface has been reduced by polishing to a level well below the fragmented layer (sketch D in Fig. 8.2) that the true lamellar structure would be observed (inset D_1).

The structures illustrated in insets B_1 and C_1 in Fig. 8.2 are false structures resulting from the presence of *preparation artifacts* or, more specifically in this case, *abrasion artifacts*. Artifact microstructures of the type illustrated in inset B_1 are rarely seen in photomicrographs published in the literature these days, but those of the type illustrated by inset C_1 sometimes are seen. The artifact structure shown in inset B_1 is the one to which Vilella[1] originally drew attention as having been the cause of much confusion when a nomenclature system was being developed for the microstructures of steels.

Vilella recommended, and it is still commonly recommended, that a surface found to have artifact structures such as those just discussed should be polished and etched repeatedly at the final polishing stage until the true structure is obtained. This is an acceptable method of removing the fragmented layer, but it is not the most effective one. The most effective thing to do is to return the specimen to a rough-polishing stage where the surface will be more rapidly lowered to the desired level. Indeed, the fragmented layer on pearlitic steels is comparatively so shallow that abrasion artifacts are not likely to be seen if a reasonably efficient rough-polishing stage is used in the preparation sequence, provided that it is realized that polishing must be continued for 2 to 3 times the period

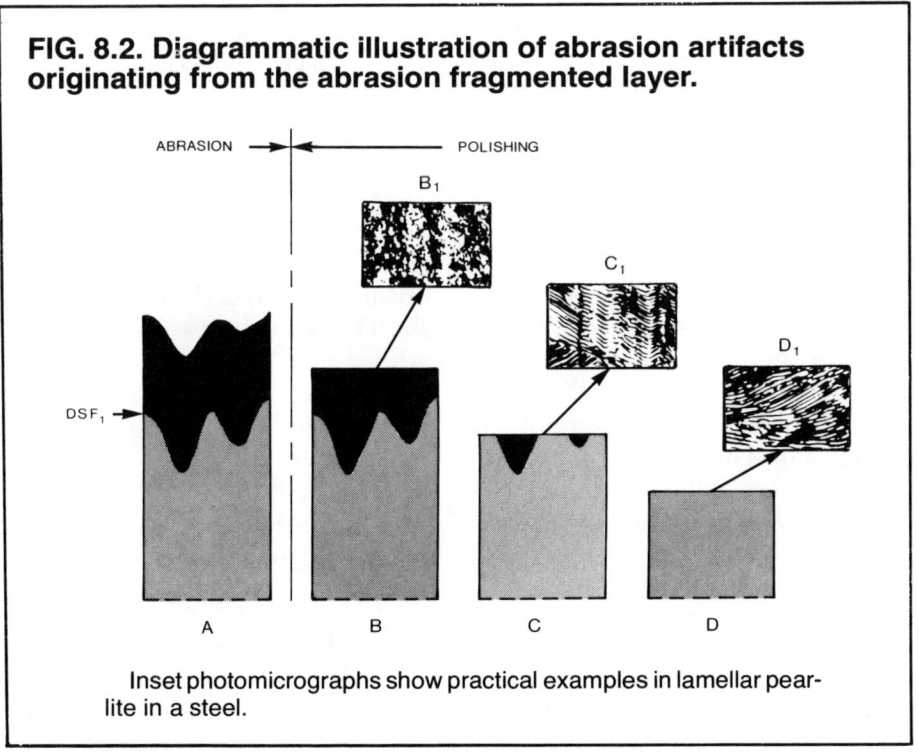

FIG. 8.2. Diagrammatic illustration of abrasion artifacts originating from the abrasion fragmented layer.

Inset photomicrographs show practical examples in lamellar pearlite in a steel.

FIG. 8.3. Abrasion artifacts in the ferrite phase in a wrought low-carbon iron.

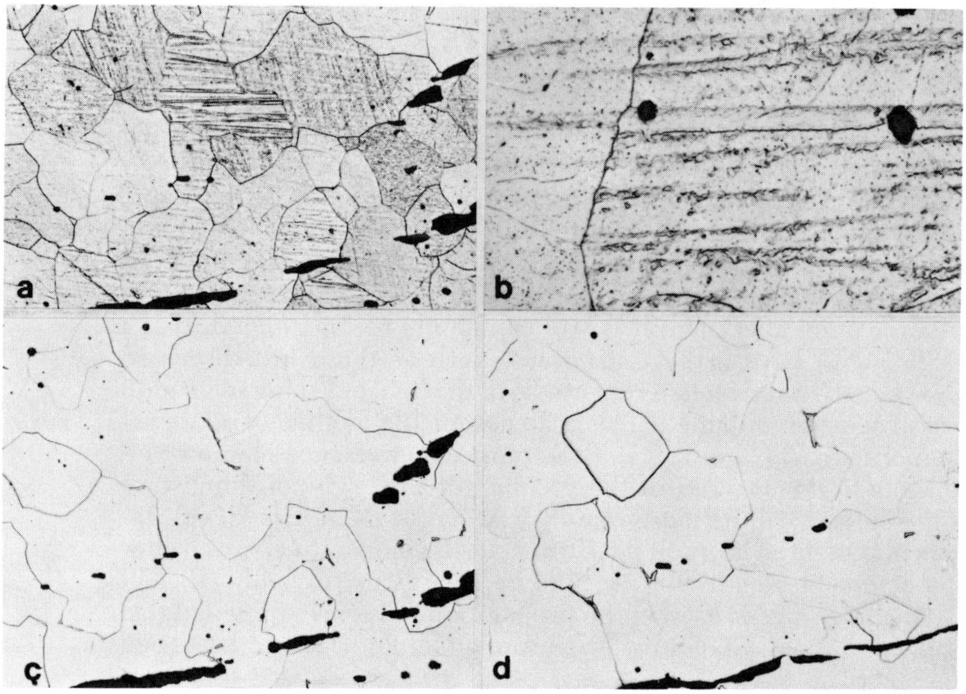

(a) Abraded on 600-grade silicon carbide paper and then final polished for a comparatively short period. Magnification, 100×. (b) As for (a). Magnification, 2000×. (c) As for (a), but returned to a rough-polishing stage and then final polished as before. Magnification, 100×. (d) As for (a), but subjected to several alternate polishing and etching cycles at the final polishing stage. Magnification, 100×. All specimens etched in nital. Compare these photomicrographs with Fig. 4.23(a) (p. 117), which shows a section of an abraded surface.

required for removal of abrasion scratches. Difficulties with artifacts of this nature in fact arise mostly when abrasion is followed by a polishing stage required to produce a final polish. This was the case with many early recommended preparation procedures and still is the case with some.

A type of artifact that generally is more likely to be found is illustrated in Fig. 8.3. The banded structures in the ferrite grains in Fig. 8.3(a), which are shown in more detail in Fig. 8.3(b), are residuals of an abrasion fragmented layer of the type illustrated in Fig. 4.23(a) (p. 117). The presence of such artifacts is possible in any material which has a structure of equiaxed grains, but their presence again would be an indication of a very poorly designed preparation sequence.

The true structure of the material illustrated in Fig. 8.3(a) is shown in Fig. 8.3(c) and (d), the artifact structures having been eliminated by the two different procedures discussed above. In Fig. 8.3(c), they were eliminated by returning the specimen to a rough-polishing stage; in Fig.

8.3(d), they were eliminated by repolishing and etching several times at the final polishing stage. The two methods have been equally effective in producing artifact-free surfaces; but considerable differences in level have developed between the ferrite grains in the second case, with the result that the ferritic grain boundaries are neither as clearly nor as uniformly delineated in Fig. 8.3(d) as in Fig. 8.3(c). The development of relief of this nature is common when the alternate etch-and-polish technique is used, and is a second reason why the technique of using an adequate rough-polishing stage is preferred. The first, mentioned earlier, is that it is less time-consuming and more reliable.

We also saw in Chapter 4 that the shapes of cavities and like features may be severely distorted when the cavities enter the abrasion fragmented zone (Fig. 4.26 and 4.27, p. 121). The shapes of the cavities will thus be misrepresented if the fragmented layer is not removed during polishing, the degree of misrepresentation depending on the extent to which the fragmented layer is left in the final surface. An example is given in Fig. 8.4 (the corresponding section of an as-abraded surface is shown

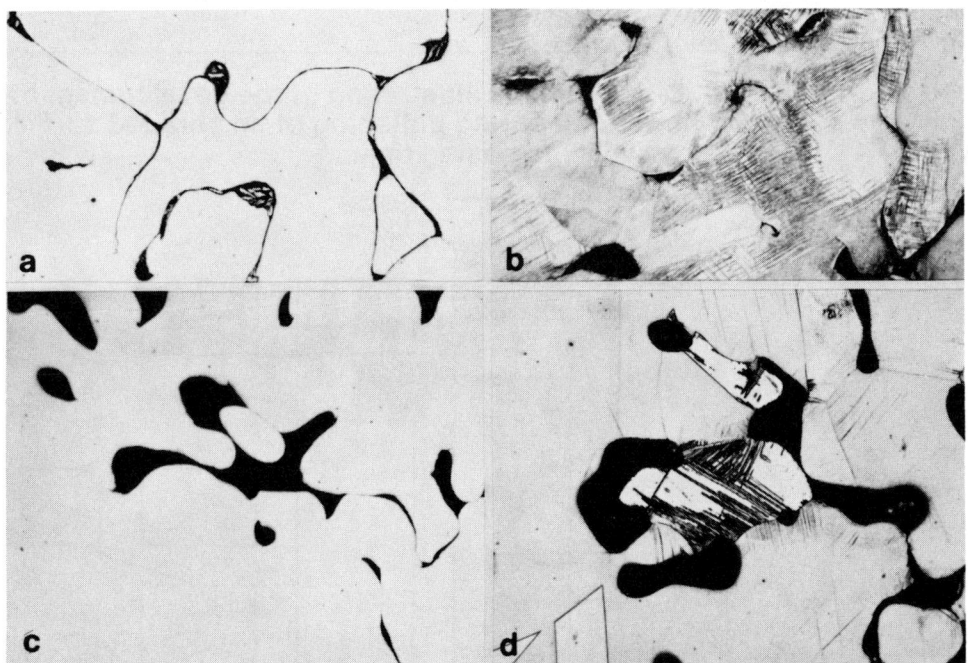

FIG. 8.4. Abrasion artifacts resulting in misrepresentation of the shapes of shrinkage cavities in a tin bronze.

(a) Abraded on 220-grade silicon carbide paper and then rough polished for a comparatively short period. Unetched. (b) As for (a), but etched in a ferric chloride reagent. (c) Abraded on 220-grade silicon carbide paper and then rough polished for a longer period. Unetched. (d) As for (c), but etched in a ferric chloride reagent. Magnification (all), 250×. Compare these photomicrographs with Fig. 4.27 (p. 121), which shows a section of an abraded surface.

in Fig. 4.27). The discontinuities in the casting appear to be of a filamentary morphology when rough polishing is continued for only a short period (Fig. 8.4a), and these discontinuities might be mistaken for oxide inclusions. The true morphology of the cavities is seen when rough polishing is continued for a longer period (Fig. 8.4c). For this particular alloy it might be suspected that artifact structures were being observed if the specimen had been examined after etching, because structures that should be recognized as abrasion artifacts (see below) are also seen (twin strain markings in Fig. 8.4b and d). However, there are many alloy systems for which this clue would not be available.

The morphologies of the lead constituent in leaded alloys and the graphite in cast irons, to take two further examples, are similarly susceptible to misrepresentation. Because, however, the distortions in shape are confined to the abrasion fragmented layer, problems of this nature again should not arise during any well-designed preparation procedure.

Artifacts Originating in the General Deformed Layer

The sketches in Fig. 8.5 illustrate possible sources of abrasion artifacts in materials in which manifestations of plastic deformation can be seen after comparatively small strains. Two etchants are considered: one that reveals manifestations of deformation with medium sensitivity (etchant

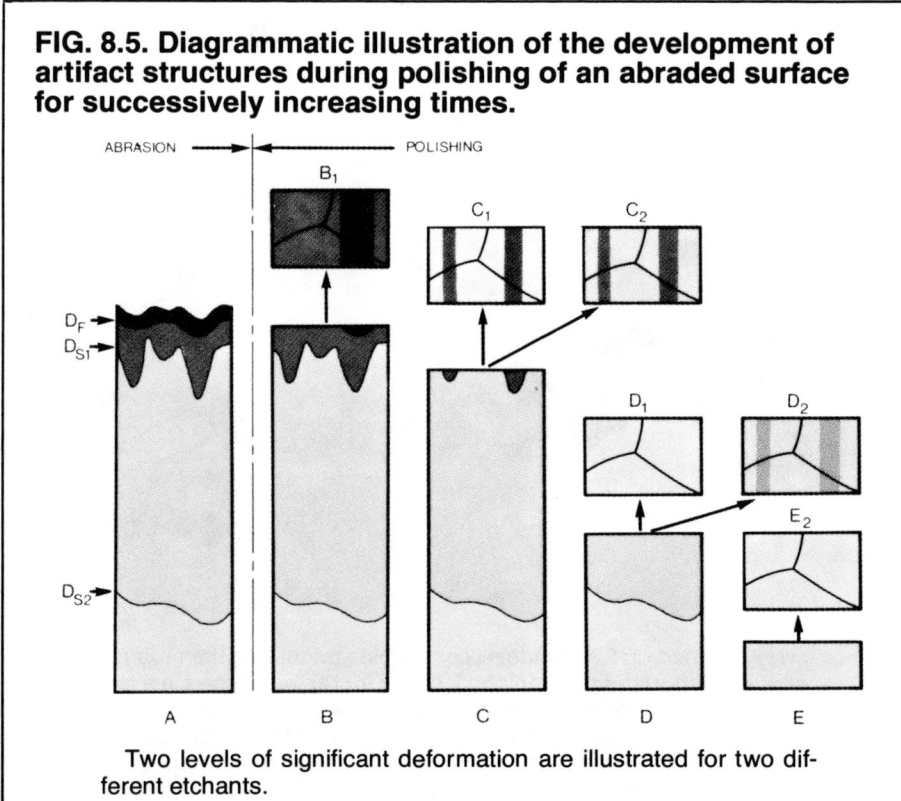

FIG. 8.5. Diagrammatic illustration of the development of artifact structures during polishing of an abraded surface for successively increasing times.

Two levels of significant deformation are illustrated for two different etchants.

1) and one that does so with high sensitivity (etchant 2). The bases of the abrasion deformation layers in which the two develop manifestations of deformation are labeled D_{S1} and D_{S2}, respectively, in sketch A. The fragmented layer, which would be revealed by both etchants, is labeled D_F. A practical example would be 70:30 brass using the ferric chloride reagent as etchant 1 and the high-sensitivity sodium thiosulfate etching procedure as etchant 2. Sections of an abraded surface etched by these two methods are illustrated in Fig. 4.10(a), 4.10(d) and 4.14 (pp. 101 and 105).

Suppose that the abraded surface has been polished until the abrasion scratches have been removed but the surface level has been reduced only to that indicated in sketch B in Fig. 8.5. The polished surface after etching by either method would then contain bands whose structure would be that of the fragmented layer originating from the deeper abrasion scratches, and other bands with a structure characteristic of the less highly strained regions of the shallower scratches (sketch C_1 in Fig. 8.5). An example for brass etched in the ferric chloride reagent is given in Fig. 8.6(a); the level to which this surface has been reduced during polishing can be deduced from Fig. 4.11 and 4.14.

Now suppose that the surface has been reduced to a lower level than that just discussed, such that the abrasion fragmented layer has been removed but not all of the D_{S1} layer (sketch C in Fig. 8.5). When etched in reagent 1 the surface would contain bands of deformation etch markings appropriate to this reagent, but the intermediate regions would be free from spurious etch markings, as indicated in sketch C_1. If this surface were etched in reagent 2, however, the same bands would be developed but the intermediate regions now would contain deformation etch markings characteristic of smaller strains (sketch C_2). These two conditions are illustrated for brass in Fig. 8.6(b) and (c); the approximate level of this surface can be deduced from Fig. 4.10 and 4.14.

Now suppose that the surface has been reduced to a still lower level such that the abrasion fragmented layer has been removed but not all of the D_{S1} layer (sketch D in Fig. 8.5). When etched in reagent 1 the surface would be free from artifact etch markings (sketch D_1 in Fig. 8.5), but deformation etch markings would be visible after etching in reagent 2 (sketch D_2 in Fig. 8.5). These two conditions are illustrated for brass in Fig. 8.6(d) and (e); the approximate level of this surface can be deduced from Fig. 4.10(a) and (d). It is only when the surface has been reduced to a considerably lower level again — to below D_{S2} — that an artifact-free result is obtained with etchant 2 (sketches E and E_2 in Fig. 8.5; Fig. 8.6f).

Clearly, artifacts originating from the deformed layer are much more difficult to eradicate than those originating from the fragmented layer. Moreover, the difficulty of eradicating these artifacts depends on the sensitivity to deformation of the etchant/specimen combination. The higher this sensitivity, the greater the thickness of material that needs to be removed from an abraded surface before an artifact-free result will be obtained. Indeed, the thickness that has to be removed varies considerably with this sensitivity because of the exponential manner in which the abrasion strains vary with depth (Fig. 4.13). It is with only a limited num-

222 / Metallographic Polishing by Mechanical Methods

FIG. 8.6. Typical abrasion artifacts in specimens of annealed 70:30 brass etched in a ferric chloride reagent (left column) and by the high-sensitivity sodium thiosulfate method (right column).

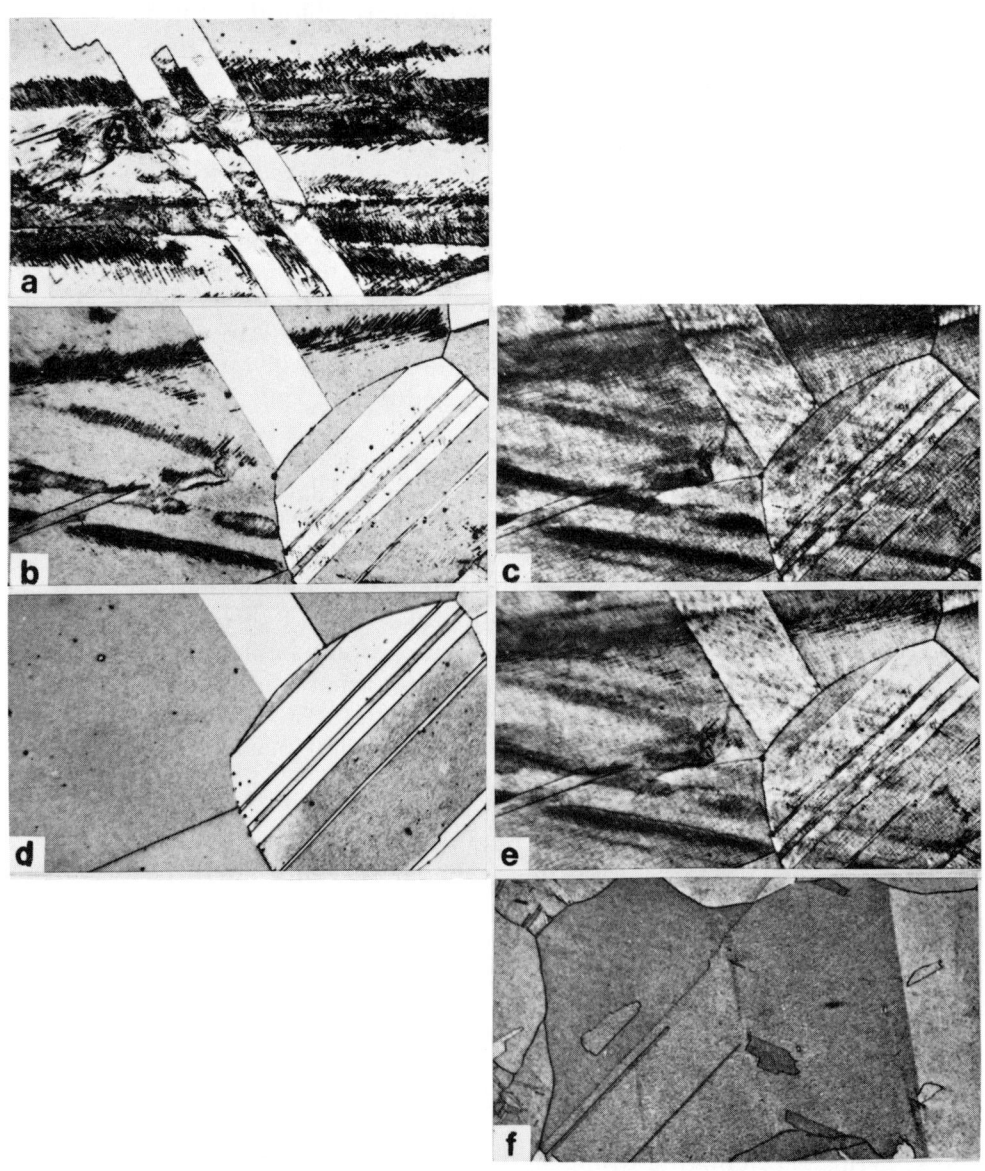

The surfaces were abraded on 220-grade silicon carbide paper and then polished until the following thicknesses had been removed: **(a)** about 1 μm (as for B_1 in Fig. 8.5); **(b) and (c)** about 5 μm (as for C_1 and C_2 in Fig. 8.5); **(d) and (e)** about 15 μm (as for D_1 and D_2 in Fig. 8.5); **(f)** about 80 μm (as for E_2 in Fig. 8.5). Magnification (all), 250×. Compare these photomicrographs with those in Fig. 4.10, 4.11 and 4.14, which show sections of abraded surfaces.

ber of systems that this sensitivity is sufficiently high to pose a serious problem. Nevertheless, the design of a basic polishing system must ensure that the system will be able to cope with the most sensitive cases. It is then easy enough to downgrade the system for less sensitive cases, but the reverse is much more difficult to achieve. Patching treatments, such as alternate polishing and etching at the final polishing stage, certainly will not suffice.

The above discussion has concentrated on the deformed layer left in the surface by the last abrasion stage and not removed adequately by the first polishing stage. As implied in the earlier discussion of Fig. 8.1, deformation that could give rise to abrasion artifacts might be left in the surface at any machining or abrasion stage of the preparation sequence. For example, the deformed layer introduced in cutting the section may not be adequately removed by the first abrasion stage. It will be appreciated that faults of this nature, once introduced, will be particularly difficult to correct at later stages. The stage at which a system of artifacts has originated can be established by introducing systems of scratches of different orientations at each stage. The direction of the artifact banding can then be related to the direction of a particular system of abrasion scratches (see discussion of Fig. 8.9). The specimen must then be returned to the stage after the source stage for a more thorough treatment.

Sufficient data are available in Chapters 3, 4 and 5 to permit quantitative assessments to be made of the likelihood that the abrasion deformation layers will be removed in practice. Some of this data for the sensitive case of 70:30 brass is assembled in Table 8.1. The treatment times required to remove the two levels of significant deformation discussed in connection with Fig. 8.6 are given, using the less rigorous of the two criteria discussed in the introductory paragraph of this chapter. Although these figures should be taken as being indicative only, a number of general conclusions can be drawn from them.

First, as qualitatively predicted earlier, the two most critical steps in the sequence are those from the final machining to the first abrasion stage and from the final abrasion to the first polishing stage. Difficulties in eliminating abrasion artifacts can therefore be reduced if the depths of the deformed layers produced by the final machining stage and the final abrasion stage are minimized, and if the material-removal rates of the first abrasion stage and the first polishing stage are maximized. The following precautions need to be taken if these criteria are to be met reasonably.

1. The cutting of a section, no matter how smooth the surface produced, needs to be followed by a machining stage or stages chosen to produce the minimum depth of deformation (see Tables 4.1 and 4.2, p. 112).
2. The first abrasion stage must be designed to achieve maximum abrasion rate, the factors which control this having been discussed in Chapter 3 (p. 55). Even so, abrasion must be continued for the time required to remove all the significant damage, which can be

TABLE 8.1 Time Required To Remove the Abrasion Deformed Layer in Annealed 70:30 Brass

Treatment stage	Preceding stage	Treatment time(a), min D_{S1}	D_{S2}
P240 silicon carbide paper	Hacksaw; abrasive cut-off wheel	0.6	8.4
	Filed	0.5	3.8
	Ground; turned	0.4	0.9
P800 silicon carbide paper	P240 silicon carbide paper	0.01	0.4
P1200 silicon carbide paper	P800 silicon carbide paper	0.02	0.5
Aluminum oxide – wax lap	P800 silicon carbide paper		
6 μm diamond, polish	P800 silicon carbide paper	0.8	5.5
	P1200 silicon carbide paper	0.6	2.8
	Aluminum oxide lap	0.4	2.0
1-μm diamond, polish	P1200 silicon carbide paper	2.0	8.8
1-μm aluminum oxide, polish		2.0	8.8
0.3-μm aluminum oxide, polish		5.0	22
0.1-μm aluminum oxide, polish		6.3	28

(a) Time required after the pre-existing scratches have been removed to reduce the depth of the deformed layer produced by the preceding stage to that of the deformed layer being produced by the new stage.

several minutes in the more critical cases.
3. If the general principles outlined in item 2 above are followed, there should be little difficulty in removing the pre-existing deformed layer at each succeeding abrasion stage. However, abrasion will have to be continued for a period much more than twice that required to remove the pre-existing scratches, which is a common recommendation.
4. Abrasion should be continued to the finest practicable stage with the objective of ending abrasion with the minimum depth of deformation (see Table 4.1, p. 112).
5. Abrasion should be followed by a polishing stage which is designed to produce the maximum possible material-removal rate consistent with a reasonable quality of polish. The factors which control polishing rate were discussed in Chapter 5 (p. 169).

It is of course possible to remove the pre-existing deformed layer by any abrasion or polishing stage, however poorly it is chosen, if it is continued for a long enough time. Requirements of efficiency alone demand, however, that this time be reduced to the minimum. Moreover, it is only when comparatively short times are needed that it can be expected that operators will consistently continue the treatment for long enough to produce artifact-free results. This applies above all to the step from the last abrasion stage to the first polishing stage, and the concept of using a polishing stage designed to have maximum polishing rate and operated primarily with the intention of removing the significant abrasion deformation is a critical one. Admittedly, such a polishing process often may not produce a polish of a quality adequate for final examination. In this event, it must be regarded as a *rough-polishing stage* and be supplemented by a *final-polishing stage* that can be designed to produce a polish of adequate quality. The presence of abrasion artifacts most commonly results from attempts to combine these two functions.

Although we have mainly been discussing one of the more sensitive cases, there are a number of other alloy systems for which abrasion artifacts cause considerable difficulties in practice. One includes alloys which are susceptible to strain-induced transformations, such as the austenitic steels (the structure of the abrasion deformed layer for these steels is discussed on p. 121). Abrasion artifacts of the type that may be found in finish-polished surfaces are illustrated in Fig. 8.7. They are similar in appearance to those illustrated in Fig. 8.3(a) but are much more difficult to eliminate because D_s is so much larger.

Another group includes most metals with noncubic crystal structures, in which massive lenticular twins form at comparatively small amounts of strain. The structures of the twin-containing deformed layers produced during abrasion of these metals were discussed on p. 123, where it was also noted that the surface portions of the deformed layer may recrystallize spontaneously in metals of comparatively low melting point, whether cubic or noncubic in crystal structure.

FIG. 8.7. Abrasion artifacts in an AISI type 304 austenitic stainless steel.

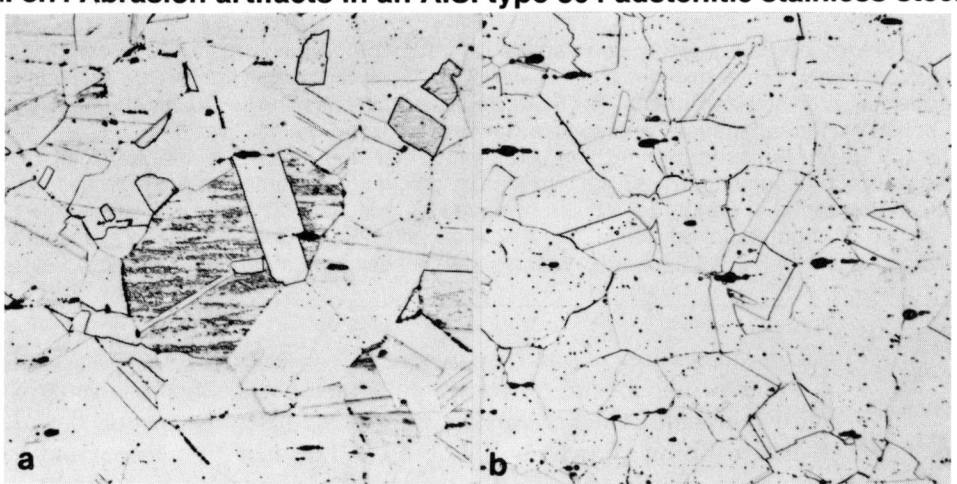

Abraded on 600-grade silicon carbide paper and then polished on 6-μm diamond, but for an inadequate time in **(a)**. The true structure is shown in **(b)**, the surface in this case having been polished for a longer time at the rough-polishing stage. Etched in aqua regia – glycerol reagent. Magnification (both), 250×. Compare these photomicrographs with Fig. 4.28(a) (p. 122), which shows a section of an abraded surface.

226 / Metallographic Polishing by Mechanical Methods

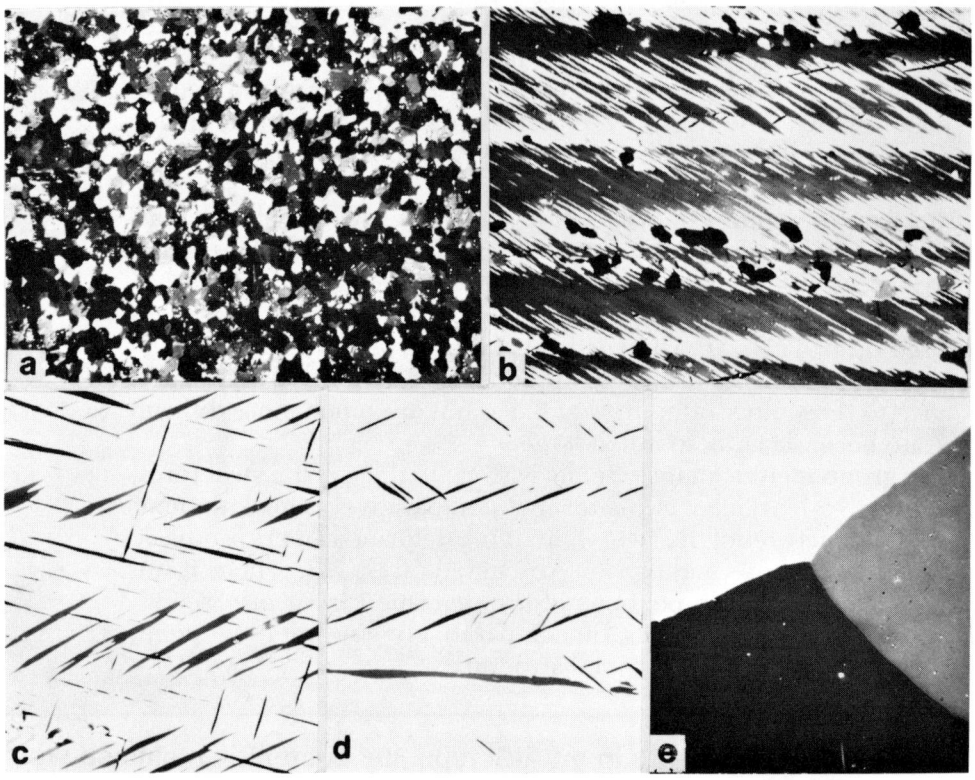

FIG. 8.8. Abrasion artifacts in polycrystalline zinc, viewed in polarized light.

Abraded on 220-grade silicon carbide paper and then electropolished to remove successively greater thicknesses of material. **(a)** 2 μm removed by polishing. Artifacts of small recrystallized grains. **(b)** 5 μm removed by polishing. Artifacts comprising bands of twinned material and some small recrystallized grains.
(c) 15 μm removed by polishing. Artifacts comprising massive twins, with only a suggestion of banding. **(d)** 30 μm removed by polishing. Artifacts comprising massive twins arranged in bands. **(e)** 45 μm removed by polishing. The true structure, consisting of large equiaxed grains. Magnification (all), 250×. Compare these photomicrographs with Fig. 4.29 (p. 125), which shows a section of an abraded surface.

The succession of microstructures that might be observed when an abraded surface of one of these materials is polished to remove successively increasing thicknesses of material is illustrated in Fig. 8.8, the true structure being shown in Fig. 8.8(e). The fact that the structures observed are false structures introduced during abrasion might be suspected in Fig. 8.8(b) and (d) because the twins are obviously aligned in bands in the direction of the pre-existing abrasion scratches. This is much less obvious in Fig. 8.8(c), and the structure might be interpreted as indicating that the material had been deformed in bulk. It is even less apparent in Fig. 8.8(a), and a completely erroneous estimate might be made of the grain size of the material if a section were examined in this condition. Difficulties with twin-type artifacts of the type illustrated in Fig. 8.8(c) and (d) are possible for titanium, zirconium, niobium, beryllium, cadmium, bismuth, uranium and antimony, among others. The difficulties are likely to be greatest when the specimen material is soft (when the depth of the twin-containing layer introduced during abrasion is greatest) and has an intrinsically low polishing rate. Recrystallized grain artifacts are a common problem with low-melting-point metals such as tin, cadmium, and lead.

Note that abrasion artifacts of the type that we have been discussing are observed when polishing has removed insufficient material from the abraded surface, irrespective of how this polishing is carried out — that is, be it mechanical, chemical or electrochemical. In fact, the surfaces illustrated in Fig. 8.8 were polished by electrochemical methods. On the other hand, no artifacts will be present if polishing is continued for long enough by any of these methods, including any mechanical method. (The photomicrograph shown in Fig. 8.20c was, for example, polished entirely by mechanical methods.) The question is whether the polishing time required is acceptably short in normal practice.

It follows that it is essential for metallographers to be able to recognize abrasion artifacts when present. Any structure which is banded in the direction of a set of pre-existing abrasion or machining scratches or grooves should be regarded with suspicion. A number of the examples already discussed can also be used as a general guide, but the illustrations given in Fig. 8.9 demonstrate the constant vigilance that must be exercised.

This aluminum bronze had been heat treated so that it contained nodules with a lamellar structure distributed in a martensitic matrix. As originally prepared, the structure appeared to be a valid one when the section was examined at high magnification (Fig. 8.9b). However, banding in the structure was apparent at low magnification, the banding being related to machining marks produced when the section surface was faced by lathe turning (Fig. 8.9a). When this surface was returned to the first abrasion stage, for a longer treatment at that stage, and then polished as before, the martensite needles in the microstructure were seen to be more sharply delineated and the lamellae in the nodules to be clearly resolved (Fig. 8.9c and d).

A common source of artifacts in quench-hardened steels originates from the deep rays of tempered material that may be introduced by preliminary abrasive cutting and machining (Fig. 4.30, p. 130). An example is given in Fig. 8.10. The specimen in (a) has been abraded dry on a belt surfacer and then treated for only a comparatively short time at the first abrasion stage; that in (b) has been treated for a longer time at this abrasion stage. The dark-etching longitudinal bands in Fig. 8.10(a) might well be mistaken for segregation bands and it might also be concluded that the specimen has been tempered; Fig. 8.10(b) indicates that neither of these conclusions would be valid. Tempering artifacts may also originate in cutoff operations that use high-speed abrasive disks, or in grinding operations, but not in any of these operations if the cutting regions are kept cool by a copious flow of water (see p. 128).

228 / **Metallographic Polishing by Mechanical Methods**

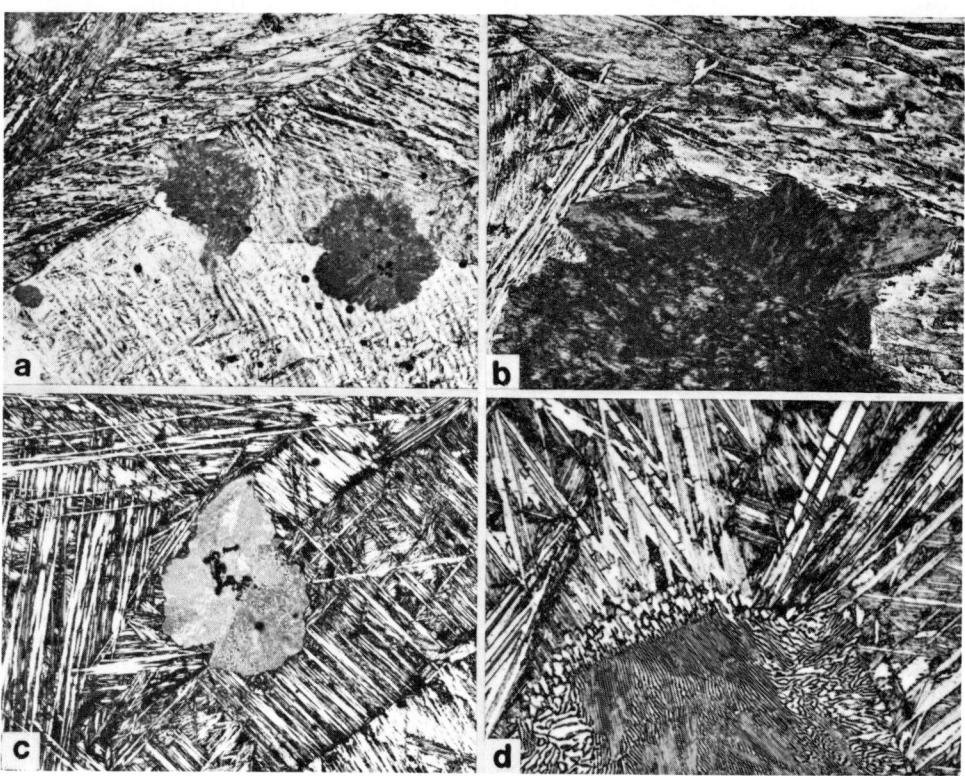

FIG. 8.9. Abrasion artifacts in a cast aluminum bronze (11.8% Al) water quenched from 800°C.

(a) Incorrectly prepared. Magnification, 25×. (b) As for (a). Magnification, 500×. (c) Correctly prepared. Magnification, 25×. (d) As for (c). Magnification, 500×. All etched in a ferric chloride reagent.

Another type of artifact worth describing is that originating from the cracking produced during the abrasion of brittle materials or brittle phases (see Chapter 7). An example of the arrays of abrasion-crack artifacts in the surface of a brittle solid is given in Fig. 8.11, the origins of this cracking having been discussed on p. 203. The cracks usually are thin, and may be difficult to see in an as-polished surface because physiologically it is difficult to see a thin black line against a bright background. Cracks widen significantly during etching and then become more readily visible. In some cases, and germanium is one of these cases, arrays of dislocations extend beyond the cracks and so also become visible as additional artifacts after appropriate etching. Noncrystalline materials (e.g., glass) would be susceptible to crack artifacts only. An example of crack artifacts in a brittle constituent in a metallic alloy is given in Fig. 8.12. As with other types of artifacts, crack artifacts can always be eliminated if a rough-polishing treatment is continued for a long enough time.

FIG. 8.10. Tempering artifacts in a quench-hardened steel (0.6%C; water quenched from 850°C).

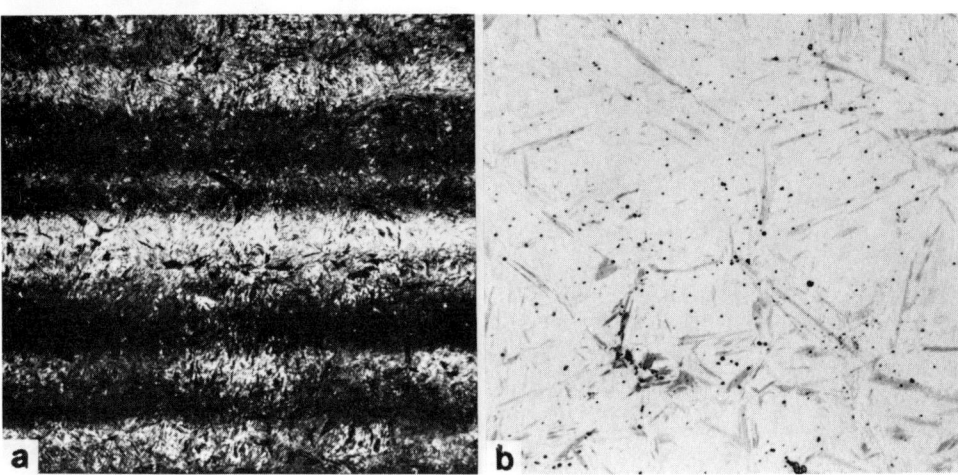

(a) Dark-etching bands of tempered martensite originating from a dry belt-surfacing operation on 100-grade silicon carbide. (b) True structure. Both etched in picral. Magnification, 250×. In (a), not only is the impression gained that the material has been tempered, but the bands might be mistaken for segregation bands. Compare these photomicrographs with those in Fig. 4.30 (p. 130), which show sections of abraded surfaces.

FIG. 8.11. Crack artifacts in germanium that were introduced at an abrasion stage, and deformation artifacts introduced at a polishing stage.

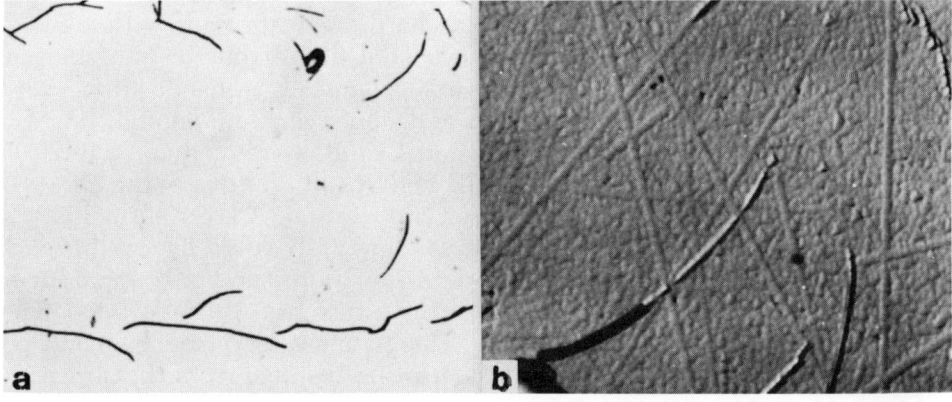

The surface was abraded on 220-grade silicon carbide paper and then polished on 1-μm diamond until the surface irregularities produced by abrasion had been removed completely. (a) Etched lightly in a ferricyanide reagent to show more clearly the cracks remaining in the surface. Magnification, 250×. (b) Etched heavily in a ferricyanide reagent, showing cracks (broad dark lines) and arrays of dislocation etch pits extending from these cracks. Note also that arrays of small etch pits have developed along the sites of polishing scratches. Magnification, 1500×. Compare these photomicrographs with those in Fig. 7.6 (p. 210), which show sections of abraded surfaces.

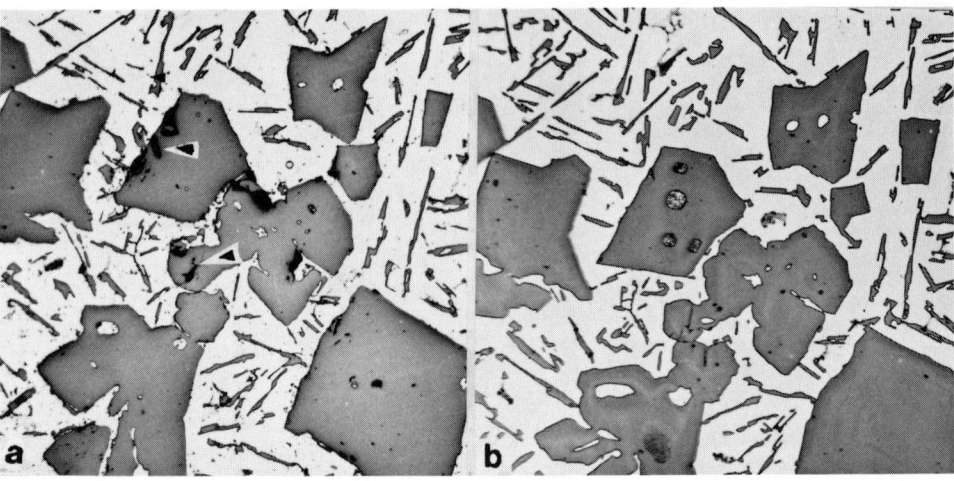

FIG. 8.12. Crack artifacts in the primary silicon constituent of a cast aluminum – 13% silicon alloy. Magnification, 250×.

(a) Abraded on P800-grade silicon carbide paper and then polished on 1-μm diamond until all surface irregularities had been removed. (b) Same area as in (a), but polished for a longer time. Note the cracks and pits in the primary silicon present in (a) but not in (b). Compare these photomicrographs with Fig. 7.7 (p. 212), which shows a section of an abraded surface.

Although not strictly concerned with micrography, several other important situations where abrasion deformation left in a surface gives rise to erroneous observations are worth mentioning briefly.

The most important concerns hardness tests made with low indenting loads, when it is frequently reported that erroneously high results are obtained in tests carried out on mechanically polished surfaces if the indentations are small enough and hence shallow enough (for example, Vickers indentations with diagonal lengths of less than 10 to 20 μm).[2] It is generally recognized that the high results are due to the effects of work hardening of the surface during preparation. A layer of thickness several times the depth of the hardness indentation would have to be work hardened to affect the hardness determinations, and it is apparent by now that a work-hardened layer of this thickness could be introduced only during abrasion or machining. That is, any layer on mechanically polished surfaces that affects hardness must be residual from the abrasion stages of preparation and its presence must be the result of failure to remove enough of the abrasion deformed layer during polishing. Moreover, plastic strains of several percent reduction are needed to cause detectable changes in hardness, so that the depth of significant deformation from this point of view is comparatively small. Nevertheless, it would be large enough in soft materials for its elimination to require reasonably good polishing procedures, which clearly have not always been used by investigators of the phenomenon.

Another important example may be encountered in the chemical analysis of multiphase alloys by x-ray fluorescence methods. This technique examines a prepared flat surface and investigates a layer that is from 1 to 100 µm thick. Consequently, the results are likely to be affected if the volume fraction of a second phase is changed in a layer of this order of thickness during preparation of the analytical surface. We have seen that changes of this nature can occur during machining or abrasion of some alloys, perhaps the most notable case being that of lead-containing alloys where the lead is always present as a separate constituent of the metal itself.

Determination of the lead content of tin bronzes has been shown to be significantly affected by the method by which the analytical surface is prepared (Table 8.2). Sections of some of the surfaces referred to in Table 8.2 are shown in Fig. 8.13, and it is apparent that the volume fraction of lead in the surface layers has been altered during preparation in all but Fig. 8.13(d); the vertical bars drawn on these photomicrographs indicate the thickness of the layer which effectively was analyzed by the x-ray fluorescent method. The turned surface following a large depth of cut (Fig. 8.13a) gave a slightly high result for lead content because some lead had been smeared across the surface. The ground and abraded surfaces (Fig. 8.13b and c) gave low results for lead, and complementary high results for tin, because lead had been extruded out of its cavities in the surface layers. The metallographically polished surface (Fig. 8.13d) gave the correct result within experimental error, but this occurred because, and only because, the polishing treatment had removed all of the lead-depleted layer remaining after abrasion. Similar effects are known to occur with leaded steels and seem possible with other alloy systems. A pos-

TABLE 8.2 Effect of Surface Preparation on the Results of Chemical Analysis, by X-Ray Fluorescence Methods, of a Leaded Tin Bronze *(Manners et al, Ref 3)*

Surface-preparation method	Analysis	
	Lead, wt%	Tin, wt%
Turned:		
Small depth of cut	10.9	6.0
Large depth of cut	16.6	5.7
Milled	9.3	6.1
Machine surface ground	10.2	6.0
Hand abraded:		
220 SiC paper	12.0	5.8
400 SiC paper	12.2	5.8
600 SiC paper	13.7	5.8
Metallographic polish (6-µm diamond)	15.9	5.7
Composition as determined by bulk chemical analysis	16.0	5.7

FIG. 8.13. Taper sections of surfaces of a 16%-lead bronze. *(Manners, Craig and Scott, Ref 3)*

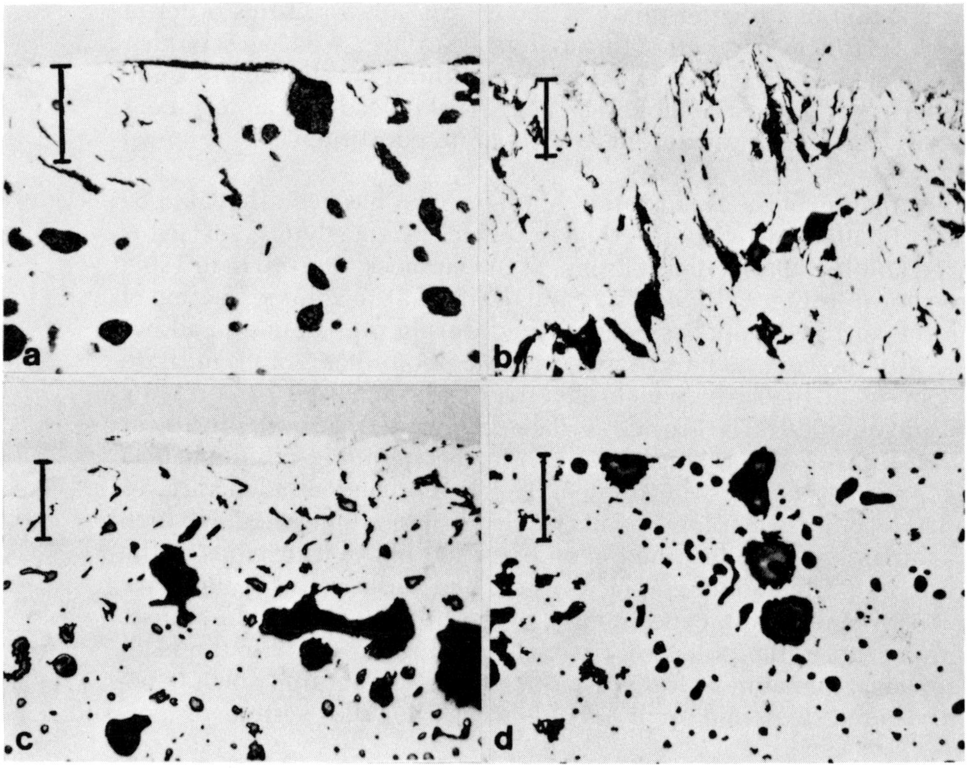

(a) Lathe turned; large depth of cut. (b) Machine surface ground. (c) Abraded on 220-grade silicon carbide paper. (d) Metallographically polished; finished on 6-μm diamond. In each instance, the vertical bar represents the depth of the layer that would be investigated by the x-ray fluorescence method of analysis used to obtain the results given in Table 8.2. Taper ratio, approx. 10. Magnification, 250×.

sible example is the determination of silicon content of aluminum-silicon alloys, where the surface layers can be depleted of silicon by abrasion (Fig. 7.7, p. 212).

Difficulties with the surface deformed layer may also occasionally be encountered during macroscopical examinations, which commonly are carried out on sections prepared by machining. The etching procedure used to reveal the macrostructure has to remove the surface deformed layer before it starts to develop the structure of interest. This usually is possible because vigorous etchants are used and deep etching is acceptable. One occasion when it is not possible, however, arises when the objective is to reveal localized zones of deformation (Lüders bands) in sections of low-carbon steels. The etchant used for this purpose is an acidified solution of ferric chloride (Fry's reagent) and is effective only if the etching time does not exceed 20 s.[4] An etch of this duration does not remove the surface deformation normally introduced during cutting of sections. Consequently, information on the presence and dispositions of Lüders bands can be obtained only if the deformed layer introduced during cutting of the section is removed by an appropriately long treatment on abrasive papers before etching (Fig. 8.14).[4]

Investigations of crystal structure by x-ray diffraction methods also examine surface layers which are on the order of several micrometres in depth. Many of the diffraction effects observed are sensitive to deformation, although usually not highly so, and the more highly strained regions of an abrasion deformed layer may cause artifact diffraction patterns. The problem of preventing these artifact patterns, however, is no more severe than that of preventing microstructural artifacts in a materials/etchant system which is moderately sensitive to deformation.

POLISHING ARTIFACTS

Let us assume that a rough-polishing stage has removed effectively the artifact-containing layer introduced during abrasion, and that it has been followed by a fine-polishing stage which has produced an adequate final polish. By "adequate polish" we mean, at this stage, a surface on which polishing scratches cannot be seen by optical microscopy using vertical bright-field illumination. Remember however that the surface might contain undetected polishing scratches (p. 141) and, moreover, that a plastically deformed layer with its own special characteristics might be present beneath these scratches, the depth and nature of this layer varying considerably beneath different scratches (p. 195). These features of a polished surface can give rise to artifact structures (*polishing artifacts*) which

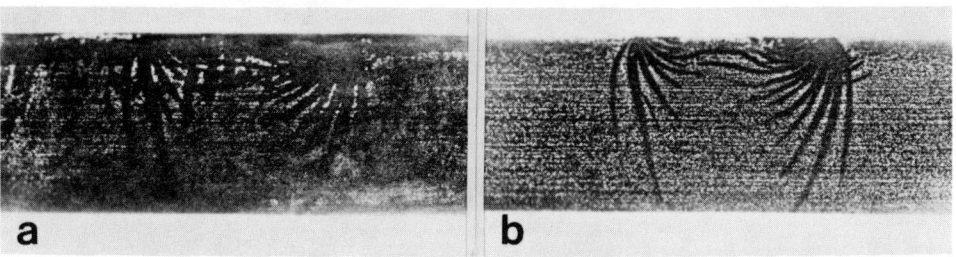

FIG. 8.14. Section of a mild steel plate subjected to local deformation.

(a) Surface prepared by removing by abrasion a thickness of 0.4 mm from the as-sawn surface. (b) As for (a), but after removal of 1.0 mm from the as-sawn surface. Both etched in Fry's reagent. Magnification, 1×. The bands in which the deformation has been concentrated (Lüders bands) are sharply delineated in (b) but not in (a). They have been partly obscured in (a) by residuals of the surface deformed layer introduced during cutting of the section.

are quite different from abrasion artifacts, and it is these polishing artifacts that we shall now discuss.

Enhancement of Polishing Scratches by Etching

To their constant disappointment, metallographers find that surfaces they thought to be scratch-free or nearly scratch-free after polishing can be seen to be profusely scratched after etching. In fact, the scratches were there all the time. They were not visible when the as-polished surface was examined, because bright-field illumination is comparatively insensitive to surface irregularities, but were made visible by etching (Fig. 8.15). The phenomenon is more marked when the combination of specimen and etching method is sensitive to deformation, when etching produces color contrast, and when etching removes little material. The apparent exaggeration of the scratches also depends on the severity of etching. Fine scratches tend to be most noticeable after light etching (Fig. 8.16a) and medium scratches after slightly heavier etching (Fig. 8.16b). Moreover, the scratches disappear progressively with increasing severity of etching as increasing amounts of material are removed from the surface (Fig. 8.16). The surface may, however, be etched to an unacceptably great depth by the time that most of the scratches have been obliterated.

The increased visibility of the scratches is due to two phenomena acting either singly or in combination. First, the scratch grooves may be

FIG. 8.15. An area on the surface of a wrought low-carbon iron polished on 1-μm diamond abrasive.

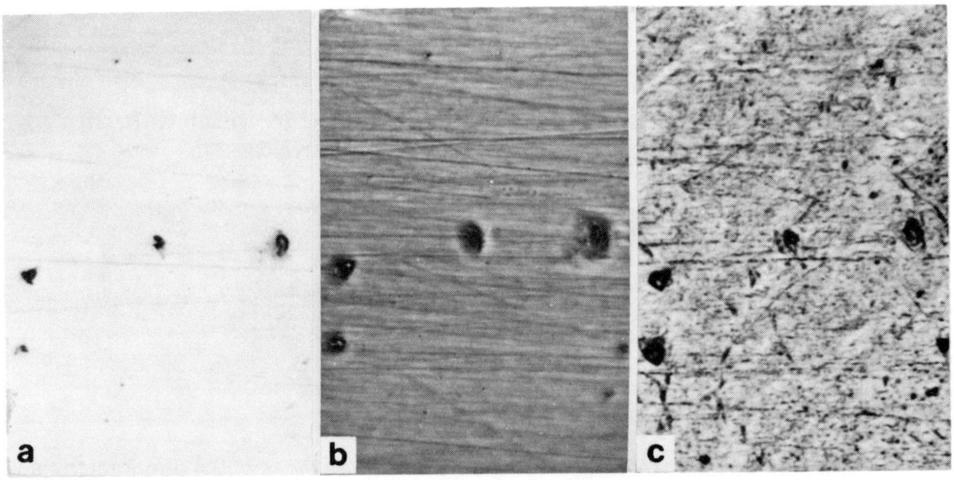

(a) As-polished; bright-field illumination. (b) As-polished; phase-contrast illumination. (c) Etched in nital; bright-field illumination. Magnification (all), 100×. Etching has made obvious in (c) many fine scratches which are not visible in (a) but which nevertheless can be seen in (b). Compare the polishing artifacts in (c) with the abrasion artifacts illustrated in Fig. 8.3(a) and (b) (p. 218).

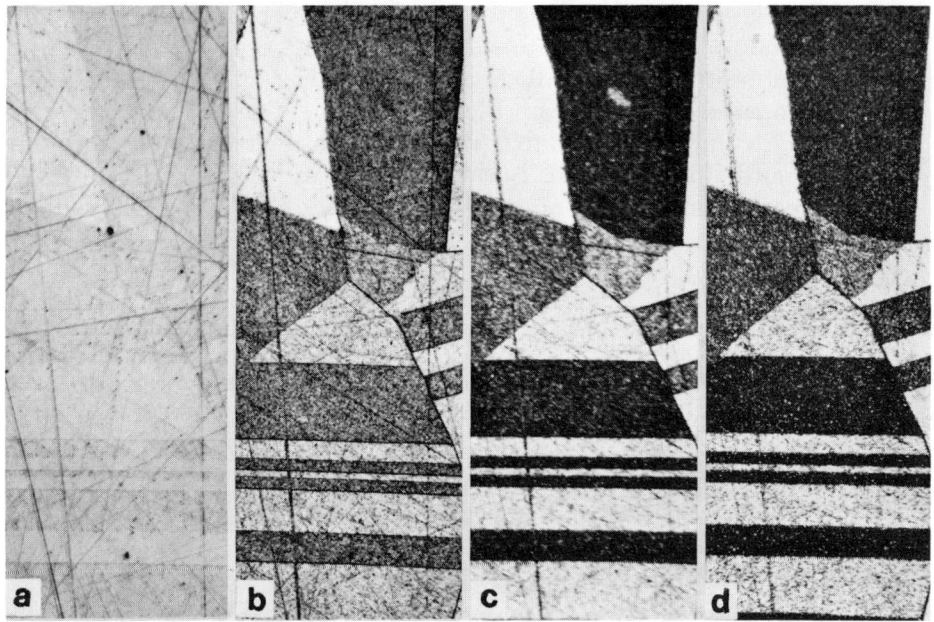

FIG. 8.16. An area on the surface of annealed 70:30 brass polished on fine aluminum oxide on Selvyt cloth and then etched in increments in a ferric chloride reagent.

The fine polishing scratches are most visible after the lightest etch, and progressively widen and then disappear as etching is continued. Magnification, 250×.

widened because the deformed region associated with them is anodic with respect to the matrix[5] and so etches preferentially. An example of this type of scratch enlargement is given in Fig. 8.11, in which a series of etch pits can be seen to have developed along the scratches; the etchant used in this case was capable of developing pits at the sites of dislocations and so the scratches are enlarged as a row of pits. Secondly, etching contrast may be developed between the deformed regions associated with the scratch which are misoriented with respect to the matrix. For example, the scratch markings have been made more obvious in Fig. 8.16(b) as a result of some areas being etched to a different color from the matrix grain (see also Fig. 8. 17d).

Degradation of Grain Color Contrast

In those systems where the etching contrast is due to differential coloring of differently oriented crystals, the grain color contrast observed will be that of the fragmented layer produced on the surface of the various grains during polishing and not that of the grains themselves. We saw in Chapter 6 that the orientations of the grains of the fragmented layer, even allowing that some of this layer will be removed during etching, will be generally different from those of the parent grains. These orientations will differ less from grain to grain than will those of the parent grains themselves. This can cause the etching color contrast to be degraded.

An example is given in Fig. 8.17. The color contrast obtained with the particular etch when no polishing deformed layer is present is illustrated in Fig. 8.17(a), that for a surface which is very finely polished but which contains a polishing deformed layer in Fig. 8.17(b), and that for a somewhat more coarsely polished surface in Fig. 8.17(c). The color contrast, compared with that in Fig. 8.17(a), decreases progressively in Fig. 8.17(b) and (c); grains that should be white are gray, and all grains have a mottled appearance. Both of these features are due to etching of discrete areas, strung along the polishing scratches, to a different color than that of the matrix grain (Fig. 8.17d). These areas are patches of the polishing fragmented layer that remain in the surface.

FIG. 8.17. Variation in the grain contrast obtained by etching annealed 70:30 brass in a ferric chloride reagent.

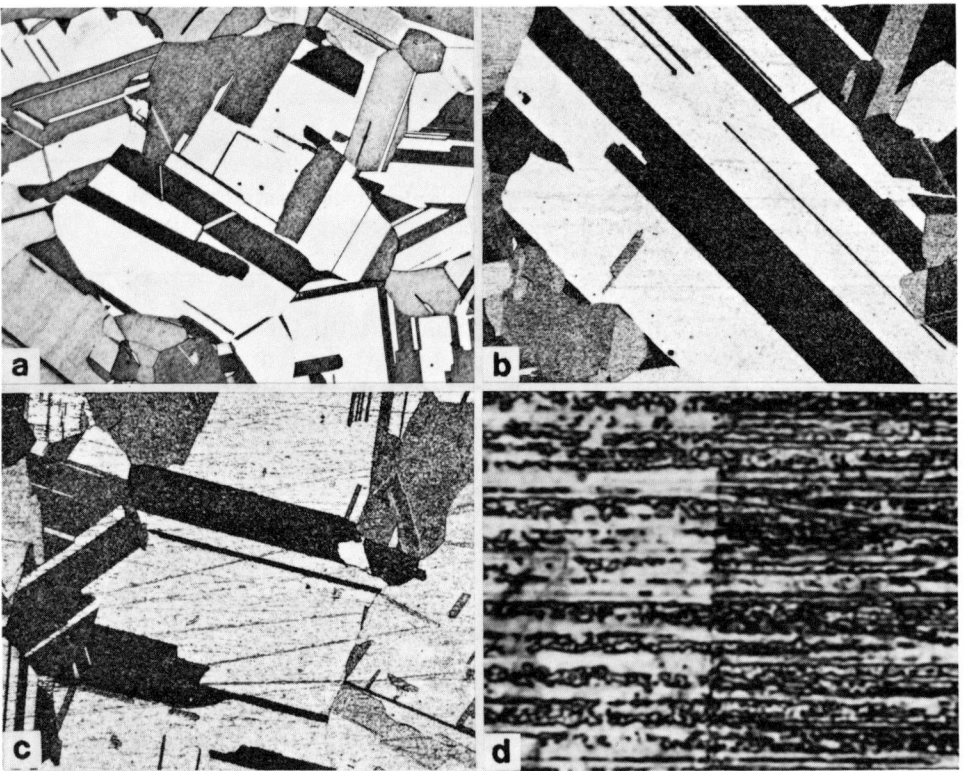

(a) Finish polished by a skidding technique on magnesium oxide abrasive. Magnification, 250×. (b) Finish polished on 0.1-μm-grade aluminum oxide abrasive. Magnification, 250×. (c) Finish polished on 1-μm-grade diamond abrasive. Magnification, 250×. (d) As for (c). Magnification, 2000×.

The effects that have just been described will be found after polishing by any process that operates with a substantial component of mechanical cutting. It can be eliminated completely only by using a final polishing stage that operates essentially by a chemical mechanism, one such process having been used for the preparation of Fig. 8.17(a). Processes that meet this requirement are discussed later (pp. 267 and 270).

Development of Scratch Traces by Etching

Suppose that a surface has been produced in which all of the significant abrasion deformation has been removed but which is roughly polished (i.e., a surface such as that illustrated in Fig. 8.17c). Suppose further that this surface has then been polished by a process such as that used to prepare Fig. 8.17(a) and that the finer polishing treatment has been continued only until all of the rough-polishing scratches have just been removed. Some of the rough-polishing scratches will seem to reappear when this surface is etched, as is illustrated diagrammatically in Fig. 8.18. This is because portions of the rays of polishing deformation which extended beneath the rough-polishing scratches remain in the surface and because these lines of deformed material are etched preferentially. They are another type of polishing artifact.

This phenomenon of the apparent reappearance of scratches is one which we have mentioned earlier as having been first observed by Beilby. He supposed, erroneously, that it was due to the original scratches themselves being re-exposed by etching (p. 144). The features may be thought of as ghosts of the original scratches but certainly are not the scratch grooves themselves.[6] We shall call them *scratch traces*.

We can explore this phenomenon in the same way that we explored abrasion artifacts. Sketches A and B in Fig. 8.18 illustrate diagrammatically the deformation produced in a surface by a typical rough polishing process (cf. Fig. 6.1, p. 196). Two levels of significant deformation are illustrated: one, the fragmented layer, can be revealed by any etchant, here referred to as etchant 1; the other can be revealed only by an etchant that is comparatively sensitive to deformation, here referred to as etchant 2. Considering again our model system for 70:30 brass, the ferric chloride reagent can be taken as an example of etchant 1 and the high-sensitivity sodium thiosulfate etch as an example of etchant 2.

If the rough-polished surface being considered is etched, the result indicated by sketch A_1 in Fig. 8.18 is obtained. A practical example of this was illustrated in Fig. 8.17(c) and was discussed in connection with that figure as an example of degraded etching contrast. If this surface had then been polished by a finer process down to the level of sketch B in Fig. 8.18, banded remnants of the fragmented layer would be left in the surface and these bands would be revealed by any type 1 etchant, as suggested in sketch B_1 in Fig. 8.18 (see also Fig. 5.4, p. 147). An example of the practical result for brass is illustrated at low magnification in Fig. 8.19(a), where the lines might be mistaken for reincarnated scratches, and at higher magnification in Fig. 8.19(b), where it is obvious that the bands are not grooves but are regions that have been colored differently from the matrix grain, more so in some grains than in others. In other alloys, however, the scratch traces might be etched as grooves (an example is given in Fig. 8.20, which will be discussed in more detail soon).

Note also that the base grain contrast in Fig. 8.19(a) is considerably better than that in Fig. 8.17(c). This is because the surface portion of the fragmented layer of fairly uniform thickness visible in Fig. 6.1(a) has been removed; it is the deeper rays of deformed material visible in Fig. 6.1(a) that have not been removed completely.

If, on the other hand, fine polishing were to be continued to the level suggested by sketch C in Fig. 8.18, an artifact-free result would be obtained with etchant 1 (cf. sketch C_1 in Fig. 8.18 and Fig. 8.19c), because all of the rays of deformation of the fragmented layer would have been removed. Scratch traces would be developed in this same surface, however, by etchant 2 (cf. sketch C_2 in Fig. 8.18 and Fig. 8.19d). In the particular case being considered, the scratch traces can be resolved at higher magnifications into bands of slip etch markings (Fig. 8.19e) identical to those developed at the appropriate level in a section of the rough-polished surface (cf. Fig. 6.1c). It is only when the surface has been reduced to a level beneath the deepest rays of significant deformation for etchant 2 that an artifact-free result is obtained with both etchants, such as those illustrated in Fig. 8.19(c) and 8.6(f), respectively.

The scratch-trace artifacts that develop in noncubic metals, illustrated in Fig. 8.20, have slightly different characteristics in detail. We saw earlier that discrete twins are produced beneath some scratches during polishing of noncubic metals, the example used being zinc (Fig. 6.3, p. 200). These twins can be seen on the polished surface when it is examined under polarized light (Fig. 8.20a). The twinned bands are less numerous

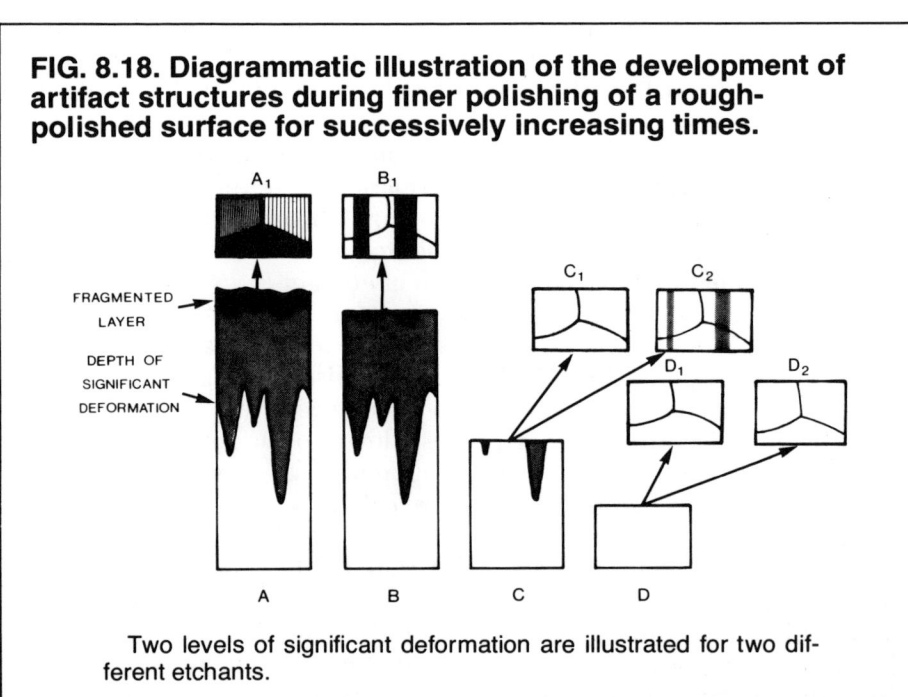

FIG. 8.18. Diagrammatic illustration of the development of artifact structures during finer polishing of a rough-polished surface for successively increasing times.

Two levels of significant deformation are illustrated for two different etchants.

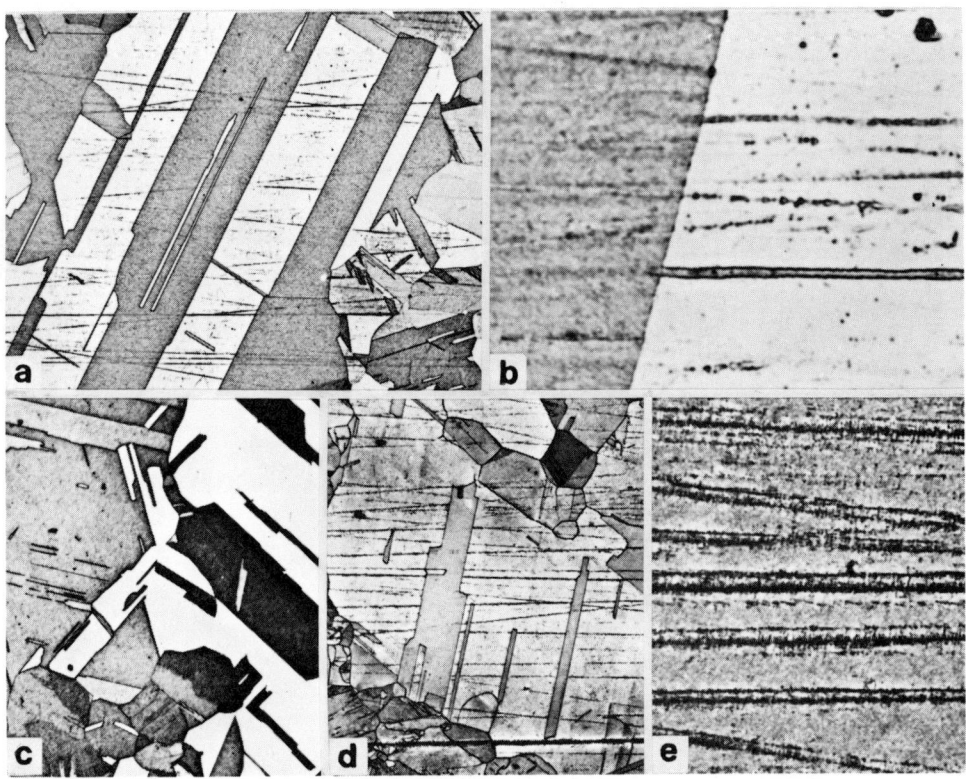

FIG. 8.19. Polishing artifacts in annealed 70:30 brass.

(a) Surface similar to that shown in Fig. 8.17(c) from which a thickness of about 0.3 μm has been removed by a finer polishing process. Etched in a ferric chloride reagent (cf. B_1 in Fig. 8.18). Magnification, 250×. (b) As for (a). Magnification, 2000×. (c) As for (a), but after removal of 1.5 μm (cf. C_1 in Fig. 8.18). Magnification, 250×. (d) As for (c), but etched in a high-sensitivity sodium thiosulfate reagent (cf. C_2 in Fig. 8.18). Magnification, 250×. (e) As for (d). Magnification, 2000×.

when the surface has been finely polished just to remove the rough-polishing scratches (cf. Fig. 8.20a and b) but are not eliminated until polishing has been continued for a much longer period of time (Fig. 8.20c). Scratch traces in the form of grooves are developed at the sites of the twins if the surface is etched and viewed in bright-field illumination (cf. Fig. 8.20b and d).

Scratch traces in the form of recrystallized grains have not been reported and are not likely.

Thus a critical requirement of any fine polishing stage is to remove not only the scratches produced by a preceding rough polishing stage but also that portion of the deformed layer introduced by the rough polishing stage that might affect significantly the observations to be made on the surface. This implies that the polishing rate of a fine polishing stage is of some importance so that the requirement just mentioned can be achieved

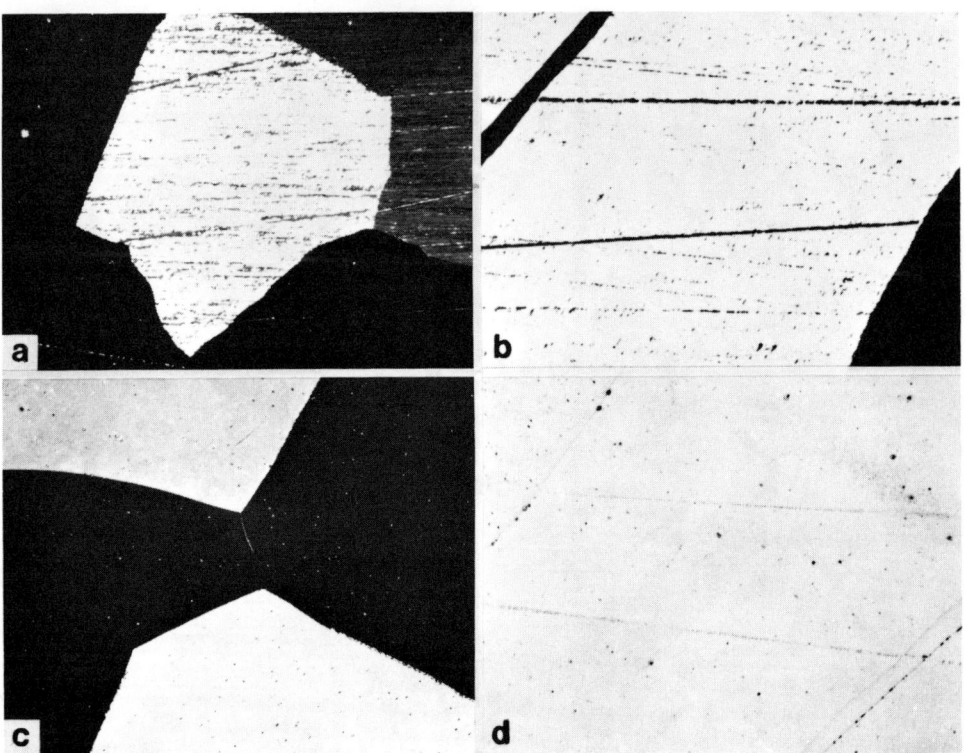

FIG. 8.20. Polishing artifacts in polycrystalline zinc.

(a) As polished on 1-μm diamond; polarized light. (b) As for (a), but polished briefly by skid polishing on magnesium oxide; polarized light. (c) As for (b), but polished for a longer time; polarized light. (d) Same field as in (b), but etched in 1% nital; bright-field illumination. Magnification (all), 250×.

in a practically acceptable polishing time. Moreover, it implies that polishing at that stage may need to be continued for a period many times that required to remove the pre-existing scratches. No information is available on the polishing rates of common fine polishing processes, so that the control of fine polishing from this point of view has to be left to experience and trial. It is consequently of special importance that a metallographer be able to recognize polishing artifacts for what they are.

We have so far discussed only final polishing processes which do not introduce new scratches or new deformed layers of their own; but most fine polishing operations do both. If so, the objective has to be to reduce the polishing damage to a level that does not affect the observations that are to be made on the final surface. The acceptable level of polishing damage varies greatly. It depends on the sensitivity of the specimen/etchant system to the effects that we have been discussing, and also on the nature of the examination that has to be made. For example, the result shown in Fig. 8.17(b), and perhaps even that in Fig. 8.17(c), would have been adequate if only the grain size of the brass had to be determined, but neither would have been adequate if the aim had been to determine whether small precipitates were present. It can well be imagined that the artifact structures of the type visible in Fig. 8.17(d) would confuse a search for small second-phase particles. A practical example is given in Fig. 8.21. The small precipitates so clearly visible in Fig. 8.21(a) cannot be discerned among the confusing background of polishing artifacts in Fig. 8.21(b). They could just be discerned in a visual microscopical examination, but might easily have been missed even then.

PRACTICAL PREPARATION PROCEDURES

Mounting, abrasion and polishing processes have so far been discussed individually in some detail, but we need to consider how these stages can be integrated to form practical preparation procedures. It will not be our aim to provide cut-and-dried recipes. Rather, we shall try to elucidate principles that will aid the metallographer in selecting the most suitable sequence for a particular situation.

Basic Preparation Sequences

Three basic preparation sequences of progressively increasing complexity are outlined in Table 8.3. It is assumed in this table that the section

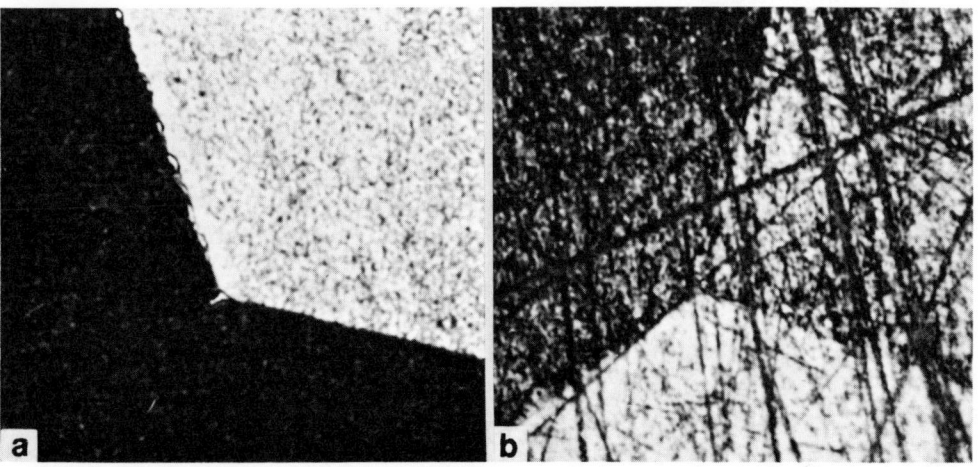

FIG. 8.21. A beta brass containing small particles of gamma phase at some grain boundaries.

(a) Finish polished by a technique (skid polishing on magnesium oxide) which does not introduce any polishing damage of its own. Precipitate particles are clearly visible at one grain boundary. (b) Finish polished by a method which introduces fine scratches. These scratches, enlarged by etching, provide such a distracting background that the precipitates cannot be discerned in this micrograph. They were just visible by optical microscopy at the vertical grain boundary. Both specimens were etched in an ammonium persulfate reagent. Magnification, 2000×.

TABLE 8.3. Practical Mechanical Preparation Sequences Based on Diamond Abrasives

Method	Stage	Preparation sequence		
		Grade I	Grade II	Grade III
Abrasion	Silicon carbide paper, P240	Stage 1	Stage 1	Stage 1
	Silicon carbide paper, P800	Stage 2	Stage 2	Stage 2
	Silicon carbide paper, P1200	Stage 3	Stage 3(a)	Stage 3(a)
	Aluminum oxide – wax lap		Stage 3(a)	Stage 3(a)
Rough polishing	Diamond, 6 μm		Stage 4	Stage 4
	Diamond, 1 μm	Stage 4		Stage 5
Final polishing	Aluminum oxide, 0.1 μm		Stage 5	Stage 6(a)
	Magnesium oxide			Stage 6(a)

(a) Alternatives.

has already been cut, and that the surface of the section has been subjected to an appropriate preliminary machining treatment. These particular sequences are based on the use of diamond abrasives for polishing and on the use of silicon carbide papers for abrasion; aluminum oxide papers would be desirable alternatives for abrasion under some circumstances (p. 62), but the principles would remain unchanged. The salient points to remember when operating each stage are summarized below, although the relevant text in Chapters 3 and 5 should be consulted for more detail.

For the abrasive-paper stages:

1. Use papers in a slightly worn, but not severely worn, condition.
2. Use heavy abrasion pressures.
3. Flush the abrasion track with a copious flow of water and check that the paper does not become clogged.
4. Give a thorough treatment at each stage — long enough, as indicated by experience, to remove the pre-existing layer of significant deformation.

For the fine abrasive lap:

1. Substitute a fine abrasive lap for the last abrasive-paper stage whenever surface flatness or retention of nonmetallic constituents is important.
2. Ensure that no loose abrasive is present on the working surface.
3. Clean and recharge the lap as soon as it shows signs of clogging.

For rough polishing:

1. Operate principally with the aim of removing the layer of significant deformation introduced during abrasion.
2. Unless other requirements dominate, use a polishing cloth with a medium-length nap and good wear resistance.

3. Charge with a diamond-containing carrier paste spread uniformly over the polishing track.
4. Keep the working surface moist with a polishing fluid appropriate to the particular carrier paste.
5. Apply a heavy pressure to the specimen.
6. Ensure that the direction of motion of any point on the specimen changes regularly with respect to the direction of motion of the polishing track.
7. Give thorough treatments, particularly at the first stage after abrasion. The treatment time required must be determined by experience as being that necessary to remove the significant abrasion damage.

For final polishing:

1. Operate principally with the aim of producing the desired quality of polish, but also —
2. Ensure that the polishing time is sufficiently long to ensure adequate removal of the rough-polishing damage.

In spite of the apparent complexities, it is only at several points in the preparation sequence that a choice has to be made between alternative courses. These are:

1. At the final abrasion stage, a decision has to be made whether or not to use a fine abrasive lap.
2. At the first rough-polishing stage, a decision has to be made as to how best to remove the layer containing potential abrasion artifacts.
3. At the final polishing stage, the least complex technique that will give the necessary quality of final polish has to be chosen.

A guide to making the second and third of these decisions is given for a number of common metals and alloys in the notes in Chapter 11. A guide to making the first decision was given on p. 76.

It is assumed in all of the procedures so far discussed that a small specimen can be removed from a bulk specimen and that it can be handled safely. This is the norm. There are occasions, however, when it is necessary to prepare a small area of a bulky specimen for examination *in situ*. There are other occasions where the specimen material is either so radioactive or so toxic that it cannot be handled safely by ordinary methods. These situations can raise considerable manipulative problems the solutions to which can be found only by consulting the specialized literature. However, this should not be allowed to hide the fact that the overall principles that we are discussing still apply.

Final Polishing

The one basic process that has not yet been considered is final (finish) polishing. It has already been noted that metallographic preparation is often regarded as being an art and not a science. This statement can by now be refuted for the abrasion and rough-polishing stages of preparation because we have seen that they can be based on sound principles. The same cannot be said to quite the same extent for final polishing. Final polishing still can demand a good deal of personal skill and touch from the operator, particularly when high-quality results are desired, although we shall see that some techniques eliminate the need for much of this skill.

The problems of final polishing are greatly increased by the range of the standards required for various applications, and by the tremendous variation in the difficulties encountered in meeting these standards in different specimens and alloys. A range of processes of increasing quality (and unfortunately of increasing complexity) must therefore be available from which a selection can be made as each occasion arises. Nevertheless, the difficulty of finish polishing by mechanical means should be kept in perspective. A finish of adequate quality can be produced on the vast majority of specimens by comparatively simple methods. These are the methods that we shall now consider. More advanced methods will be considered in Chapter 9.

The choice of an abrasive for final polishing lies mainly between magnesium oxide and aluminum oxide. Although other abrasives (such as rouge and chromic oxide) have achieved limited popularity, they will not be considered here, because little systematic information on their performance, and none to indicate that they have significant advantages over either aluminum oxide or magnesium oxide, is available. Between aluminum oxide and magnesium oxide, magnesium oxide produces better results (for example, cf. Fig. 9.1c and e), but gives lower polishing rates and is more difficult to use. Aluminum oxides, on the other hand, are simpler to use, give higher polishing rates, and are available commercially in a number of grades excellently suited to metallographic purposes. They are the first choice, the use of magnesium oxide tending to be confined to situations where aluminum oxide does not produce a polish of fully acceptable quality.

Magnesium oxide for metallographic polishing preferably should be purchased commercially as a dry calcined grade, because it is extremely difficult to prepare in the laboratory without serious contamination. A source must be selected which is adequately free from carbonate and other gritty particles. Even then, it is desirable to screen the powder onto the polishing pad through a 200-mesh sieve. The particles of magnesium oxide appear visually to be comparatively large, but actually consist of agglomerates of very much smaller particles (Fig. 8.22a). The agglomerates soon break up in use into the smaller particles visible in Fig. 8.22(b). These small particles are, nevertheless, considerably larger than the working particles of a 0.1-μm grade of gamma aluminum oxide (Fig. 8.22d) which, as we have already noted, produce a rather coarser polish. This cannot be explained.

In use, the magnesium oxide powder is moistened with water to form a thick slurry on the polishing pad. This slurry can be used for only a limited period because moist magnesium oxide carbonates fairly rapidly in air and the resultant carbonate grits cause severe scratching. Pastes using propylene glycol as the suspending fluid are, however, nearly as

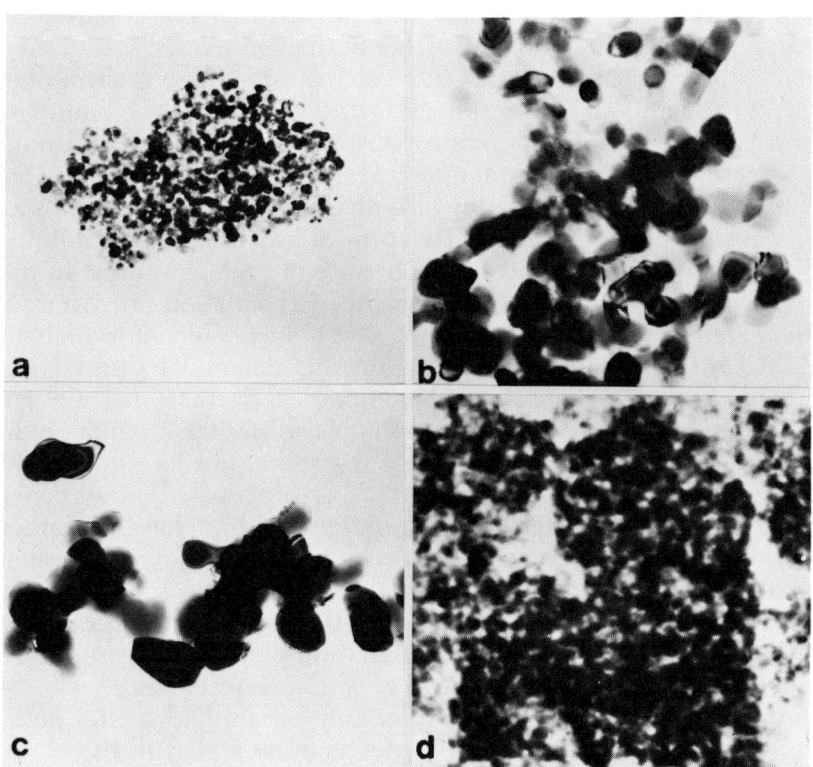

FIG. 8.22. Shadow electron micrographs of fine polishing abrasives. **(a)** A single agglomerate of calcined magnesium oxide. **(b)** Individual idiomorphic particles of calcined magnesium oxide. **(c)** Agglomerates of 0.1-μm-grade gamma aluminum oxide. **(d)** Individual particles of 0.1-μm-grade gamma aluminum oxide. Magnifications: **(a)** and **(c)**, 20,000×; **(b)** and **(d)**, 80,000×.

effective as those based on water, and can be used for many days without serious formation of carbonates. Care must be exercised in selecting other alternative fluids, however, because the polishing rate may be drastically reduced; presumably, a component of chemical-mechanical mechanisms needs to operate with this abrasive. Polishing cloths used with magnesium oxide should be washed thoroughly before and after use. They should be treated with, and stored in, a 1:1 hydrochloric acid solution to prevent the possibility of contamination by carbonates.

Aluminum oxides suitable for final polishing can be prepared from crushed grades of oxide using, for example, the methods developed by Rodda (see Appendix 5-A, p. 191). This is a very tedious process. They can also be produced in the laboratory by calcining an aluminum oxide,[7] when the process can be controlled to produce a definite maximum particle size and, if desired, to produce a preponderance of either alpha or gamma phases. It is widely thought that, for final polishing, gamma aluminum oxide produces better results and higher cutting rates than does the alpha form. However, alpha aluminum oxide can be prepared to have the same platelike morphology as that of the gamma form, and then has similar polishing characteristics. The calcination process is a troublesome one, and in small-scale operations it is difficult to avoid contamination.

Preparation of aluminum oxide polishing abrasives in the laboratory thus is scarcely to be recommended. This is particularly so now that excellent and closely sized grades of both alpha and gamma types are so readily available commercially. Two proprietary brands of aluminum oxide, known as Linde A* and Linde B*, are of special interest in this regard. Both are produced by the controlled calcination of hydrated ammonium aluminum sulfate and do not have to be graded after calcination.[8] Linde A contains about 90 vol. % alpha alumina and 10 vol. % gamma alumina; it has an average equivalent spherical particle size of about 0.3 μm. Linde B contains about 10% alpha phase and 90% gamma phase; it is composed of particles ranging up to 0.1 μm equivalent diameter with an average size of about 0.05 μm. The powders as supplied consist of agglomerates of small particles (Fig. 8.22c and d). These agglomerates usually, but not always completely, break up in use. Nevertheless, it has been suggested that sometimes it is desirable to break up the agglomerates before use by running the powder through a pulverizer mill;[9] tests need to be carried out on a particular supply of abrasive to determine whether this or some similar course of action is necessary.

Aluminum oxide abrasive powders may be applied to the polishing cloth in several ways. The dry powder may be added to the cloth and mixed there with the polishing fluid. Alternatively, it may first be made up as a slurry in the polishing fluid. A slurry may be prepared simply in the laboratory, in which event it will need to be shaken into a uniform dispersion immediately before application. Premixed permanent suspensions are available commercially.

The polishing rates obtained with both types of aluminum oxide — but particularly with the gamma form, to which the following discussion applies specifically — are sensitive to the condition of the abrasive slurry. First, the polishing rate decreases significantly as the pH of the slurry increases. Second, further drastic reductions in polishing rate occur, particularly at medium pH values, if mildly soapy agents are added; examples of such agents are triethanolamine and most mild alkalis that might be used for pH control. Corresponding reductions in polishing rate also result when many detergents are added, causing formation of a characteristic fine froth on the pad. Considerable improvements in cutting rate are obtained, on the other hand, when viscous soap solutions are added. For example, a soft coconut-oil soap mixed with water in a 1:1 ratio (by weight) has this beneficial effect.

Over-all, it is desirable that the pH of the abrasive slurry be kept in the range 6 to 7. The pH of a simple water-base slurry depends on the

---

*Trademark, Union Carbide Corporation.

quality of both the abrasive and the water. When accurate control of pH is not essential, the desired pH is best obtained simply by suitable selection of the two ingredients rather than by addition of pH-adjustment agents which might adversely affect the polishing rate.

The fluid in which the abrasive is suspended can affect the performance of both magnesium oxide and aluminum oxide abrasives in other ways. The fluid most commonly used is water. The water used for this purpose must be of good quality. Filtered, demineralized or distilled water should be used when the mains supply is of doubtful quality. This is particularly necessary for polishing of the more reactive metals, the most reactive of which may even require the use of a non-ionizing fluid such as propylene glycol, kerosine, or a light oil. Remember again, however, that changing the fluid may significantly change the polishing rate.

There are some specimens (e.g., galvanized steels, clad aluminum alloys, and leaded steels) in which there are considerable electrochemical differences among various areas. These electrochemical differences may result in severe etching of the anodic areas of the specimen if the polishing fluid is ionizable (Fig. 8.23a). The problem usually can be overcome by using a slurry of aluminum oxide abrasive which has a pH of exactly 7 (Fig. 8.23b). This can be achieved by very carefully selecting the components of the slurry, adjusting the pH to exactly 7 if necessary by the addition of a mild alkali.[10] An alternative and simpler method is to use a standard pH 7 buffer solution as the polishing fluid; this was the technique used in preparing the specimen shown in Fig. 8.23(b).

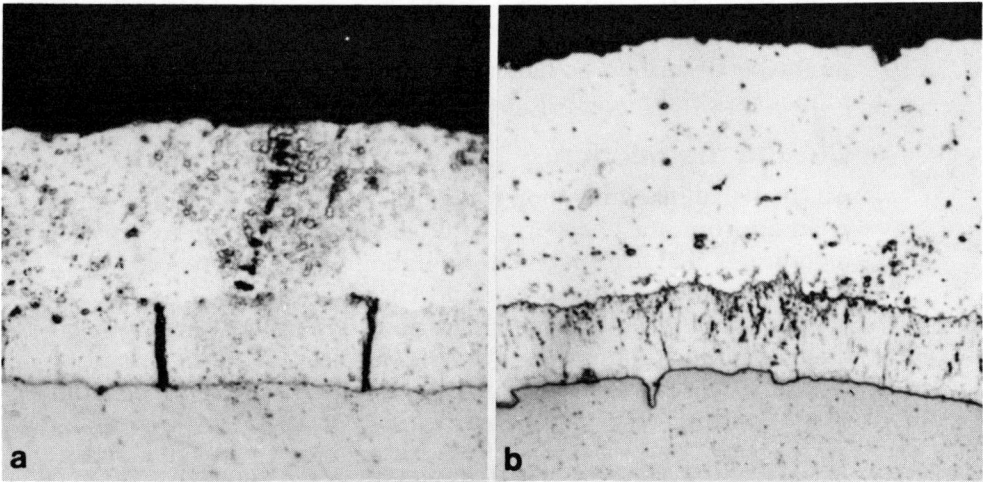

FIG. 8.23. Section of a sheet of galvanized steel finish polished with aluminum oxide abrasive suspended in two different polishing fluids.

(a) Abrasive suspended in good-quality mains water. (b) Abrasive suspended in a pH 7 buffer solution. The severe etching of the coating in (a) is due to the electrochemical differences between the zinc coating and the steel base. Magnification (both), 750×.

The same range of polishing cloths might be considered for use in final polishing as for rough polishing. Now, however, the degree to which the cloth itself scratches the surface of the specimen becomes of major importance. It is clearly counterproductive to use a cloth that scratches the surface more severely than the abrasive. The tendency, therefore, particularly with softer specimen materials, is to use a napped cloth and the softest possible cloth at that. However, a compromise may have to be made in some circumstances, particularly when flatness of the surface is important. The use of a harder cloth will improve surface flatness with some sacrifice in quality of finish. Alternatively, a short polishing time may be possible with a soft cloth, with some sacrifice in flatness. In the second case, the quality of finish achieved at the rough-polishing stage will need to be good if the rough-polishing scratches are to be removed successfully.

STANDARD PROCEDURES

There is a veritable plethora of devices, either available commercially or described in the literature, which can be used for the various stages of specimen preparation. Selection from among these devices depends on the particular requirement of the laboratory concerned; and, as we have mentioned previously, these requirements vary enormously (a) from laboratories that handle a few specimens a week to those that handle hundreds per day, (b) from those that handle a single type of specimen material to those that handle a wide variety, (c) from those performing routine quality-control tests to those involved in sophisticated research, and (d) from those handling the easiest to those faced with the most difficult specimen materials. It clearly will be possible for us to discuss only a selection of the available equipment and processes, selected to give an indication of the range that might be considered. Moreover, we shall discuss in the immediately following sections only those processes that might be described as basic processes. Specialized processes designed to meet special needs will be discussed in Chapter 9.

Cutting the Section

All the usual machine-shop practices are available for cutting the section at an appropriate plane in a bulk specimen. But there are widely used techniques that are specific to metallographic practice.

The first is the abrasive cutoff wheel (Fig. 8.24a) which is especially appropriate for sectioning of materials that are too hard to be cut by normal machining techniques. The critical variables in the operation of these machines are the type of abrasive wheel, the supply of coolant, and the rate of cutting. The wheels are usually of a vitreous-bonded aluminum oxide, supplied in a range of grades to match particular types of materials; suppliers' recommendations should be adhered to when selecting among various grades. Steel disks, the peripheries of which are impregnated with diamonds, are also available for cutting very hard materials. The vital matter, however, is that it must be possible to keep the actual

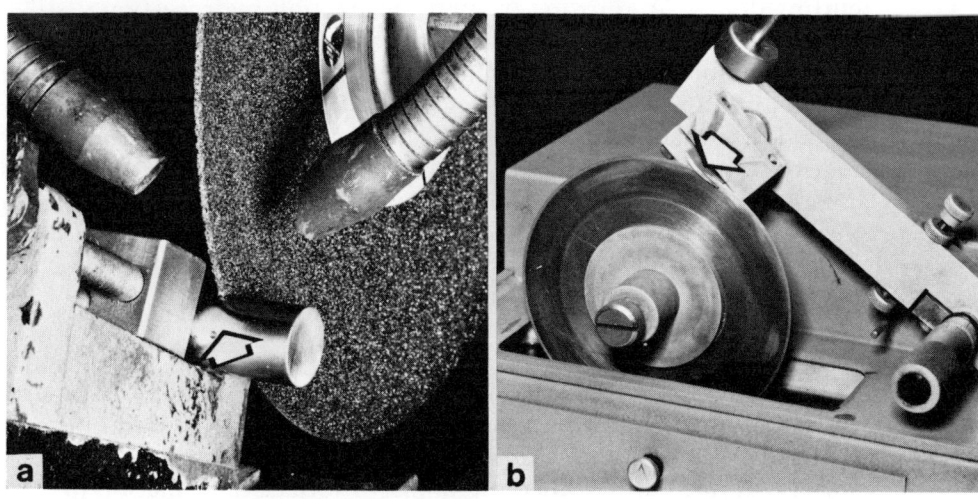

FIG. 8.24. Representative abrasive wheels used for cutting metallographic sections.

(a) A vitreous-bonded abrasive wheel which rotates at high speed. In actual use, cooling fluid is pumped through the two adjustable nozzles. The specimen (arrow) is clamped, and the wheel can be oscillated in a vertical plane on a radius arm. The specimen clamp can be advanced horizontally in small increments between oscillations of the wheel. (b) A thin metal wheel with a diamond-impregnated periphery. The specimen (arrow) is clamped in a radius arm and rests against the cutting edge of the wheel under an adjustable dead load.

cutting interface flooded with a copious supply of water. The cutoff wheels operate at comparatively high speeds (approximately 3000 rpm; peripheral speeds of 20 to 40 m/s), and severe thermal heating of the cut surfaces will occur unless the cutting region is kept flooded with a coolant (see p. 126). Likewise, the depth of cut for each traverse of the abrasive wheel must be kept within limits; otherwise, the specimen may be distorted or surface heating may be accelerated.

In commercial machines, either the wheel or the specimen is swung manually on an arm so that the two can interact with one another. In addition, the specimen can usually be advanced in increments into the wheel. Essential features are that it should be possible (a) to clamp rigidly specimens of a variety of shapes; (b) to direct a copious supply of coolant into the slot that is being cut; and (c) to control the depth of cut of the abrasive wheel.

The second characteristic metallographic sectioning machine is a more delicate one (Fig. 8.24b). It uses a thin metal disk, the periphery of which is impregnated with diamonds. The disk runs at comparatively low speeds (up to 400 rpm; peripheral speeds of up to 3m/s) and is designed to run truly. The specimen is clamped in a radius arm and is fed against the wheel under a variable dead weight. Cutting fluid is supplied to the cutting area by arranging that the wheel dip into a tank of the liquid. The objective of the fluid is to increase the cutting rate and to flush away the cutting debris. A kerosine or medium-fraction hydrocarbon distillate to which 10 vol % of a liquid detergent has been added is a suitable cutting fluid. Suitable fluids are available commercially. A range of machines which differ in mechanical arrangements rather than in principle is available. These machines cut slowly. They have application for small specimens, for fragile or brittle specimens, and where thin cuts or thin slices are required. A very fine finish is produced, but the depth of the plastically deformed layer on the cut surface is not greatly different from that produced by a normal abrasive cutoff machine (Table 4.1, p. 112). There is no risk of significant surface heating.

Preliminary Machining

The range of tools available in a standard machining shop can be used for the preliminary machining of a cut surface, but it needs to be recalled that the objective of preliminary machining is not only to produce a flat surface. It must also reduce the depth of surface deformation to the minimum. It is also necessary to avoid significant surface heating.

Files are commonly used for softer materials,* but produce comparatively deep deformed layers (Table 4.2, p. 113). Hand grinding is also commonly used, and produces a reasonably shallow deformed layer, but care is necessary to ensure that serious surface heating does not occur; either a coolant has to be directed at the grinding wheel, which is not often done, or the specimen has to be immersed frequently in a tank of coolant.

Abrasive belt surfacers are also commonly used in metallographic laboratories. They run at linear speeds of 10 to 20 m/s. This is sufficiently high to cause significant surface heating if abrasion is carried out dry for more than a few seconds, but not if the surface of the belt is flooded with water. The latter therefore is essential. Moreover, the abrasion characteristics of the belts will be the same as those of any other coated abrasive product, as discussed in Chapter 3. From this it follows that there are no advantages, and that there are disadvantages, in using belts coarser than P240 to 150 grades for many materials (Fig. 3.16, p. 58, and Fig. 3.20, p. 64). Moreover, the total thickness of metal that can be removed on one track on the belt is strictly limited for some other materials, the most important examples being ferrous alloys (Fig. 3.19). Belt life will be longer for belts coated with aluminum oxide than for those coated with silicon carbide abrasive, but will still be limited to less than 1000 traverses. This may represent only a few minutes' operation. Exceptions to this limit are found only with very coarse grades of belts (80 grade or coarser; see pp. 69 and 70). Such belts achieve comparatively low abrasion rates but will continue to cut for comparatively long times. This advantage must be balanced against the poor finish that is obtained.

---

*Extremely soft materials are discussed on p. 313.

Another device characteristically used in metallography for coarse finishing is a horizontal rotating disk to which a coated abrasive paper is attached. These devices are similar to those described below for abrasion except that the disks are of greater diameter (12 in., or 30 cm) and rotate at higher speeds to give linear speeds on the abrasion track of up to 50 m/s. Consequently, the same remarks apply as for belt surfacing. Vitreous-bonded abrasive disks may also be used in this way (in machines similar to that illustrated in Fig. 8.27), but the characteristics of these disks have not been established in detail.

Diamond abrasive disks might be considered for preliminary machining when the limitations of conventional abrasives become restrictive. They may be cost-effective for a range of conditions (see p. 71).

Metallographic Abrasion

At the simplest level, a sheet of coated abrasive paper can be laid on a flat backing (such as a sloping glass plate), the working surface of the paper flooded with a stream of water, and the specimen rubbed by hand against the working surface with an oscillating motion (Fig. 8.25a). The paper will be held down more satisfactorily if its back surface is coated with a contact adhesive, and papers coated in this way are available. Strips of double-sided adhesive tape can also be laid on the back surfaces of normal papers (dispensers are available for this purpose; see Fig. 8.25b). Flushing the abrasion track with water generally is essential to prevent it from becoming clogged with abrasion debris (see p. 59). It is desirable to

FIG. 8.25. Basic techniques for hand abrasion.

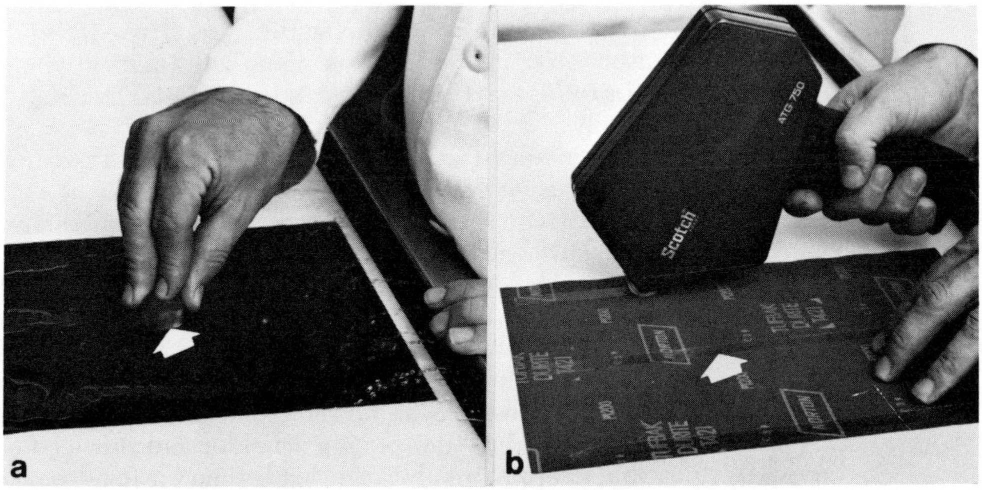

(a) The specimen (arrow) is rubbed by hand on an abrasive paper over which a stream of water is flowing. The paper rests on a glass plate. (b) Strips of double-sided adhesive tape (arrow) are applied to the back surface of an abrasive paper. Several strips will hold the paper flat against a backing plate.

FIG. 8.26. Techniques for mechanized abrasion.

(a) The specimen is held by hand against an abrasive paper attached to a rotating wheel. (b) An attachment designed to automate abrasion on the wheel in (a). Both specimen and wheel are driven mechanically. Mounted specimens are clamped in a collar chuck (A) which is driven by a spindle in (B). The collar chuck is loaded by a spring in (B), with a screw adjustment (C). Power is taken off the main wheel through (D). Water is supplied from (E). The specimen-holder collar chuck is similar to that in Fig. 8.27(b).

apply pressure to the specimen only during the away stroke if the surface is to be kept reasonably flat. The section plane will then tilt slightly in the direction of abrasion, but this tilting can be obviated by rotating the specimen periodically.

Various degrees of mechanization of the abrasion process are possible. As a first step, the abrasive paper may be attached, by use of a contact adhesive, to a power-driven rotating disk or wheel and the specimen held against the abrasive surface by hand (Fig. 8.26 a). The wheel must run in a bowl to which drainage is provided so that the surface of the abrasive paper may be kept flooded with water, either from a constant-head device (as in Fig. 8.26a) or directly from the mains supply.

As a second step, a group of specimens may be clamped in a collar chuck (Fig. 8.27b) in which they can be rotated against the abrasive surface. A load can be applied to the specimen holder and this load can be varied; it is possible to apply much larger loads than for hand-held abrasion. Attachments to standard power-driven abrasion wheels are available for the purpose; one is illustrated in Fig. 8.26(b). More sophisticated machines are available in which the specimen holder is driven independently (Fig. 8.27a), and in some of which the holder can be made to oscillate across the surface of the abrasive paper. The abrasion parameters can be adjusted and reproduced, and long abrasion times are acceptable, but long abrasion times may not even be necessary if high abrasion pressures are used.

Automated machines need, however, to be operated with some thought to the characteristics of the abrasive papers, which are still the heart of the machine. The papers will behave in exactly the same way as they would under any other condition of use: that is, they will still have the limitations that were discussed in Chapter 3 (p. 55). These limitations are not likely to be directly apparent to the operator. The fact that an abrasive paper has been used beyond its useful limit is certainly not likely to be as apparent as in hand abrasion.

The rotational speed of an abrasion wheel needs to be several hundred rpm; 200 rpm corresponds to a linear speed at a typical abrasion track of about 5 m/s. These relatively low speeds are desirable to ensure good control of the specimen and hence retention of adequate flatness of the section surface, and also to eliminate the risk of adverse effects due to heating of the specimen surface. Flushing the abrasion track with water is still essential, however, to ensure that clogging with abrasion debris does not occur easily (see p. 59). Thus a water supply and a drain are necessary.

Polishing

Polishing can be carried out most simply on a disk-shape block over which the polishing cloth is stretched taut (Fig. 8.28a). The specimen is then rotated by hand over a section of cloth that has been charged with abrasive and polishing fluid. The disk may be made of a metal (usually a

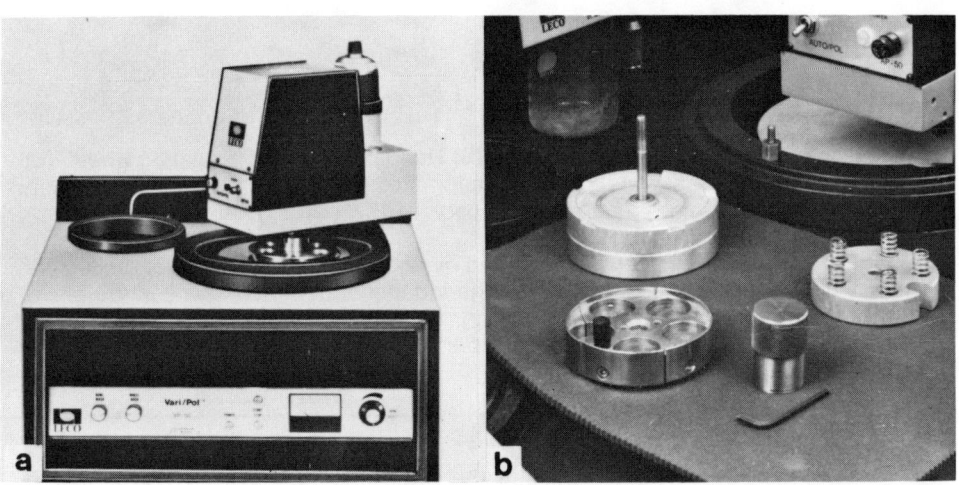

FIG. 8.27. A special-purpose fully automated abrasion/polishing machine.

(a) General view, showing the head used for driving the specimen collar chuck. (b) A more detailed view of the specimen collar chuck, and ancillary devices for loading the collar chuck so that the section surfaces are approximately coplanar before abrasion is commenced.

254 / Metallographic Polishing by Mechanical Methods

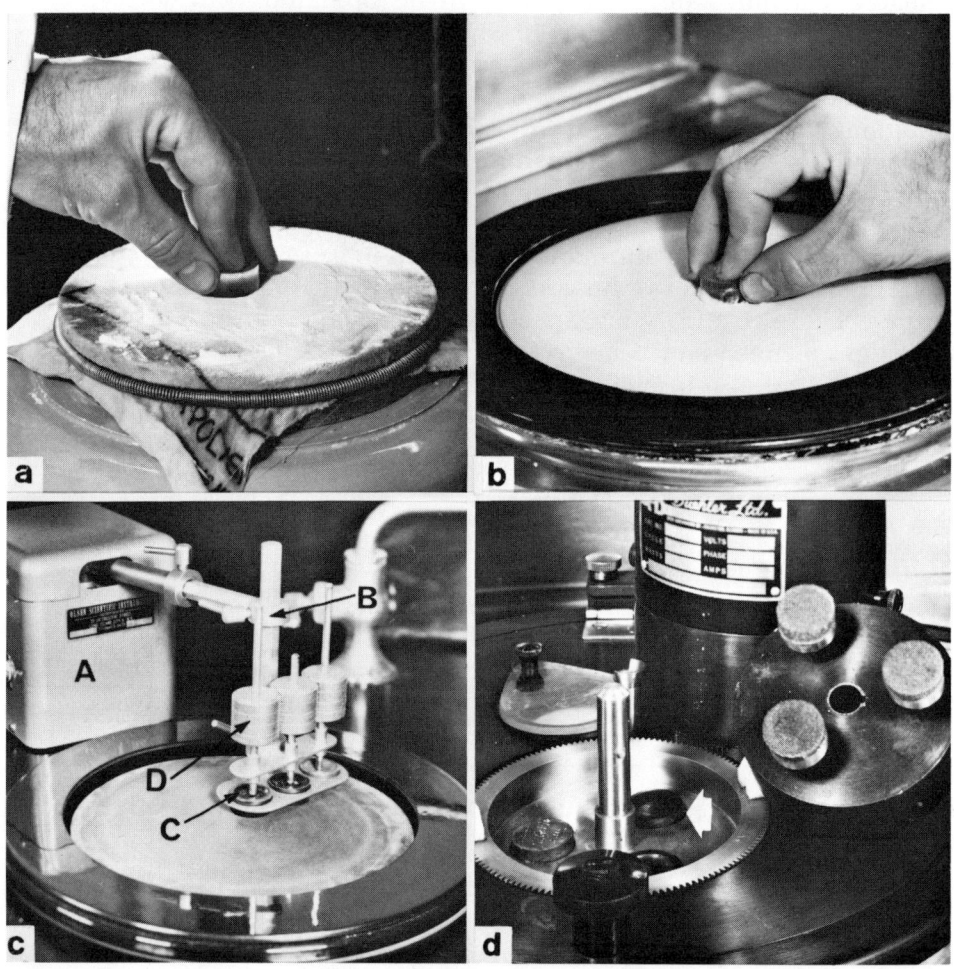

FIG. 8.28. Techniques for hand and mechanized polishing.

(a) Hand polishing on a stationary pad. (b) Hand polishing on a rotating wheel. (c) A device which can be set on a bench to automate polishing on a standard polishing wheel (see text for a more detailed description). (d) An attachment designed to automate polishing on the wheel shown in (b).

Three specimen mounts (one is indicated by an arrow) rest in a circular guideplate which is rotated through a gear train by the independent motor visible in the background. A load is applied to the specimens through pads beneath a dead weight which normally rides on the vertical spindle; this part of the device is shown inverted to the right of the guideplate.

bronze) or a plastic, the choice normally being important only from the point of view of corrosion by the polishing fluids (but see p. 269). The polishing cloth may be clamped mechanically to the disk (as in Fig. 8.28a); alternatively, some types of cloths are available whose back surfaces are coated with contact adhesive.

At least some mechanization of the polishing process is desirable, even more so than for abrasion and particularly so for rough polishing. Faster polishing becomes possible, which reduces the probability of abrasive artifacts being present in the final surface. Moreover, the most appropriate procedure can be standardized more readily, and higher output rates become possible. The first question that might arise with mechanical polishing is whether a satisfactory quality of final polish can be achieved. At worst, however, only finish polishing by other methods would be necessary to achieve any desired improvement in the quality of finish. The main question, however, is likely to be whether the capital costs of the equipment are justified. Complex and comparatively expensive equipment tends to be needed — particularly with more advanced stages of mechanization. This is true for several reasons. First, it must be possible to apply a reasonably large load to the specimen, but in such a way that the full area of the section contacts the polishing cloth; that is, the load-applying device must not allow the specimen to tilt. Moreover, the load must be applied in such a way that the specimen does not rock if it changes direction relative to the local direction of motion of the polishing cloth. Regular changes in the relative direction of motion are in fact necessary; otherwise, unacceptable tracking grooves may develop in the section surface. These grooves first become noticeable as tails on any cavities in the surface, when they are commonly known as *comet tails* (Fig. 8.29). Cavities produced by complete or partial loss of nonmetallic constituents may also initiate comet tails.

FIG. 8.29. Cavities in an aluminum alloy casting polished by unidirectional and random polishing.

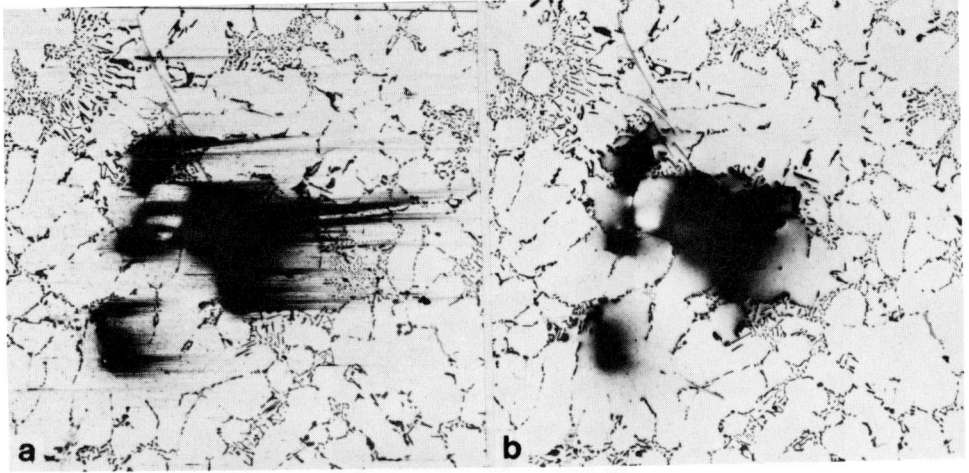

(a) Section polished unidirectionally, the abrasion motion being from left to right. (b) Same field as in (a), but polished with random relative motion between the section surface and the polishing cloth. Both polished on suede cloth charged with 1-μm-grade diamond abrasive. Magnification, 100×.

Mechanization of polishing can again be introduced in a number of stages. First, the disk can be power driven as a wheel encased in a bowl, the specimen being pressed against the polishing cloth by hand (Fig. 8.28b); usually, the specimen is rotated against the direction of motion of the wheel. This method provides good control of the specimen, particularly when results of the highest quality are required. A satisfactory degree of random motion between the specimen and the polishing cloth can be achieved.

Next, comparatively simple specimen-handling devices are available that can be added to a polishing wheel of the type just described. A typical example is illustrated in Fig. 8.28(c). The mounted specimens (C) sit loosely in a guideplate and can adjust themselves in the plate so that the full area of the abraded surface sits on the polishing cloth; the specimens can also move vertically to accommodate irregularities in the plane of the polishing wheel as it rotates. A variable dead-weight load (D) with free vertical movement is applied through point contact to the back of the specimen mount; the guideplate is made to oscillate across the surface of the polishing cloth by means of a radius arm (B) driven by an electric motor (A). The difference in the surface speed of the polishing pad between the inner and outer edges of the specimen mount causes the mount also to rotate in the guideplate about its vertical axis. The specimens are, however, likely to rock a little as the guideplate oscillates across the polishing cloth.

The device illustrated in Fig. 8.28(c) sits on the bench adjacent to the polishing wheel. Others, similar in principle, are matched as attachments

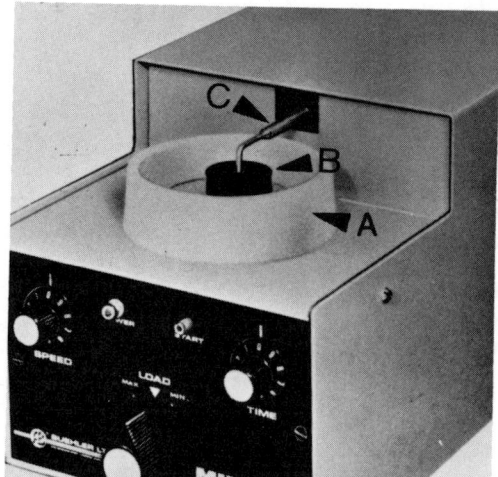

FIG. 8.30. A simple automatic polishing machine.

The polishing cloth covers the base of the rotating bowl (A). The specimen mount (B) is held against, and traversed across, the cloth by the oscillating arm (C). The pressure applied to the mount by the arm can be varied.

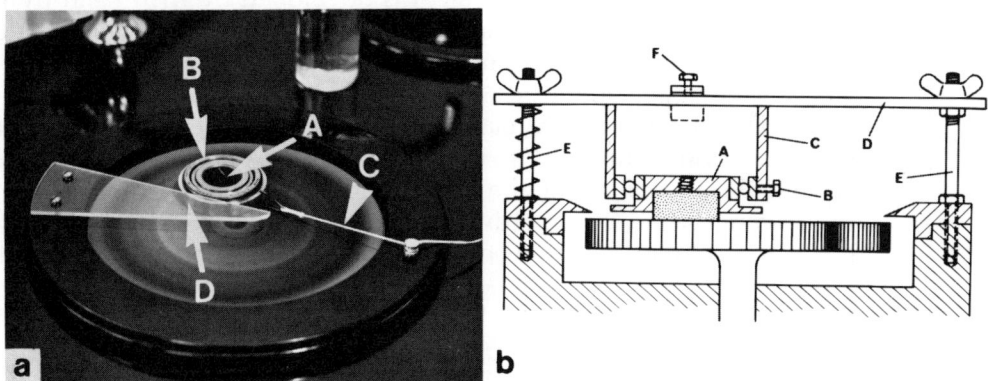

FIG. 8.31. Two simple devices for automatic polishing on a standard polishing wheel.

  (a) The specimen (A) is fixed in and protrudes below the inner race of a ball bearing (B), which is held in position by an anchor cord (C) and a stop plate (D). Additional weights can be applied to the bearing if desired. *(Kjer, Ref 14)*
  (b) The specimen (speckled) is held in a flanged cup (A) which is a press fit in the inner race of a ball bearing. The bearing is a loose fit within a cylinder (C), but can be clamped in suitable vertical location by means of a lock nut (B). The cylinder (C) is attached to a lever arm (D) which is located by two posts (E). One end of the lever arm is fixed and the other rides on a return spring. The vertical location of the free end of the arm, and hence the load applied to the specimen, can be varied by means of an adjustment screw (F). *(Evans, Ref 15)*

to particular polishing machines; one of these is illustrated in Fig. 8.28(d). An even simpler but specially manufactured device is illustrated in Fig. 8.30. The entire polishing bowl (A) rotates, and the polishing cloth is attached to a removable plate contained in its base. Point pressure is applied through a hole drilled in the back of a standard specimen mount (B) by means of a loaded lever arm (C); the arm is made to oscillate as the bowl rotates. Consequently, the specimen oscillates across the face of the rotating disk, and also rotates about its vertical axis.[12,13] The bowl is small, can readily be interchanged, and so can be dedicated to a particular grade of abrasive.

A number of devices have also been described which could be made simply enough in the laboratory. An extremely simple one credited to Kjer[14] is illustrated in Fig. 8.31(a); the specimen mount and inner race of the bearing rotate as the polishing wheel rotates. A slightly more elaborated and more permanent device based on the same principle is illustrated in Fig. 8.31(b).[15]

As a final stage of mechanization, fully automated machines of the types illustrated in Fig. 8.26(b) and 8.27 can be used for polishing. In this event, all abrasion and polishing have to be carried through in the machine without disturbing the specimens in the collar chuck.

Polishing wheels of all these types need to run at comparatively low rotational speeds (100 to 200 rpm). This alleviates problems of specimen control, minimizes throw-off losses of abrasive, and minimizes the risk of heating of the surface of the specimen (see p. 127).

An automatic polishing machine based on an entirely different principle from those discussed so far was developed by Krill[16] and by Long and Gray[17]. The polishing pad is contained in a bowl which constitutes a portion of a torsional pendulum driven approximately at resonance by an electromagnet operating on mains alternating current supply (Fig. 8.32). The specimens are clamped within cylindrical weights. The vibrating motion imparted to the bowl causes the specimens to move around the periphery of the bowl without rotating significantly about their vertical axes. The relative motion between specimen and polishing pad is thus similar to that in conventional hand polishing. The vibrational

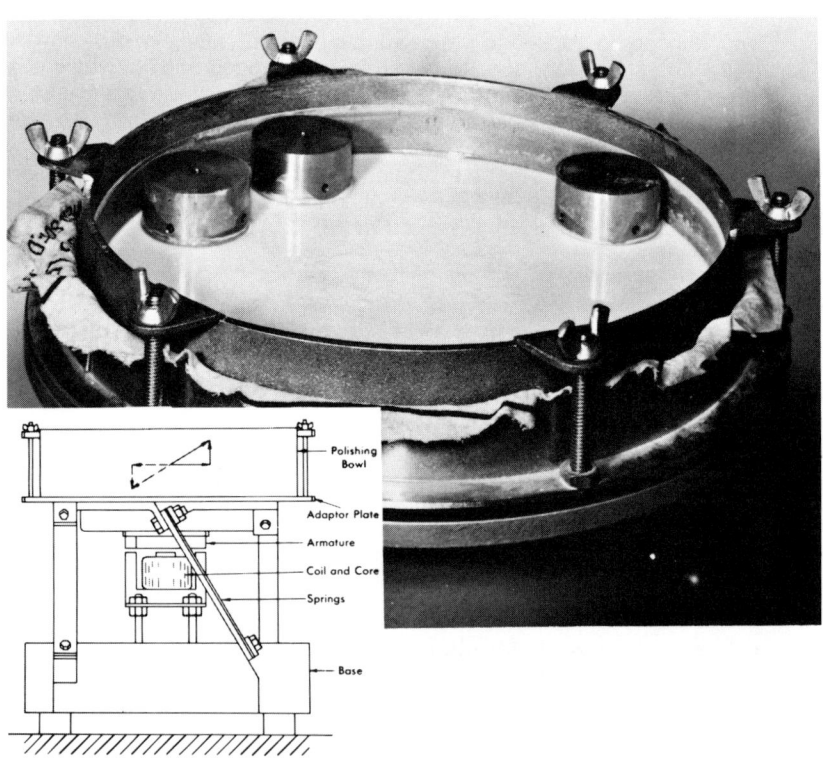

FIG. 8.32. An automatic vibratory polishing machine.

The specimens are clamped in cylindrical holders, three of which are shown here in the polishing bowl. The polishing cloth is spread across the base of the bowl and covered with a slurry of abrasive. The inset shows mechanical details of the apparatus and the nature of the motions that are imparted to the bowl. The bowl should be covered with a transparent dust cover that can be easily removed.

amplitude of the device can be controlled (Appendix 8-A, p. 262). Generally, it should be set to the lowest value that ensures smooth progression of the specimen around the bowl without bouncing. Suitable machines are available commercially which are equipped with adequate polishing bowls, and a number of special bowl designs have also been developed which facilitate changing of cloths.[17-19] Some attention to detail is necessary if these machines are to be run successfully, which is discussed in Appendix 8-A.

The first point to note about the vibratory technique is that the polishing rate obtained is roughly the same as that which would be obtained for the same relative movement by a conventional technique. The specimens track quite slowly around the periphery of the bowl. Hence the polishing rates obtained are comparatively small. The technique therefore is not suited to the removal of large thicknesses of material. It follows from the arguments developed earlier in this chapter that this technique is not a highly desirable one for rough polishing, even though it allows the machines to be left unattended for long periods. It is only in exceptional circumstances that the technique has significant advantages over conventional rough-polishing techniques. The exceptions depend on the special mechanical features of the system, such as the ease of arranging remote handling of the specimens. On the other hand, the technique is eminently suited to final polishing, because very small loads can be applied to the specimen, polishing parameters can be well controlled, and long polishing times are practicable. In fact, as discussed later (p. 279), special finishing processes can be based on this technique.

Ideally, a polishing machine of any type should be dedicated to a particular grade and type of abrasive in order to prevent cross contamination.* If this is not possible, cleaning of the machine is necessary between stages. Designs which enable this to be done both easily and effectively then are desirable.

In a sense, the distinction that has been made between abrasion machines and polishing machines is an artificial one because all of the machines described can be used for either purpose and in different combinations. For example, specimens could be abraded by any of the methods illustrated in Fig. 8.25, 8.26 and 8.27 and then polished by any of the methods illustrated in Fig. 8.28 and 8.30 to 8.32. They could be both abraded and polished by the methods illustrated in Fig. 8.26(b), 8.27, 8.28(c), 8.28(d) and 8.30. Specimens could be abraded and rough polished by any of these methods and then, if necessary, be given individual attention by hand for finish polishing.

Nevertheless, some machines are more suited to abrasion and less suited to polishing, and vice versa. For example, generally it is desirable to apply comparatively large loads to the specimen during abrasion and rough polishing. This is not easily done in some machines without causing excessive rocking of the specimen. Other machines can apply large loads

---

*A bowl can be dedicated to each grade of abrasive in machines of the type illustrated in Fig. 8.30, since the bowls are interchangeable.

with ease and produce particularly flat surfaces (e.g., the machines illustrated in Fig. 8.26b and 8.27). On the other hand, comparatively small loads need to be applied in final polishing, and some devices are not particularly suitable in this respect. Moreover, although fully automated machines become desirable and even essential when large numbers of specimens have to be handled, they are not particularly convenient when the throughput is small. Each laboratory has to consider its needs and make a selection from the range of machines available, considering effectiveness and cost.

Interstage Cleaning and Drying

Thorough cleaning of the specimen and of the operator's hands to prevent carryover of abrasive between stages is of the utmost importance. It is also necessary to clean the specimen at the end of the polishing sequence, both before and after etching.

It is usually desirable, when oily polishing fluids have been used, first to flush the surface with a nontoxic solvent. Washing with a soap or a detergent then usually is adequate. Washing may be assisted by swabbing with a suitable material, provided that this does not scratch the surface more than the preceding polishing process. The washed surface may then be flushed with a strong stream of water. The flat jet illustrated in Fig. 8.33(a) is an example of one effective way of doing this.

Ultrasonic cleaning in a bath of dilute detergent is also efficient and usually reliable — particularly so when oily polishing fluids have been used, when the specimen contains pores or cracks, or when the specimen

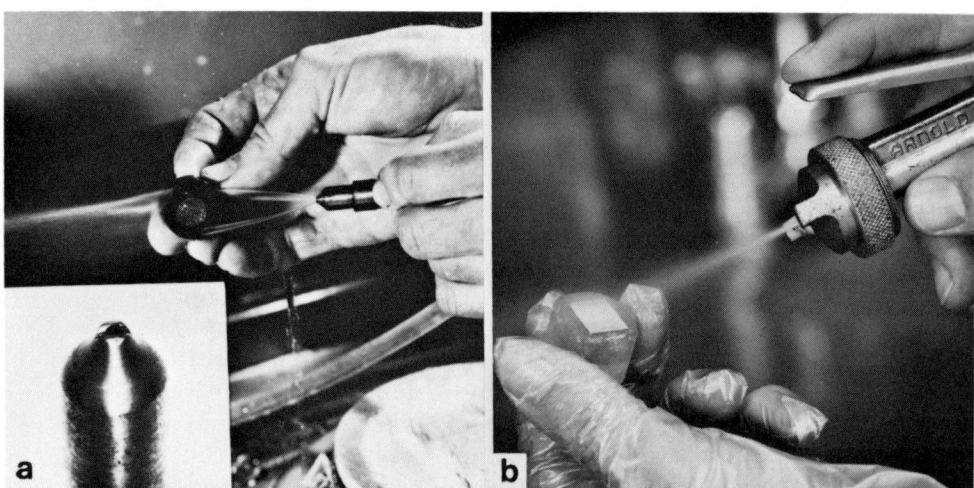

FIG. 8.33. Techniques for washing and drying polished surfaces.

(a) The surface is washed with a flat jet of water. A nozzle that produces such a jet is shown in the inset. (b) The surface is dried with a blast of atomized solvent.

is clamped in a collar chuck such as that illustrated in Fig. 8.27(b). It may also be possible to remove minor stains by ultrasonic cleaning. It will certainly be possible to remove microscope immersion oil and to dislodge any polishing debris that has accumulated in cavities and depressions on the polished surface (see p. 131).

Nevertheless, it is possible to damage polished surfaces by ultrasonic cleaning. The process works by cavitation erosion, and so cavitation pits can be produced on the surface. The damage may be serious with some materials, dental amalgams being a specific example. Ultrasonic cleaning erodes some of the phases in these alloys selectively; the phases themselves cannot then be examined and the apparent porosity content of the alloy is erroneously increased.[20] Pits may also be produced in the surfaces of comparatively hard materials, but only after long cleaning times (30 min or more). Even then, the pits are likely to be resolved only by sensitive examination methods such as phase or Nomarski contrast.[21] The resultant damage may often be noted visually, however, as patches of cloudiness when the surface is viewed at grazing incidence.

The method of drying the washed surface after final polishing (and after etching) is also important. Drying must be carried out in such a way that stains are not produced. A standard technique is to warm the surface under a stream of hot water, then to flood it with a volatile solvent, and finally to evaporate the solvent rapidly by means of an air blast. The solvent must be of such a quality that it leaves no residue; alcohol and acetone are commonly used, although acetone attacks some mounting plastics (e.g., the methacrylates). Hand hair dryers conveniently supply a stream of either warm or cool air for drying but are difficult to keep in maintenance in the rather severe operating environments involved. Specially designed dryers based on a similar principle but more robust are available from metallographic supply houses. An air line fitted with filters and with a pen-type outlet is often satisfactory. In the case of mounted specimens, the risk of staining due to minor contaminants in the air supply can be reduced by directing the primary air blast onto the surface of the plastic adjacent to the specimen.

A most satisfactory drying method is to direct a blast of atomized solvent onto the specimen surface. This can be done by means of a standard paint spray-gun head (Fig. 8.33b), or some similar atomizing head.[22] The technique is particularly effective in eliminating staining caused when etching solutions seep out of cracks, crevices and cavities in the specimen/mount system. An extensive survey[23] has shown that a solution with the following composition is most suitable for safe and effective use in this way: 48 vol % Freon, 48 vol % methylene chloride, and 4 vol % ethanol. This spray is nontoxic, nonflammable and nonexplosive. It does not attack any of the plastic mounting media and does not frost the specimen surface.

In spite of all these precautions, occasionally it is found that stain marks develop around inclusions such as manganese sulfide and graphite in ferrous alloys (Fig. 8.34a and c). The marks appear to be rust stains and can be prevented by washing and rinsing the specimen in cold water only (Fig. 8.34b and d).

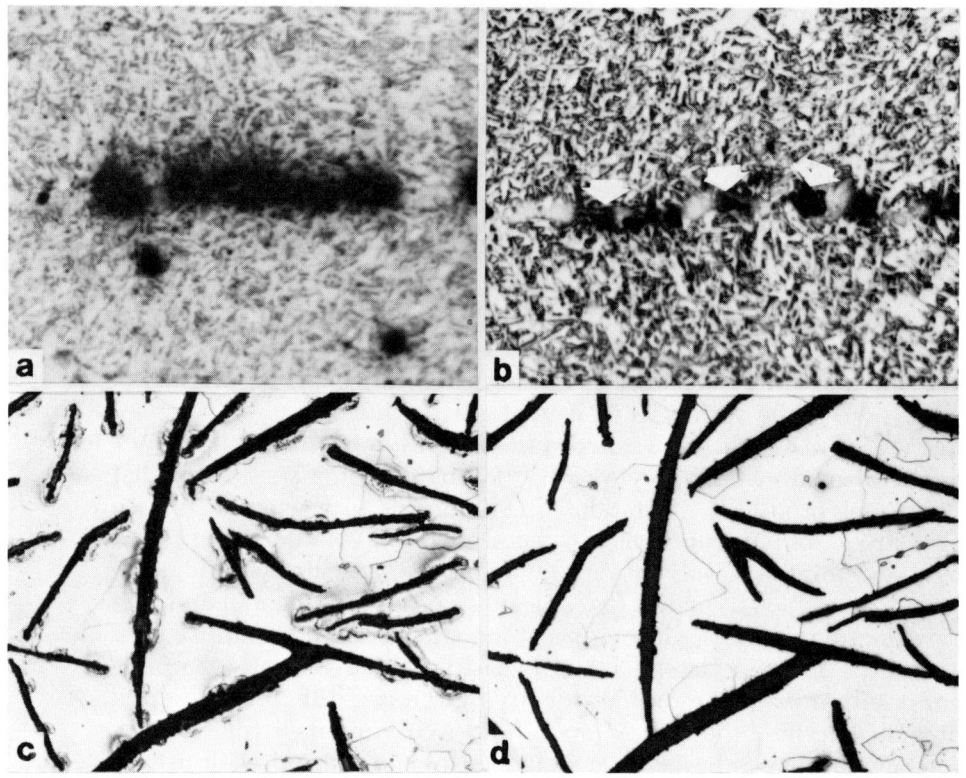

FIG. 8.34. Stains developed around inclusions in ferrous alloys as a result of the surface being washed in hot water.

(a) A heat treated steel, etched in picral and then washed in hot water. (b) Same area as in (a), but washed in cold water. Sulfide inclusions (arrows) in a segregate band can now be seen; neither is visible in (a). (c) A gray cast iron that has been polished, etched and then washed in hot water. Note stains around graphite flakes. (d) As for (c), but washed in cold water. Magnifications: (a) and (b), 250×; (c) and (d), 100×.

APPENDIX 8-A
Operating an Electromagnetic Vibratory Polisher

The natural frequency of a torsional vibratory system is described by:

$$V_o = \frac{\pi}{2}(k_t/T)$$

where V_o is natural frequency, k_t is spring constant and T is mass moment of inertia.

As Long et al.[18] have pointed out, the vibratory polishers used in metallography usually are adaptations of commercial vibratory feeding devices. The mass moment of inertia (T) of the feeder will have been matched to the spring constant (k_t) in the original design of the feeder, and it would be difficult to design a polishing bowl to have exactly the same value of T. It is best to aim at a close value and then to adjust k_t. Springs of different sizes may be available from the manufacturer for this purpose, and the number and size of the springs in one or both diametrically opposite pairs of springs can be changed; opposing pairs of springs must, however, be kept matched.

The torque of the bolts holding the springs is also important because it determines the degree of damping of the spring and the distribution of the vibratory forces between the springs. A minimum torque of 500 lb/in. is recommended by the manufacturer for one common type of machine. The air gap between the armature and the coil core is also adjustable for fine tuning. However, the armature must not strike the core during operation. A nominal gap of about 0.5 mm (0.02 in.) is usually optimum.

The frequency of the exciting system is not controllable, being that of the mains supply. The relationship between exciting frequency and vibrational amplitude of the bowl is illustrated in Fig. 8.35, which also indicates the effect of damping.[18] The characteristics of these curves in-

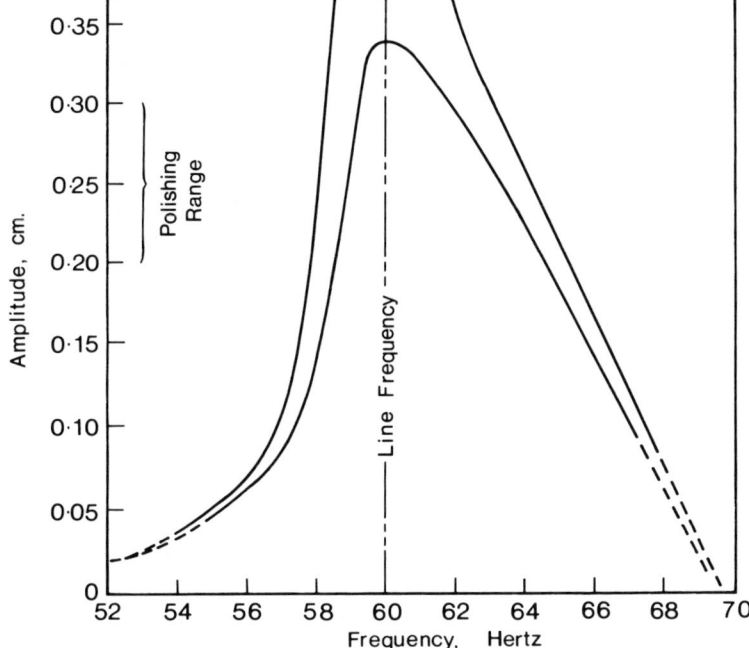

FIG. 8.35. Relationship between exciting frequency and vibrational amplitude of the bowl of a vibratory polishing machine. *(Long, Meador and Gray, Ref 18)*

The upper curve represents the ideal case of no damping, and the lower curve represents a satisfactorily damped system.

dicate that the natural frequency of the vibratory system should be adjusted to be somewhat higher than the frequency of the mains supply. The system is then more stable and the addition of specimens to the system is less likely to destroy the resonance of the vibration. Some damping of the system clearly is necessary, but excessive damping could reduce the amplitude of vibration to unacceptable levels.

Because the operation of the machine is critically dependent on a satisfactory vibrational amplitude being achieved, and because this may be seriously affected by factors such as overloading of the bowl with specimens or deterioration of the spring systems, it is desirable to have a method of checking the vibration amplitude at all times. One suitable method is illustrated in Fig. 8.36. A card marked with a vee figure is attached to the side of the bowl such that the axis of the vee is parallel to the torsional springs (this usually means that $\phi = 75°$). When vibrated, the figure appears as in the central sketch where the shaded area represents portions of the figure that appear dark because they are swept by the moving lines. The distance (x) from the base of the dark vee to the base of the bright vee is then a measure of vibrational amplitude (δ); for example, if $\tan \theta = 0.1$ (i.e., $\theta = 5°43'$), then $\delta = x/10$ to a satisfactory approximation. A figure can be constructed, as shown in the right-hand sketch in Fig. 8.36, so that the position of the base of the bright vee can be read to give the vibrational amplitude directly.

The vibrations from these machines may be transmitted to adjoining fittings and apparatus, and they may be particularly troublesome if transmitted to microscopes. The vibrations of a standard machine weighing 40 to 50 kg and operating at 50 to 60 Hz can be adequately damped by mounting the machine on a lead slab 30 × 30 × 3 cm and weighing approximately 40 kg. Sponge-rubber pads 12× 1.5 cm should be inserted between the feet of the machine and the lead slab, and a double layer of

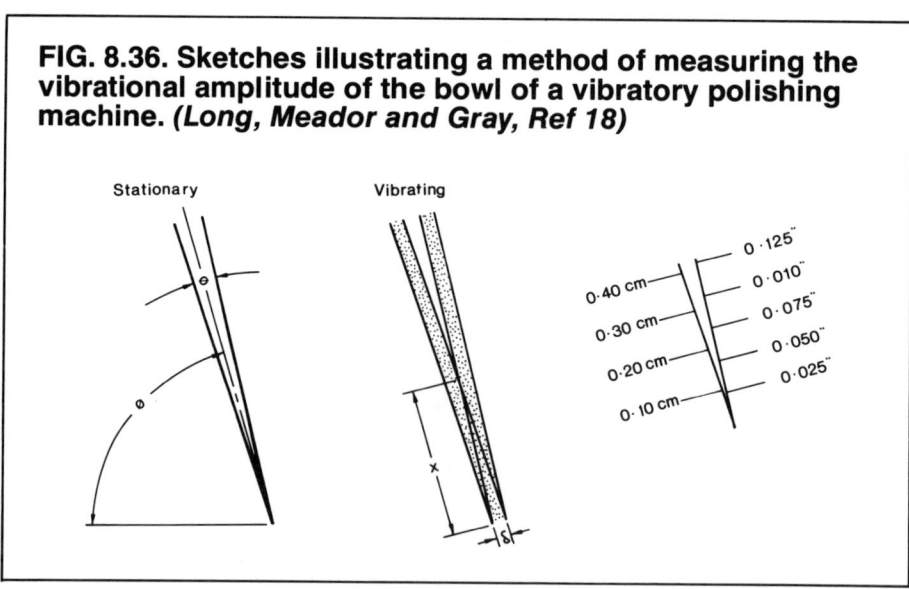

FIG. 8.36. Sketches illustrating a method of measuring the vibrational amplitude of the bowl of a vibratory polishing machine. *(Long, Meador and Gray, Ref 18)*

similar pads between the four corners of the slab and the supporting bench or stand.

Most standard polishing cloths can be used in a vibratory polisher, and a specimen carrier weighing 350 g commonly is recommended, although it is desirable to have a range of weights available. One of the practical difficulties in operating these machines is in ensuring that the specimens continue to track around the bowl when a satisfactory vibration amplitude has been achieved. This is important, because it must be possible to leave the machine unattended for hours. The first precaution necessary is to lay the nap of the polishing cloth around the polishing track in the intended direction of specimen movement; this can be achieved by rubbing a blank specimen around the track by hand under heavy pressure. Secondly, any attachments to the polishing bowl must be arranged so that no large mass accumulates at a local sector of the circumference; otherwise, the vibrational pattern may be disturbed locally and a dead spot developed at which the specimens stop. Thirdly, the abrasive slurry should be of as low a viscosity as possible.

REFERENCES

1. J. R. Vilella, "Metallographic Technique for Steels", American Society for Metals, Cleveland, 1938.
2. B. W. Mott, "Micro-Indentation Hardness Testing", Butterworths, London, 1956.
3. V. J. Manners, J. V. Craig and F. H. Scott, *J. Inst. Metals*, 1967, *95*, 173.
4. R. L. Bish, *Metallography*, 1978, *11*, 215; 1979, *12*, 147.
5. G. R. Wallwark, "Proceedings of the VIth International Conference on Corrosion", Australian Corrosion Assn., Sydney, 1981.
6. L. E. Samuels, *J. Inst. Metals*, 1956-57, *85*, 51.
7. E. C. Rollason, E. Sharratt and R. R. Roberts, *J. Iron Steel Inst.*, 1949, *162*, 265.
8. F. R. Charvat, P. C. Warren and E. D. Albrecht, "Metallographic Specimen Preparation", edited by J. L. McCall and W. M. Mueller, Plenum, New York, 1974, p. 95.
9. T. G. Gregory and D. R. Schuyler, *Metallography*, 1972, *5*, 195.
10. D. H. Rowland and O. E. Romig, *Trans. Amer. Soc. Metals*, 1943, *31*, 980.
11. E. D. Albrecht, E. H. Stearns and F. J. Wittmayer, "Proceedings of the Second Annual Meeting", International Metallographic Society, 1969, 11.
12. K. H. Roth and D. M. R. Taplin, "Metallographic Specimen Preparation", edited by J. L. McCall and W. M. Mueller, Plenum, New York, 1974, p. 129.
13. J. A. Nelson, *Practical Metallography*, 1974, *11*, 735.
14. T. Kjer, *Practical Metallography*, 1967, *4*, 365.
15. P. E. Evans, *J. Scientific Instruments*, 1962, *39*, 242.
16. F. M. Krill, *Metal Progress*, 1956, *70*, (1), 81.
17. E. L. Long and R. J. Gray, *Metal Progress*, 1958, *74*, (4), 145.
18. E. L. Long, J. T. Meador and R. J. Gray, "Symposium on Methods of Metallographic Specimen Preparation", ASTM Publication No. 285, 1960, p. 79.
19. R. Rothstein and F. R. Turner, "Symposium on Methods of Metallographic Specimen Preparation", ASTM Publication No. 285, 1960, p. 90.

20. O. F. Makinson and J. R. Abbott, *Metals Forum,* 1980, *3,* 249.
21. A. D. McLachlan and W. E. K. Gibbs, *Applied Optics,* 1977, *16,* 544.
22. J. F. Turner, *Metallography,* 1971, *4,* 187.
23. J. F. Turner and D. L. Rivenburgh, *Metallography,* 1979, *12,* 181.

CHAPTER 9

Advanced and Special Preparation Methods

IN THIS CHAPTER we shall consider advanced and special preparation methods. First, we shall discuss advanced methods of final polishing that are needed to produce high-quality results with more difficult specimen materials. Secondly, we shall consider modifications to standard procedures that are necessary to meet particular needs; examples are retention of sharp edges of specimens, retention of nonmetallic phases and inclusions, and preparation of materials with unusual physical properties.

ADVANCED METHODS OF FINAL POLISHING

Skidding Techniques

The fineness of the finish that can be achieved by standard polishing techniques is often limited by the fact that the polishing cloth itself, no matter how "soft", scratches the surface being polished. Special techniques have consequently been developed which aim at filling the nap of the polishing cloth with a thick paste, the polishing conditions then being adjusted so that the specimen skids on this paste without ever touching the polishing cloth. This presupposes that very light pressures are applied to the specimen, from which it follows that low polishing rates will be obtained. A further practical requirement usually is that the specimen be set in a cylindrical mount. The corners of the working surface of the mount must be beveled; otherwise the mount will easily break through the protective paste to the polishing cloth.

A method applicable to diamond abrasives[1] employs a cream based on the diamond-abrasive carrier paste described in Appendix 5-B (p. 192). A paste is prepared to the formula given in that appendix but without abrasive; it is whipped into a smooth cream after it has cooled. A suede cloth is charged with the standard abrasive-containing paste in the normal way and an appropriate polishing fluid added. A small amount of the cream is then worked into the nap of the polishing cloth on a wheel which is rotated at 200-500 rpm. The specimen is held lightly by hand against the cream while being rotated in the opposite direction to the wheel. The resulting improvement in polish can be illustrated by comparison of Fig. 9.1(a) and (b).

268 / Metallographic Polishing by Mechanical Methods

FIG. 9.1. Annealed 70:30 brass finish polished on three different abrasives by conventional and skidding techniques.

(a) Diamond abrasive (1 μm); conventional technique.
(b) Diamond abrasive (1 μm); skidding technique.
(c) Aluminum oxide abrasive (0.1 μm); conventional technique.
(d) Aluminum oxide abrasive (0.1 μm); skidding technique.
(e) Magnesium oxide abrasive; conventional technique.
(f) Magnesium oxide abrasive; skidding technique.

All specimens etched in a ferric chloride reagent. Magnification, 250×. The quality of polish should be judged on both severity of scratches and clarity of grain contrast.

In a similar technique for aluminum oxide abrasives,[2] a thick soap solution is prepared by adding about 25 wt % of water to soft coconut-oil soap, allowing the mixture to stand until it becomes homogeneous. The resulting thick liquid is spread over a napped polishing cloth on a rotating wheel and the abrasive sprinkled generously on top. The specimen pressure applied during polishing is adjusted so that the specimen is supported on the layer of bubbles that forms. A significant improvement in polish is again achieved (cf. Fig. 9.1c and d). Etchants may be added to increase the polishing rate, although strongly acidic solutions cannot be used because they are incompatible with the soap solution.

The best skidding technique so far developed is based on magnesium oxide abrasives (Fig. 9.1f).[3] A high-quality grade of magnesium oxide is mixed with water into a thick paste with water and forced through a 200-mesh sieve onto a polishing wheel covered with a long-napped cloth (e.g., Selvyt cloth). The paste should then have a very thick, creamy consistency. The paste is worked into the nap of the polishing cloth and the specimen is rotated by hand against the direction of rotation of the polishing wheel in such a way that the specimen skids over a bed of paste. The consistency of the paste is important and the correct consistency must be maintained during polishing by sparing additions of water; one breakthrough to the underlying cloth completely ruins the effect sought. Slurries made up with hydrogen peroxide (30%) instead of water are reported to have improved stability in this respect.[4] The polishing rate is always low but sometimes can be increased by controlled additions of an etchant. Because of the alkaline nature of magnesium oxide, the etchant must also be alkaline. Moreover, the magnesium oxide paste carbonates rapidly and is usable for only about a 5-min polishing period.

Considerable personal skill is required to take advantage of this technique, and the technique is slow. A complete 5-min polishing period, or even two such periods, may be necessary to remove the intermediate polishing damage even in favorable cases. In fact, the polishing rates for some metals are so low that they virtually preclude the use of this technique. The justification of the technique lies solely in the fact that it can produce damage-free surfaces on even the softest material. For example, polishing scratches may not be detectable by examination under phase-contrast illumination (Fig. 9.2) or even by electron microscopy. It is probable therefore that material is removed primarily by a chemical-mechanical mechanism (see p. 159). Indeed, the polishing rate becomes vanishingly small when the water in the paste is replaced by an inert liquid.

A further precaution must be taken if a metallic polishing wheel is used and if there is a considerable electrochemical difference between the metals of the specimen and the polishing wheel. Extensive etch pits may develop in the specimen when it is the cathodic member of the pair (Fig. 9.3a). This pitting can be prevented, however, by using a polishing fluid with a pH of exactly 7 (or, preferably, by using a buffer solution). But this is usually inconvenient and may not even be practicable. A more satisfactory cure is to use an electrochemically neutral material for the polishing wheel. This may be a metallic alloy with electrochemical charac-

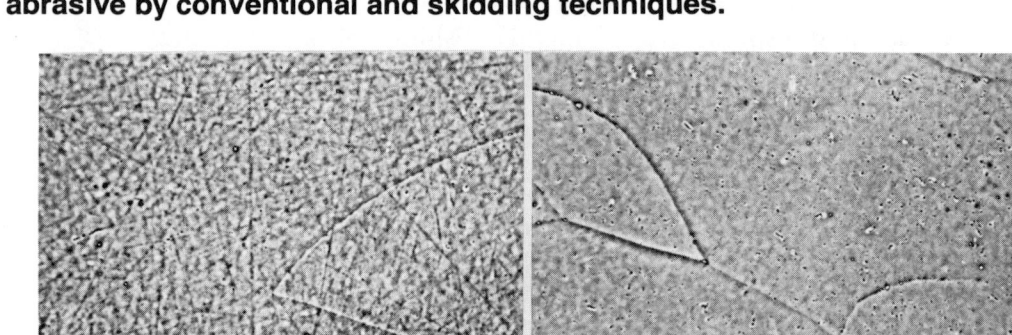

FIG. 9.2. High-purity aluminum finish polished on magnesium oxide abrasive by conventional and skidding techniques.

(a) Conventional technique. **(b)** Skidding technique. Both specimens viewed under phase-contrast illumination. Magnification, 100×.

teristics similar to those of the specimen (Fig. 9.3b). Alternatively, it may be a nonmetallic material; for example, a plastic or glass. A film of plastic may also be inserted between the polishing cloth and the wheel. Similar problems may arise with other fine polishing processes, and can be cured in the same way.

Chemical-Mechanical Techniques

We have defined chemical-mechanical mechanisms of polishing as those in which an etching reagent is deliberately incorporated in the abrasive slurry (p. 159). The purpose of the reagent is either to increase the polishing rate or to change the mechanism of polishing to one predominantly involving dissolution, or both. The technique has definite advantages in that the tendency for relief to develop between grains and constituents or for etch pits to form is considerably less than if etching were used alone. Moreover, reagents which do not ordinarily attack the specimen material may be effective because passivating surface films are removed continuously.

The increase in polishing rate is the important advantage in rough-polishing processes. Situations where this becomes particularly important are encountered, for example, with the precious metals (e.g., silver, gold and platinum) and the refractory metals of high melting point (e.g., titanium, zirconium, niobium and tungsten). These metals may be quite soft when unalloyed, and consequently the depth of the abrasion damage is of a similar order to that found in the more common metals. The polishing rates achieved by conventional rough-polishing methods, on the other hand, tend to be small (Table 5.2, p. 171). The removal of the abrasion damage consequently tends to become difficult.

In the case of final polishing, the change in polishing mechanism is an additional important advantage because it results in reduction, or even elimination, of the polishing damage. The technique then assumes considerable importance because it provides a method of obtaining completely damage-free surfaces. In a sense, the skidding techniques discussed immediately above are in this category; the difference is that a chemical component is not introduced deliberately.

If the advantages of mechanical polishing are to be retained, however, careful control of the whole process is necessary so that a proper balance is achieved between the actions of the abrasive and the etchant. High polishing rates and minimum polishing damage are achieved by using aggressive etching conditions, but aggressive etching conditions tend to result in the development of an unacceptable degree of relief between grains and constituents. They may also result in preferential attack at nonmetallic inclusions or even in the development of severe etch pitting. There is clearly an optimum concentration of etchant for any application, and this concentration depends not only on the composition of the alloy but also on the type and grade of the abrasive and the way in which it is used.

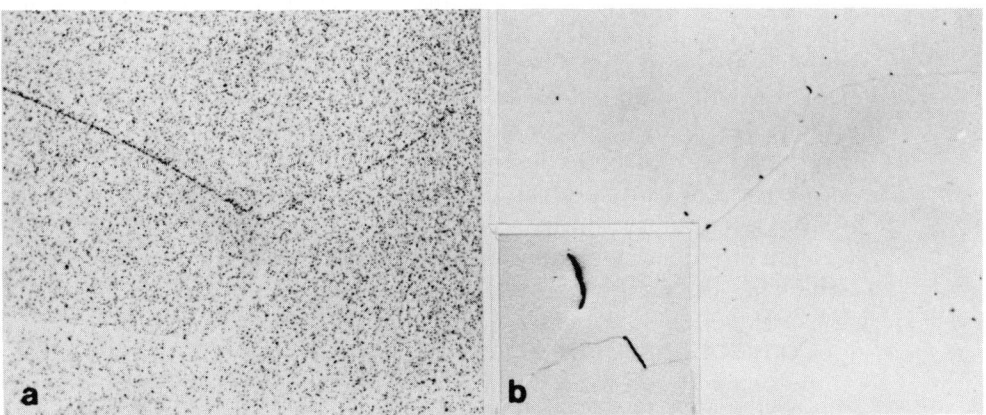

FIG. 9.3. High-purity aluminum – 0.04% iron alloy finish polished by skidding on magnesium oxide abrasive.

(a) Polished using a bronze polishing wheel. The spots are artifact etch pits developed as a result of electrochemical differences between the specimen and the polishing wheel. The pits might be mistaken for a real structure. They certainly obscure any structural detail that might be present.

(b) Polished using an aluminum alloy polishing wheel to eliminate electrochemical effects. No etch pits are present. The few spots that can now be seen are particles of an iron-rich intermetallic compound; some are shown in more detail in the inset. Magnifications: (a) and (b), 100×; inset in (b), 500×.

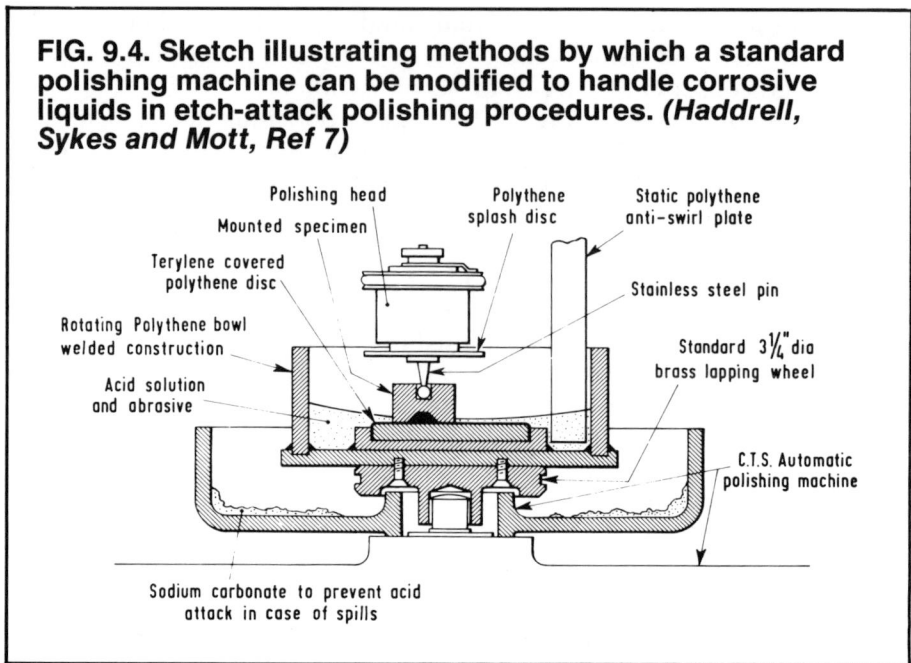

FIG. 9.4. Sketch illustrating methods by which a standard polishing machine can be modified to handle corrosive liquids in etch-attack polishing procedures. *(Haddrell, Sykes and Mott, Ref 7)*

With practice, some degree of control can be achieved by adding the etchant to the polishing pad after the abrasive slurry has been made up in the normal way. It is more satisfactory, however, to make up a solution of the etchant of controlled concentration and to use this as the polishing fluid. The volume of etching solution added to a given weight of abrasive may also be controlled with advantage. The pressure applied to the specimen is also important because it changes the etching component. The use of a controlled automatic polishing process is clearly advantageous.

As a result of the very nature of the process, however, the surfaces produced cannot be expected to be as flat as those produced by similar processes not using etch attack. Control usually means keeping the relief within acceptable limits. On the other hand, controlled etch attack in certain circumstances can improve the flatness at the edges of a specimen which has been mounted in plastic, a matter which will be discussed later (p. 295).

Conventional Etch-Attack Techniques. Details of etch-attack techniques developed for a number of metals, and their alloys, are given in Table 9.1. Note that many of the reagents used are quite corrosive, and this presents problems. The first problem encountered is rapid disintegration of the polishing cloth, but this can be overcome satisfactorily by using a cloth woven from acid-resistant synthetic fibers. The second problem is corrosion of the polishing apparatus. The precautions necessary here depend on the severity of the etchant. Vitreous enameled or stainless steel bowls and stainless steel polishing heads can withstand the milder reagents, but the more severe reagents require special apparatus constructed largely from acid-resistant plastic. The principles of the design of equipment of this nature are illustrated in Fig. 9.4. The final problem is the protection of the operator. The wearing of rubber gloves is one solution, although their continued use is unpleasant and tends to induce skin complaints. The best solution is to use a specimen-holding device.

The etching reagents listed in Table 9.1 are of two general types, namely: (a) those that form a reaction product which is soluble (e.g., the techniques listed for copper alloys), and (b) those that form a reaction product which is insoluble (e.g., hydrogen peroxide used with uranium and hydrofluoric acid used with zirconium). The second type is the more amenable to easy control because control is exercised through the abrasive component. The abrasive component has only to be made large enough to ensure that sufficient of the reaction layer is removed to permit the reaction to proceed. Control with the first type has to be exercised through the etching component, which has to be adjusted to keep the reaction within certain bounds. This is often difficult to do.

Electromechanical Polishing. A further development, known as *electromechanical* polishing, employs electrolytic instead of straight chemical methods for the etching component.[16-22] The advantages are greater flexibility, more precise control, and less corrosive reagents. Less relief is developed between grains and constituents and a more uniform polish can be achieved than by straight etch-attack techniques (see, for example, Fig. 9.5). The retention of nonmetallic constituents is also good.[19,20]

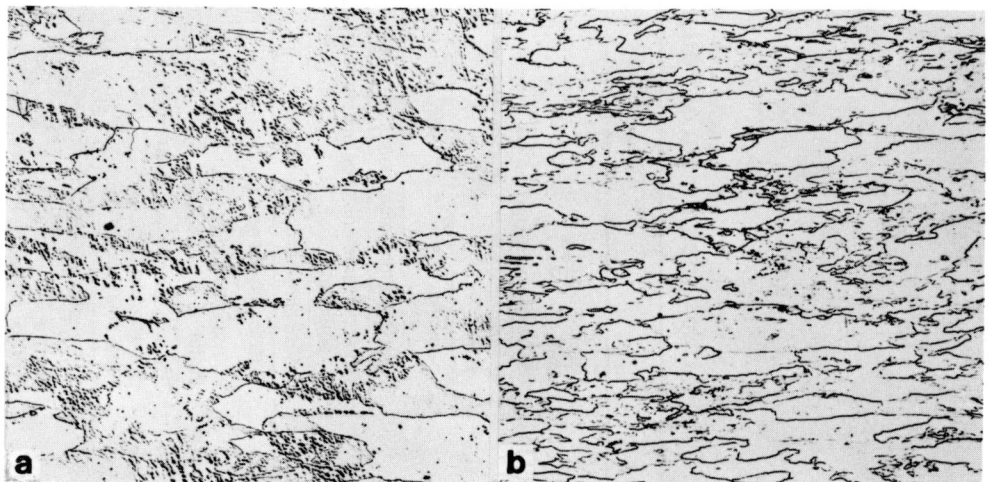

FIG. 9.5. Polycrystalline tungsten polished by a conventional etch-attack technique and by an electrochemical technique. *(Dickinson, Ref 19)*

(a) Polished by a conventional etch-attack technique. Unetched. Note the relief between the grains and the vertical system of wiping marks. (b) Polished by an electrochemical technique. Etched in 30% H_2O_2. No structure could be seen after polishing. The structure now visible was developed solely by etching. Magnification (both), 100×.

TABLE 9.1. Etch-Attack Techniques of Polishing

Metal	Reagent	Remarks	Reference
Beryllium	Oxalic acid . . . . 10 g Water . . . . 100 g Abrasive . . . . 5-8 g	Rough polishing: 0.5-μm-grade alumina on a silk cloth. Final polishing: 0.1-μm-grade alumina on a napped cloth.	5,6
Chromium	Acetic acid . . . . 15 g Water . . . . 150 ml		7
	1% oxalic acid in 1:1 ethyl alcohol-water solution.	Final polishing; Terylene cloths necessary.	8
Copper	Up to 3% ammonium persulfate solutions.	Final polishing by skidding technique.	3
	Ammonia, ammonium persulfate, cupric ammonium persulfate.	Added to polishing pad.	
	2-10% chromic acid solution.	Abrasive suspended in reagent and used on a napped cloth.	9
Gold	5-10% chromic acid solution.	Abrasive suspended in reagent and used on a napped cloth.	9
	Potassium iodide . . . . 12.5 g Water . . . . 100 ml	Drops of reagent added to normal polishing pad.	9
Magnesium	2% potassium dichromate solution . . . . 20 ml Sat. solution boric acid . . . . 150 ml Nitric acid (conc.) . . . . 15 drops	Suitable for pure magnesium, slight variations being desirable for alloys (see original reference).	10
Thorium	10% oxalic acid solution.	Rough polishing. Abrasive suspended in reagent.	9
	Nitric acid . . . . 10 ml Hydrofluoric acid . . . . 1 ml Water . . . . 98 ml	Final polishing. Reagent added to normal polishing pad.	9
Titanium	20% chromic acid solution . . . . 30 ml Water . . . . 200 ml Abrasive . . . . 15 g	Rough polishing on a high-speed wheel. Surface produced is passive but passivity can be removed by a further treatment on a normal pad.	10

Metal	Solution		Notes	#
Titanium, contd.	Hydrofluoric acid	1.5 ml	Final polishing. Add few drops to a normal pad at the end of the treatment.	9
	Nitric acid	3.5 ml		
	Water	95 ml		
	5% oxalic acid solution.		Final polishing. Suspend abrasive in reagent.	12
	Hydrofluoric acid	1.5 ml	Final polishing.	12
	Nitric acid	1.5 ml		
	Water	100 ml		
	Abrasive	10 g		
	1% hydrofluoric acid solution.		Final polishing. Suspend abrasive in reagent.	12
Tungsten	Potassium ferricyanide	3.5 g	Final polishing. Develops some grain structure.	13
	Sodium hydroxide	1 g		
	Water	100 ml		
	Abrasive	10 g		
	Copper sulfate	1 g	Final polishing. Suspend abrasive in reagent. Give alternate treatments on a normal polishing pad.	14
	Ammonium hydroxide	5-10 ml		
	Water	1000 ml		
Uranium	Chromic acid	10 g	Rough polishing. Suspend abrasive in reagent. Polythene apparatus and Terylene cloth necessary.	7
	Nitric acid	10 ml		
	Water	1000 ml		
	Hydrofluoric acid	30 ml	Final polishing. Swab surface with conc. nitric acid immediately after polishing.	5
	Nitric acid	30 ml		
	Water	60 ml		
	Alumina	5-8 g		
	5% suspension of abrasive in hydrogen peroxide (30 wt. %)		Final polishing. Polythene apparatus and Terylene cloth necessary.	15
Zirconium	Hydrofluoric acid	4-30 drops	Final polishing. Polythene apparatus and Terylene cloth necessary.	7
	Water	10 ml		
	Abrasive	5-6 g		

NOTE: Most of these solutions are intended for use with aluminum oxide abrasives. Only the alkaline solutions can be used with magnesium oxide.

FIG. 9.6. Apparatus for electrochemical polishing.

The scoop (A) continuously returns the abrasive slurry to the center of the polishing bowl (B). A polishing cloth is fixed over a stainless steel disk electrode in the base of the polishing bowl. Electrical contact is made with the disk and with the specimen; the connection to the specimen can be seen (C).

Any standard low-speed polishing machine can be modified for this purpose by providing a corrosion-resistant polishing head and an electrical contact to this head. An apparatus specially developed for the purpose is illustrated in Fig. 9.6. This equipment has a plastic bowl attached to the polishing head, so that a large volume of electrolyte can be held on the polishing wheel. A baffle scoop redistributes to the center of the wheel the electrolyte thrown to the periphery by rotation of the wheel. Electrical contact to the specimen must also be provided. The polishing cloth should be woven from acid-resistant fibers.

In its most elementary development, a low-voltage dc power supply is adequate. A polarity-reversing switch is desirable because, although the specimen is usually made anodic ("normal" polarity in Table 9.2), it is sometimes desirable to make it cathodic ("reverse" polarity in Table 9.2). In other cases, moreover, a low-voltage ac supply is desirable and, in still others, a device which reverses the polarity every 1 or 2 s ("cyclic" polarity in Table 9.2).[19]

Polishing conditions developed for a variety of metals are listed in Table 9.2. The current density listed should be used as a guide only, because the optimum value depends on several ill-defined factors. These factors include the specimen area, the pressure applied to the specimen, and the type and concentration of the abrasive. The current density does not, however, have to be confined within narrow limits, as in the case of electrolytic polishing. As a general rule, the current density is considered to be too low when the edges of the specimen are properly polished but the center is etched; and it is considered to be too high when the reverse occurs. Excessively high current densities also cause filming or uneven and excessive etching of the surface. Within these limits, the current density should be adjusted to be as high as possible. The polishing methods listed in Table 9.2 are intended for use in conjunction with fine grades

TABLE 9.2. Electromechanical Techniques of Polishing

Material	Electrolyte	Polarity	Current density, A/cm^2	Abrasive	Reference
Beryllium	1% hydrochloric acid and 2% nitric acid in ethylene glycol	Normal	1.0-1.5	None	21
Chromium	12% sodium thiosulfate 19% potassium thiocyanate solution in water	Normal	0.2-0.6	None	22
Gold	12% sodium thiosulfate 19% potassium thiocyanate solution in water	Normal	0.07	None	22
Molybdenum	3% potassium ferricyanide solution in water	Reverse	0.01	Aluminum oxide	19
Niobium	3% hydrogen peroxide	Cyclic	0.01	Aluminum oxide	19
Platinum	3% potassium cyanide solution in water	Cyclic	2.5	Aluminum oxide	19
Rhenium	Saturated sodium chloride solution in water	Cyclic	0.1	Aluminum oxide	19
Ruthenium	1% sodium hydroxide solution in water	ac	0.04	Aluminum oxide	24
Silver	12% sodium thiosulfate 19% potassium thiocyanate solution in water	Normal	0.07	None	22
Tungsten	Saturated sodium chloride solution in water	Cyclic	0.04	Aluminum oxide	19
	Saturated potassium ferricyanide solution in water	Normal	0.04	Aluminum oxide	19
Vanadium	12% sodium thiosulfate 19% potassium thiocyanate solution in water	Normal	0.5-0.6	None	22

FIG. 9.7. Annealed 70:30 brass rough polished on 1-μm diamond and finished by vibratory polishing.

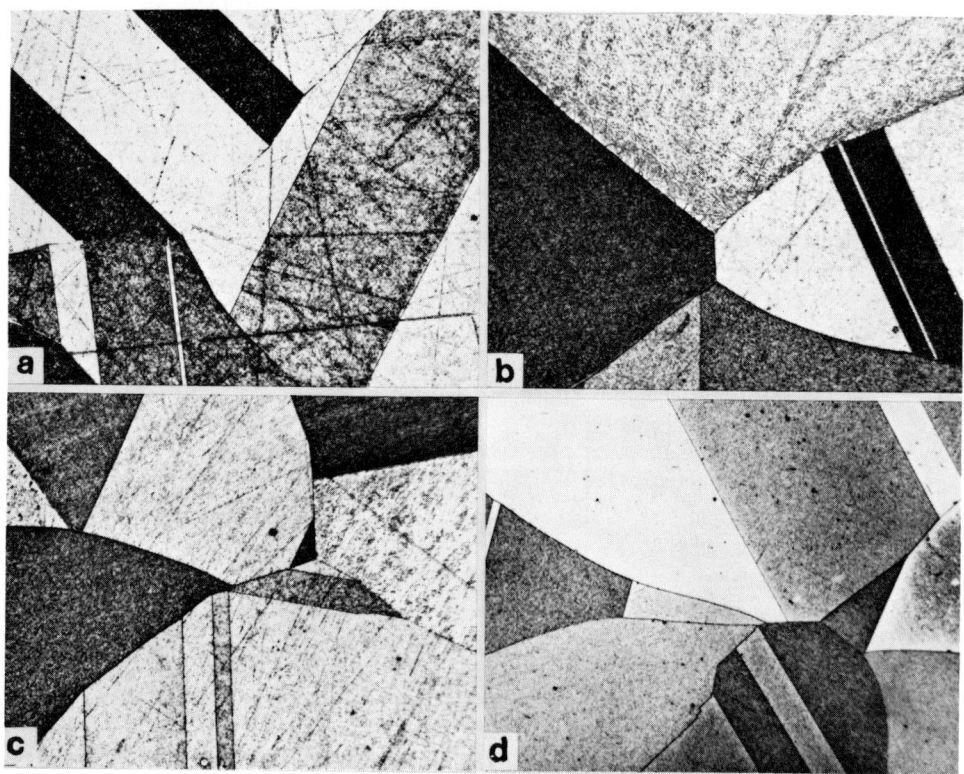

Conditions for vibratory polishing were as follows:

Micrograph	Abrasive	Specimen load	Etchant added to abrasive slurry	Polishing time*, h
a	Aluminum oxide (0.1-μm)	350 g	...	1
b	Aluminum oxide (0.1-μm)	0	...	8
c	Magnesium oxide	350 g	...	1
d	Magnesium oxide	0	Ammonium persulfate†	2

*Time required for removal of rough-polishing scratches.
†Abrasive slurry made up with a 0.5% solution.

All specimens were etched in a ferric chloride reagent. Magnification, 250×.

of aluminum oxide abrasive. However, some do not require the use of an abrasive at all. It is then a matter of arbitrary definition whether the process should be described as mechanical or chemical or electrolytic polishing.

Advanced and Special Preparation Methods / 279

Vibratory Polishing. We have discussed earlier (p. 250) the use of a vibratory polishing machine as a conventional method of final polishing. A final polish comparable to a hand polish employing the same abrasive and polishing cloth is obtained if the specimen is loaded with a comparatively heavy weight (cf. Fig. 9.7a with 9.1c, and Fig. 9.7c with 9.1e). Rather better results than those for hand polishing will be obtained, however, if only a light load is applied (cf. Fig. 9.7b and 9.1c). The latter results often are quite acceptable in many applications (e.g., harder specimens). In this respect, however, vibratory polishing is simply an improvement on standard techniques. We shall now discuss its use in special ways for producing scratch-free and damage-free surfaces.

First, note in Fig. 9.7 that a much improved result was obtained when the specimen was polished without an additional load and when an etching reagent was added to the abrasive slurry (Fig. 9.7d). The result was in fact comparable to that obtained with the most advanced skidding technique (cf. Fig. 9.1f) in the sense that it was nearly scratch-free and damage-free. Much less skill is, of course, required to produce this result by vibratory polishing. When vibratory polishing techniques that produce results of this high standard are considered in detail, it becomes apparent that a chemical-mechanical mechanism of polishing must be dominant. The nature of the liquid in which the abrasive is suspended, the weight applied to the specimen, and the polishing time are all important. The ratio of abrasive to liquid is also important on occasions.

TABLE 9.3. Vibratory Polishing Techniques

Material	Polishing fluid	Approximate optimum load(a)	Abrasive concentration(b)	Polishing time required
Aluminum alloys	Na hydrogen phosphate: 59 g Citric acid : 3.39 g Water(c) : 1000 ml (See footnote d)	10-15 g	100-200 g/l	6 h
Copper alloys	Ammonium persulfate : 5 g Water(c) : 1000 ml	10-20 g	100-200 g/l	2 h
Magnesium alloys	Propylene glycol : 3 parts Water(c) : 1 part	250-300 g	50 g/l	10-15 min(e)
Steels, low-alloy	Propylene glycol : 3 parts Water(c) : 1 part	30-50 g	100-200 g/l	3-6 h
Steels, corrosion-resistant	Water(c)	30-60 g(f)	100-200 g/l	3-6 h

(a) Applied to a 1-in.-diam mount, and includes the weight of the specimen and mount. (b) Expressed as grams of 0.1-μm gamma alumina abrasive per liter of suspending liquid. (c) In all cases, "water" refers to distilled water saturated with carbon dioxide. The variability of even good-quality mains water may have drastic effects on the results. (d) This is a pH 7 buffer solution, and standard solutions and tablets supplied commercially for making up buffer solutions may be used. The pH should, if anything, be on the acid side of pH 7; otherwise etch pitting may develop. The solution should be reasonably fresh; it can be stored for a week at most.
(e) Some alloys must be washed immediately in alcohol on removal from the polishing bowl to avoid development of staining. (f) Austenitic steels require loads and polishing times towards the high end of the range; loads and times towards the lower end of the range are satisfactory for martensitic steels.

FIG. 9.8. Specimen of an annealed low-carbon steel rough polished on 1-μm diamond and successively finished by vibratory polishing in three different slurries.

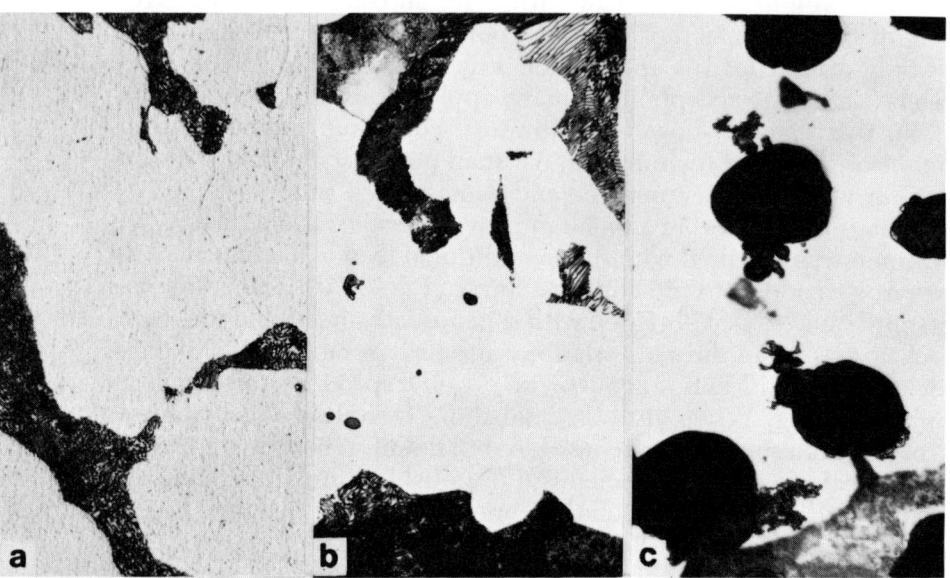

Finished by vibratory polishing for 4 h on 0.1-μm gamma aluminum oxide made into slurries with: **(a)** propylene glycol; **(b)** propylene glycol (2 parts) and water (1 part); and **(c)** water. The total load carried by the 1-in.-diam. specimen mount was 50 g. The polishing rate in **(a)** was too low to remove the rough-polishing damage. Severe etch pitting has occurred in **(c)**. The optimum result has been developed in **(b)**. All specimens were etched in nital. Magnification, 500×.

These factors have been investigated systematically for only a few alloys, the results being summarized in Table 9.3. This table is based on the use of fine aluminum oxide abrasive and on the concept that it is desirable to keep the time necessary to produce a good polish within reasonable limits. A good polish here is defined as one for which (a) scratches cannot be detected by optical microscopy either before or after etching and (b) relief between constituents is acceptably low, and excessive etching has not occurred.

A characteristic of many successful polishing fluids is that they contain agents which complex with the fresh metal surface. Propylene glycol is the complexing agent in two of the solutions listed in Table 9.3, but it has one adverse characteristic: it tends to make the suspension more viscous than is desirable if present in high concentrations. In addition, some mixture of the complexing agent and a reactive material (even water) is usually required. An example is given in Fig. 9.8. Use of propylene glycol alone resulted in such a low polishing rate that it would have taken an unacceptably long time to remove the rough-polishing damage from this specimen (Fig. 9.8a); use of water alone resulted in excessive corrosive attack (Fig. 9.8c); but use of an appropriate mixture of the two resulted in an adequate polishing rate and a good finish (Fig. 9.8b).* The complexing agents listed in Table 9.3 for aluminum are sodium hydrogen phosphate and citric acid. The solution listed happens to be a standard buffer solution with a pH of about 7, but it is not the control of pH that is significant in this application: it is the combination of complexing agents, which may have to be tried in different dilutions for some alloys (Fig. 9.9). Sometimes, addition of inhibitors is also useful; for example, addition of sodium nitrate (1 g per 300 ml) to a 1:1 alcohol – ethylene glycol solution minimizes pitting at inclusions in beryllium. Inhibitors should be added with caution, however, because they may passivate the surface to such an extent that the subsequent etching characteristics of the surface are drastically affected. On the other hand, some specimen materials may be so reactive to any water-base solution that a more inert polishing fluid has to be used. An example is uranium carbide, for which oils must be used; a silicone oil has been found to be effective.[25]

Another important factor is the load applied to the specimen. The polishing rate may be unacceptably low with small applied loads, but excessive scratching may be produced at high loads. The applied load also

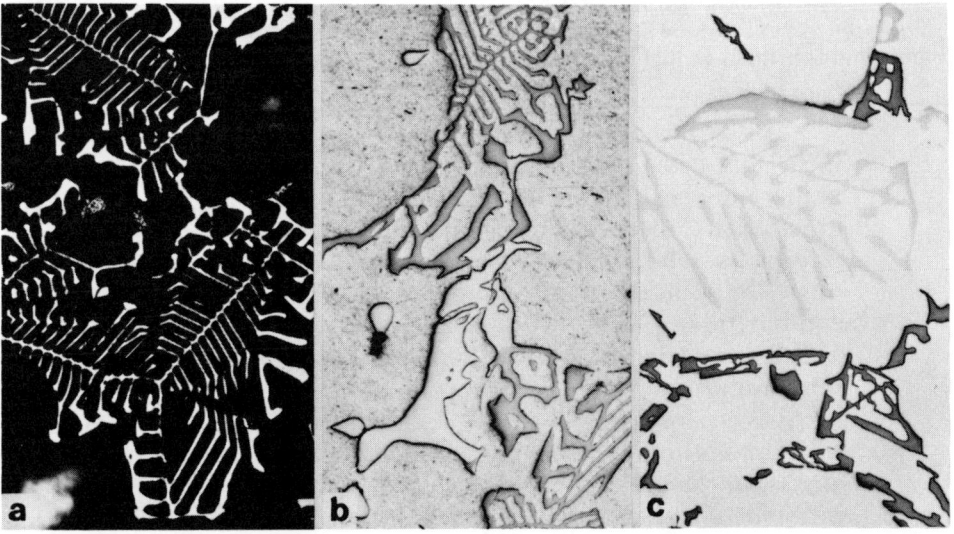

FIG. 9.9. A cast aluminum – 7.5% copper – 2.5% silicon alloy finished by vibratory polishing. Not etched after polishing.

Finished by vibratory polishing on magnesium oxide abrasive made into slurries with three different concentrations of the phosphate – citric acid polishing fluid in Table 9.3: **(a)** undiluted solution; **(b)** solution diluted 2:1 with water; and **(c)** solution diluted 1:1 with water. Both **(a)** and **(b)** have etched during polishing, **(a)** excessively so; **(c)** has etched slightly but the result is acceptable. Specimens were not further etched after polishing. Magnification, 250×.

---

*A method of making propylene glycol – water mixtures which can be stored for long periods is given in Appendix 9-A.

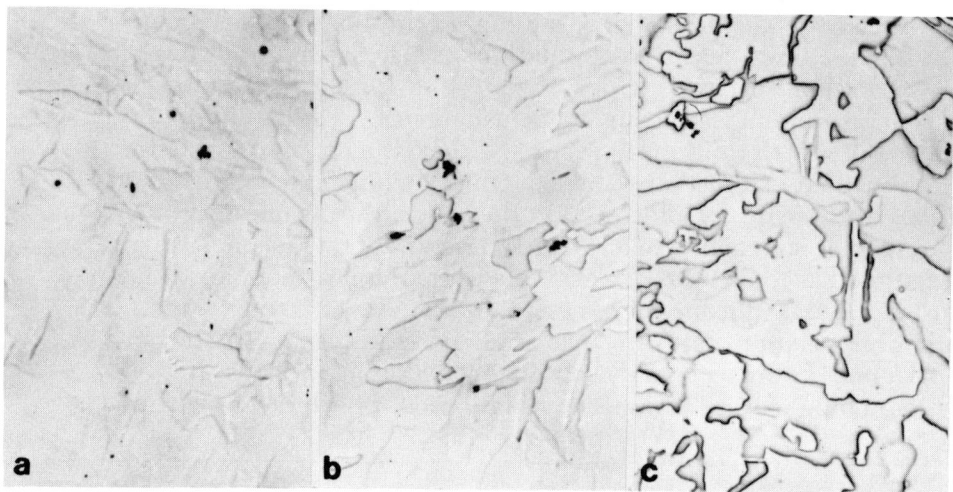

FIG. 9.10. Specimen of an annealed low-carbon steel successively finished by vibratory polishing using three progressively increased loads.

Finished by vibratory polishing on 0.1-μm gamma aluminum oxide made into a slurry with propylene glycol (2 parts) and water (1 part). Total load carried by 1-in.-diam specimen mount: **(a)** 40 g; **(b)** 70 g; **(c)** 380 g. Increases in applied load have resulted in increases in the relief between the ferritic and pearlitic areas of the structure. This is the same specimen as that shown in Fig. 9.8. Unetched. Magnification, 500×.

invariably has an influence on the degree of relief developed between grains and constituents. An example is given in Fig. 9.10, where the degree of relief increased markedly and progressively with increases in applied load. The choice of load clearly involves something of a compromise, there being an optimum load range for each metal/liquid/abrasive combination.

Polishing time is less critical but still has to be controlled. A certain minimum time is necessary to remove the rough-polishing damage, and excessively long times usually result in the production of unacceptable relief and other associated etching effects. The liquid:abrasive ratio usually is not particularly critical either, although excessively viscous suspensions are undesirable in that they make it more difficult to ensure continuous movement of the specimen in the bowl of the polishing machine.

Abrasives other than aluminum oxide can be used for vibratory polishing, both chromic oxide[23] and magnesium oxide having been applied with success. The optimum abrasive varies with the specimen material. For example, aluminum oxide produces an acceptable result with ferrous alloys (Fig. 9.8), but magnesium oxide is more satisfactory for most aluminum alloys (Fig. 9.9 and 9.11) and most copper alloys (Fig. 9.12). Magnesium oxide is satisfactory only if used in solutions which inhibit the formation of carbonates.

FIG. 9.11. A solution-treated cast aluminum – 4% copper alloy finished by vibratory polishing on two different abrasives.

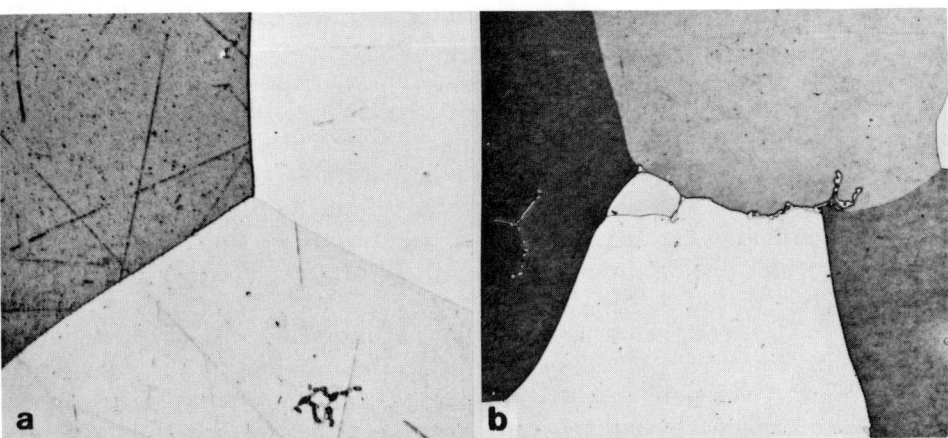

(a) Polished on 0.1-μm gamma aluminum oxide. (b) Polished on magnesium oxide. The polishing fluid used for both specimens was the phosphate – citric acid solution in Table 9.3, diluted in the proportion of 1 part solution to 2 parts water. Etched in a mixed acid reagent (0.5% HF, 1.5% HCl and 2.5% HNO$_3$ in water). Magnification, 250×.

FIG. 9.12. A cast alpha-beta brass rough polished on 1-μm diamond and finished by vibratory polishing.

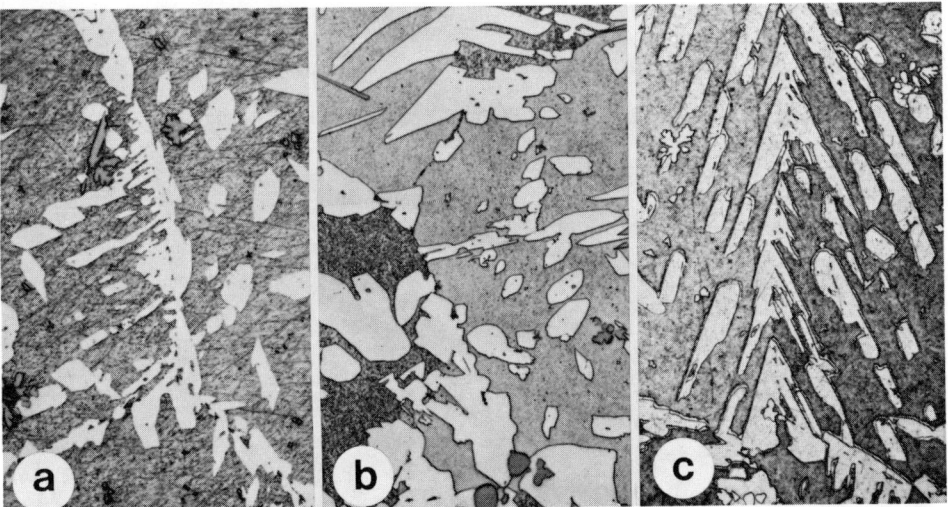

Finished by vibratory polishing on magnesium oxide made into a slurry with propylene glycol (3 parts) and water (1 part). An etchant composed of ammonium hydroxide, ammonium persulfate and water was added to the polishing fluid in increasing amounts. **(a)** Insufficient amount of etchant added. Excessive scratching has occurred. **(b)** Optimum amount of etchant added. An acceptable result, even though etched. **(c)** Excessive amount of etchant added. Excessive relief has developed between the alpha and beta phases. The double lines outlining the alpha phase are a result of this. All unetched. Magnification, 500×.

With some alloys, such as copper alloys, it is necessary to supplement the chemical attack deliberately by adding an etching reagent to the polishing fluid. (Copper alloys are good examples because of the problems that arise owing to the sensitivity with which etching of these alloys reveals polishing damage.) Table 9.1 can be used as a guide to selection of suitable reagents. Once again, however, optimum results will be obtained only if the etching conditions are carefully controlled and balanced with the other polishing parameters that we have just discussed. An insufficient addition of etchant will result in an excessive mechanical component in the polishing mechanism and hence in the development of scratches (Fig. 9.12a). Excessive additions, on the other hand, will result in the development of unacceptable relief between grains and phases (Fig. 9.12c; cf. Fig. 9.12b).

Thus a very complex interplay of factors determines the result obtained by vibratory polishing. An optimum combination for a particular alloy must be established by intelligent experimentation, using the parameters that we have discussed as a guide. Nevertheless, the process is highly reproducible once the optimum parameters have been established. Moreover, a quality of finish can be obtained that would be extremely difficult to achieve with many alloys by other methods. Examples are given in Fig. 9.7(d), 9.9(c), 9.11(b) and 9.12(b).

Vibratory polishing does, however, have several fundamental limitations. The most important limitation is that the technique is not applicable at all to some very soft materials because it causes extensive embedding of abrasive in the surfaces of these materials (Fig. 9.13). This presumably occurs during the drop portion of the vibratory cycle. The

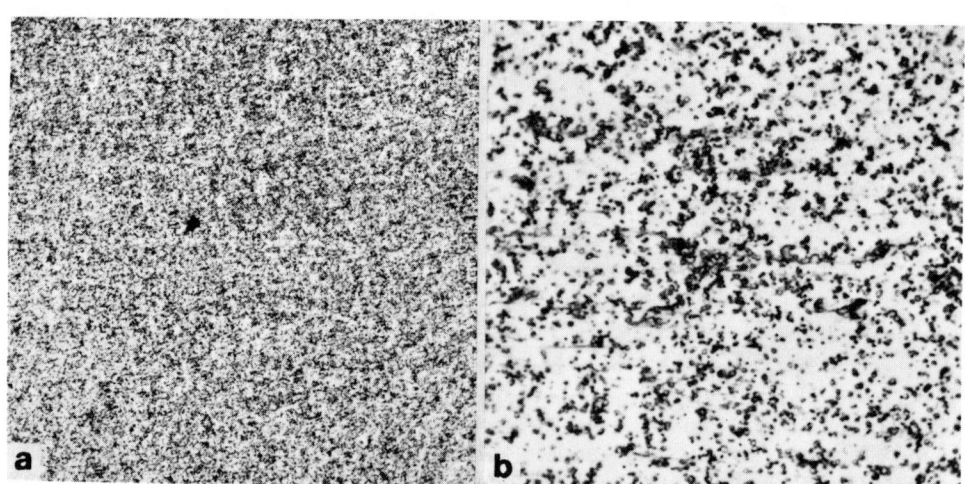

FIG. 9.13. High-purity aluminum (20 HV) finished by vibratory polishing.

Polished on 0.1-μm gamma aluminum oxide. Unetched. Magnifications: **(a)** 100 ×; **(b)** 500×. In both, all spots are particles of abrasive embedded during vibratory polishing; they can be identified as such by electron probe microanalysis.

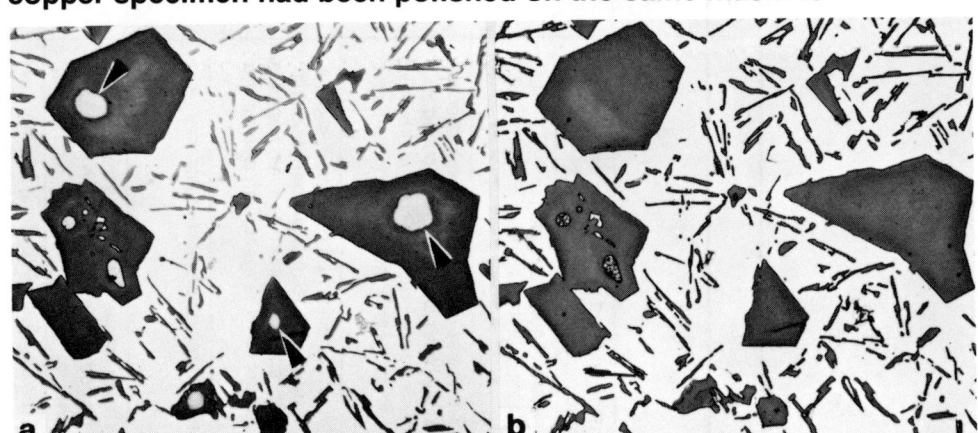

FIG. 9.14. An aluminum-silicon alloy vibratory polished immediately after a copper specimen had been polished on the same machine.
Polished on magnesium oxide abrasive made into a slurry with ethylene glycol. **(a)** As polished. **(b)** Subsequently treated in concentrated nitric acid. The features indicated by arrows in **(a)** are films of copper deposited on the silicon particles. Magnification (both), 250×.

embedded particles just might be mistaken for etch pits or structural features, but this is not likely because they are so numerous. They certainly destroy the effectiveness of the polishing treatment.

Another limitation of vibratory polishing is a consequence of the chemical activity. A film of a more noble metal may be deposited on the surface of a less noble metal or phase if the more noble metal was polished previously using the same abrasive slurry (Fig. 9.14). A third limitation is that the development of some undulations and some relief between phases is inevitable. These polishing defects may tend to be more severe than those produced by the polishing methods discussed earlier. The aim is to optimize the process so that they are kept within acceptable limits.

Alternate Etch-and-Polish Technique

We discussed earlier the application of alternating an etching treatment with conventional polishing (see p. 217) and saw that it is a technique not to be recommended for elimination of abrasion artifacts. A different, but sounder, use for the technique is to apply it to specimens with areas so different in polishing characteristics that they tend to polish to very different levels. This tendency can be reversed by interspersing etching treatments which preferentially attack the area that is polished to the higher level. Examples are coarsely graphitic cast irons[26] (for which nital and picral are suitable etching reagents) and copper-lead alloys containing extensive areas of free lead (for which swabbing with a chromic-hydrochloric-sulfuric acid mixture* is required[27,28]). The successful ap-

---

*Chromic acid (AR), 25 g; hydrochloric acid (conc.), 2 ml; sulfuric acid (conc.), 2 ml; water, 500 ml. Apply by swabbing hard with cotton wool soaked in the reagent.

TABLE 9.4. Chemical Polishing Techniques

Base metal	Polishing solution(a)		Remarks	Reference
Aluminum	Sulfuric acid Orthophosphoric acid Nitric acid	25 ml 70 ml 5 ml	Use solution at 85°C.	31,32
Cadmium	Nitric acid (fuming) Water	75 ml 25 ml	Cycle dipping for a few seconds followed immediately by washing in a stream of water. Can also be used for zinc.	33
Copper	Nitric acid Hydrochloric acid Orthophosphoric acid	30 ml 10 ml 10 ml	Use solution at 70-80°C. Specimen should be agitated.	34
Gadolinium	Lactic acid Phosphoric acid Acetic acid Nitric acid Sulfuric acid	5 g 5 ml 10 ml 15 ml 1 ml	Important that no water enters the solution. Gently swab specimen with solution for 10-15 s. Swabbing time must be reduced to 2-3 s if many nonmetallic inclusions are present. Rinse in ethyl alcohol. Also suitable for other rare-earth metals in the lanthanide series.	43
Germanium	Nitric acid Acetic acid Hydrofluoric acid Bromine	50 ml 30 ml 30 ml 0.6 ml	Make up 200 ml solution without bromine. Add bromine in slight excess, stir gently, and let stand for 10-15 min. Decant 50 ml into a container, use for polishing, and then return to main container. Solution lasts 5-6 h. Known as CP-4 reagent. Can also be used for silicon.	
Indium	Hydrochloric acid Boric acid Ethyl alcohol	200 ml 4 g 400 ml	An increase in the alcohol and hydrochloric acid contents may be desirable with some alloys.	35
Iron	Hydrogen peroxide (30%) Hydrofluoric acid Water	100 ml 14 ml 10 ml	Maintain solution at or below room temperature. Use a fresh solution. Wash promptly. Rinse in hydrogen peroxide (30%). A good standard of pre-polishing is desirable.	36

Lead	A Hydrogen peroxide (30%)
	   Acetic acid
	B Molybdic acid
	   Ammonium hydroxide
	   Water
	   Nitric acid (add before use)
		20 ml
		80 ml
		10 g
		140 ml
		240 ml
		60 ml
			Establish optimum concentration of solution A by trial; add more peroxide if finish is dull with microscopic facets; add more acetic acid if surface is bright but pitted. Condition solution first with a dummy specimen. Immersion time must not exceed 10 s, to avoid specimen heating. Alternate immersion in solutions A and B is desirable.(b)	37,33

Lead, contd.	Lactic acid Nitric acid Hydrogen peroxide (30%)	50 ml 30 ml 7 ml	For bismuth-containing alloys. Apply by light swabbing for 30-60 s; rinse in running water. Repeated treatments may be required, but take care not to overetch duplex alloys.	38
Nickel	Nitric acid Sulfuric acid Orthophosphoric acid Acetic acid	30 ml 10 ml 10 ml 50 ml	Use at 85-95°C for 30-60 s.	34
Silver	Chromic acid, sat. soln. Hydrochloric acid, 5% soln.	180 ml 5 ml	Apply by swabbing with ball of cotton wool, rinsing frequently with water. A film giving interference colors will form if hydrochloric acid content is incorrect. This film can be removed by swabbing with orthophosphoric acid or by wiping the surface with a soft tissue under running water. Removes material at ~ 4 μm/min.	39
	Chromic acid, sat. soln. Hydrochloric acid, 10% soln. Water	100 ml 45 ml 800 ml	Gives better results than above, particularly in material with a low density of dislocations, but must be used in special apparatus (see reference). Surface must finally be swabbed lightly with the polishing solution for 30 s and then rinsed in water while swabbing. Removes material at approximately 25 μm/min.(c)	11
Zinc	Chromic acid Sodium sulfate Nitric acid Water	200 g 15 g 50 ml 950 ml	Dense layer formed on the surface is soluble in water. Removes material at approximately 7 μm/min.	40
Zirconium	A Acetic acid Hydrofluoric acid	45 ml 8-10 ml	Alternative solutions. Apply by swabbing or immersion.	41

	B Hydrogen peroxide (30%)
	  Nitric acid
	  Hydrofluoric acid
		45 ml
		45 ml
		8-10 ml		

	C Nitric acid
	  Hydrofluoric acid
	  Water
		45 ml
		8-10 ml
		45 ml		

(a) All acids are concentrated unless otherwise stated. Polish by immersion at room temperature unless otherwise stated. (b) Alternatively, the solution may be poured rapidly over the surface, so that it runs over the surface randomly. Control of the process is then less difficult.[42] (c) See also p. 365.

plication of this technique requires considerable experience and skill. In the case of cast irons at least, it has been replaced by simpler and more effective techniques based on diamond abrasives,[29,30] which we shall discuss later in this chapter (p. 307).

Chemical Polishing

There are cases in which chemical polishing can be considered to give a final touch to specimens that can readily be rough polished by mechanical methods but which are difficult to finish polish by mechanical methods. It is possible to remove a small thickness of material by chemical polishing — just enough to remove the rough-polishing damage. If the polishing time is kept short, the advantages of the preceding mechanical stages may not be destroyed. Electrolytic polishing, on the other hand, inevitably removes a considerable thickness before polishing conditions are established. There is no particular point under these conditions in using any preliminary mechanical polishing stage.

Some suitable methods of chemical polishing are presented in Table 9.4.

RETENTION OF EDGES

A major advantage of mechanical over other available methods of specimen preparation is that they allow the edges of a specimen to be retained sufficiently well to permit critical examination of the surface regions. The structure of the surface layers of a specimen is frequently a matter of considerable interest, and may even be the whole point of the metallographic examination. It is important, therefore, to establish the factors that will enable the best possible standards of edge retention to be achieved with the greatest ease.

Many articles have been published which offer solutions to this problem. No doubt they do offer satisfactory solutions in the specific cases discussed, but we shall see that the general problem is a complex one

TABLE 9.5. Depths of Field of Typical Microscope Objectives

Final magnification, diameters	Objective		Area of field(a), μm	Depth of field, μm
	Magnification, diameters	Numerical aperture		
100	5.6	0.20(b)	1000	20
250	8.0	0.40(b)	400	3
500	21.0	0.65(b)	200	1
750	41.0	0.85(b)	135	0.4
1000	58.0	0.95(b)	100	0.1
1000	50.0	1.0(c)	100	0.6
1500	75.0	1.4(c)	65	0.2

(a) For a final projected image 10 cm in diameter. (b) Dry objective. (c) Oil-immersion objective.

and that there is no simple panacea. We shall discuss the problem of retaining an edge corresponding to the outer surface of the specimen. The retention of the junction region of a composite specimen is a special case of the same general problem.

The basic requirement is that the maximum variation in level in the specimen over a useful area of the field of view should not exceed the depth of field of the objective being used in the microscopical examination. The depth of field of an objective is not a precise quantity because the figure ascribed depends on the sharpness of focus that is acceptable. Nevertheless, approximate comparative values can be ascribed to microscope objectives; values for typical objectives used in metallographic microscopes are listed in Table 9.5. Although they must be interpreted with some caution, these figures do illustrate the point that what might be considered to be acceptable edge retention under some conditions of examination may not be so regarded under others. Note also that the problems of edge retention can sometimes be eased by an appropriate choice of microscope objective.

We shall concentrate, therefore, on elucidating those factors which tend to improve edge retention rather than making absolute recommendations. We shall base these considerations on quantitative comparisons of the contours of the surface in the edge region as determined by the "Talysurf" instrument. Micrographs made at a magnification of 500×, which is a moderately severe case (Table 9.5), are used for more graphic but qualitative comparisons. We shall consider first what adjustments can be made to the standard polishing techniques discussed in Chapter 8, assuming that the specimen is mounted in plastic. We shall then discuss special precautions that may be adopted to improve edge retention.

Effects of the Abrasion Stages

The abrasion rates obtained with the mounting plastics used in metallography are greater than those of most metals. Plastics which do not contain a filler, or for which the filler is not abrasion resistant, behave during abrasion in a manner similar to that of group 1 metals (see p. 56) — that is, they cause no noticeable deterioration of the abrasive paper. Consequently, they can be characterized by a simple abrasion rate. The values of abrasion rate indicated by asterisks in Table 9.6 are for plastics of this type; these rates can be compared directly with those for metals listed in Table 3.3 (p. 57).

It follows that the surface of such a plastic will tend to be reduced to a level below that of the metal specimen during abrasion. The surface contours of the metal and the plastic will then blend into one another at the interface between the two and the contour of the metal will become rounded in the edge region. Herein lies one of the basic problems in obtaining sharp edge retention in the finish-polished surface. The factors that first have to be understood when considering techniques of edge retention are therefore those that determine the difference in level that develops between the metal and plastic during abrasion.

The first factor to be considered is the nature of the abrasion process. Finer abrasion processes produce smaller differences in level and result

TABLE 9.6. Abrasion and Polishing Rates of Specimen-Mounting Plastics

Mounting plastic		Abrasion		Maximum polishing rate(b), μm/100 m
Type	Filler	Rate(a), μm/m	Maximum thickness removable, μm	
Phenolic	Wood	* 11.5		18
	Mineral(c) (7 wt%)	* 11		15
Acrylic	Nil	* 10		22
Epoxy, casting	Nil	* 20		14
	Aluminum oxide(d) (20 wt%)	* 1.3	2850	1.3
	Aluminum oxide(e)	0.1	550	0.5
Epoxy, molding	Mineral(f) (72 wt%)	0.1	450	1.0
Allyl	Mineral(g) (55 wt%)	3.0	2500	3.5
Formvar	Nil	* 5.0		2.6
Polyvinyl chloride	Nil	* 0.9		6

(a) Determined by the same methods and under the same conditions as for Table 3.3; under the particular conditions, μm/100 m ≡ μm/min. The figures listed are for P240 grade silicon carbide paper; those for P240 grade aluminum oxide paper are not greatly different. The asterisk indicates that the plastic has Type 1 abrasion characteristics; the remainder have Type 3 abrasion characteristics. (b) Determined by the same methods and under the same conditions as for Table 5.2; under the particular conditions, μm/100 m ≡ μm/min. 6-μm polycrystalline diamond used on a synthetic suede cloth. (c) Probably mica and asbestos. (d) Added as 600-mesh aluminum oxide abrasive. (e) Added as sintered pellets. (f) Probably silica. (g) Probably a mixture of mica and glass.

in less edge rounding (Fig. 9.15a). Fine abrasive laps are a considerable improvement over abrasive papers (Fig. 9.15b). The best edge retention is obtained on very hard laps, such as a diamond-plastic lap (Fig. 9.15b), but such laps produce a comparatively coarse finish. The aluminum oxide – wax lap described in Appendix 3-D (p. 83) produces both a reasonably sharp edge and a reasonably good finish (Fig. 9.15b). Its use contributes greatly to improved edge retention whatever other precautions are taken in addition.

The second factor of importance is the nature of the plastic used for mounting. The abrasion rates of plastics vary over a considerable range (Table 9.6). As is then to be expected, plastics with the lowest abrasion rates produce the smallest degrees of edge rounding (Fig. 9.15c). Three groups of plastics may be distinguished on these grounds: (a) formvar and polyvinyl chloride plastics, which are the most satisfactory (formvar is slightly superior to polyvinyl chloride); (b) allyl plastics, which are slightly less satisfactory than formvar and polyvinyl chloride plastics; and

(c) epoxy and phenolic plastics, which are much inferior to those in (a) and (b). One plastic from each group is represented in Fig. 9.16. All of the specimens in Fig. 9.16 were prepared in a similar manner by the simple grade III procedures listed in Table 8.3 (p. 242), using an abrasive-wax lap for stage 3 and magnesium oxide for stage 6. The edge retention of this steel specimen improves noticeably from phenolic to allyl to formvar plastic, but is by no means perfect even with formvar plastic.

The third factor to be considered is the nature of the specimen material. The tendency toward edge rounding, even assuming optimum mounting and abrasion conditions, will be more severe the lower the abrasion rate of the specimen material (Fig. 9.15d and 9.17; see also Table 3.3, p. 57). It will also tend to be more severe with specimen materials which cause rapid deterioration of the abrasive device; in these cases, abrasion of the metal will eventually stop but that of the plastic will continue. The most important group of materials in the latter category is the ferritic steels. They have been used in this discussion as examples of materials that are highly susceptible to edge rounding.

The fourth factor is the method by which the abrasion is carried out. Considering conventional techniques first, the most satisfactory results by a small margin are obtained in fully automatic machines of the types illustrated in Fig. 8.26(b) and 8.27. These machines produce randomly

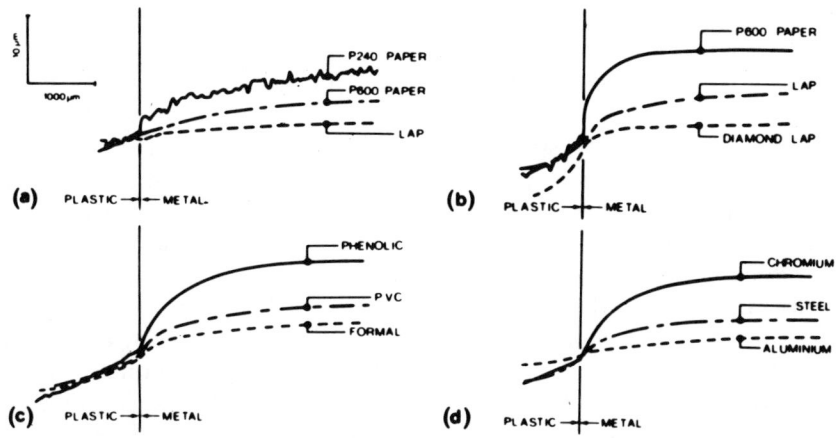

FIG. 9.15. Contours at the plastic/metal interface following abrasion.

(a) Annealed low-carbon steel (130 HV) mounted in formvar plastic; abraded as indicated. (b) Annealed low-carbon steel (130 HV) mounted in phenolic plastic; abraded as indicated. (c) Chromium mounted in plastics indicated; abraded on 15-μm aluminum oxide – wax laps.

(d) Metals indicated mounted in phenolic plastic; abraded on 15-μm aluminum oxide – wax laps. For (a), (b) and (d), the steel specimen was chosen as being representative of metals with a polishing rate lower than that of the mounting plastic.

292 / Metallographic Polishing by Mechanical Methods

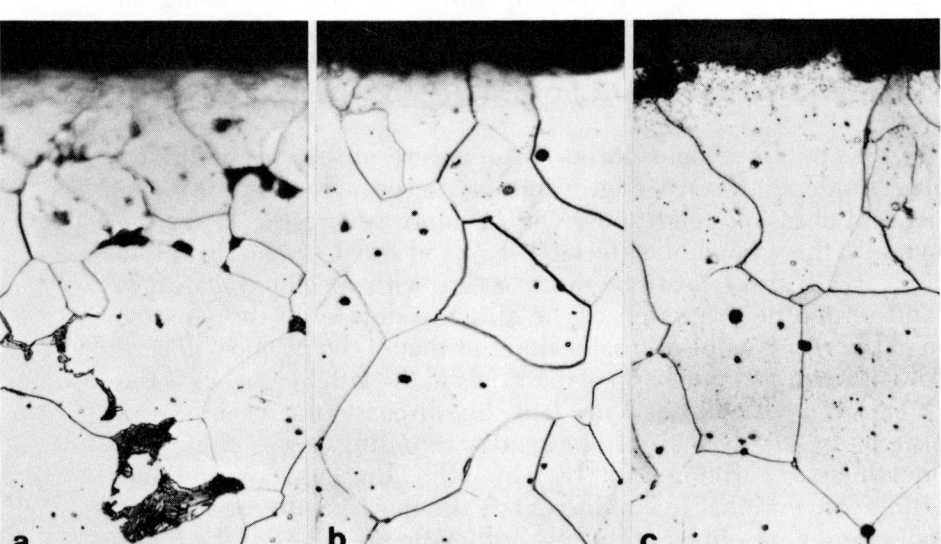

FIG. 9.16. Effect of mounting plastic on edge retention of steel specimens polished by the same standard technique.

(a) Specimen mounted in phenolic plastic (also representative of epoxy). (b) Specimen mounted in allyl plastic. (c) Specimen mounted in formvar plastic (also representative of polyvinyl chloride). All specimens etched in nital. Magnification, 500×. Steel is representative of metals with a polishing rate lower than that of the mounting plastic.

oriented scratches, and the degree of edge rounding is uniform around the circumference of the specimen. The next most satisfactory result is obtained when the specimen is held stationary against a rotating wheel, such as that illustrated in Fig. 8.26(b). The least edge rounding then is obtained along an axis perpendicular to the abrasion scratches, and somewhat more rounding in the perpendicular direction. The most severe edge rounding occurs with full hand abrasion, but again it is less along an axis perpendicular to the abrasion direction than parallel to the scratches. The magnitude of these differences is, however, small when differences in specimen-holding arrangements are considered. This suggests that the local elasticity of the abrasive device largely determines the degree of edge rounding for a given potential difference in level between metal and plastic. The comparatively small degree of rounding obtained with fine abrasive laps (Fig. 9.15b) is in agreement with this conclusion.

There is, however, one method of manual treatment by which edge rounding can be reduced to a low level, at least at one edge: the technique known as *trailing the edge*. The specimen is abraded only in a direction perpendicular to the edge of interest. Pressure is applied to the specimen only during a stroke in which this edge is the trailing edge, and the specimen is lifted perpendicularly to the surface of the abrasive paper at the end of the stroke. It is returned to the start of the stroke without contacting the paper surface. This technique requires practice and is tedious. Its use is to be considered only when other methods of improving edge retention cannot be applied. It does greatly improve edge retention (Fig. 9.18a) but at the expense of degradation of edge retention at the leading edge.

FIG. 9.17. Effect of specimen material on edge retention of specimens mounted in epoxy resin and polished by comparable standard techniques on napped cloths.

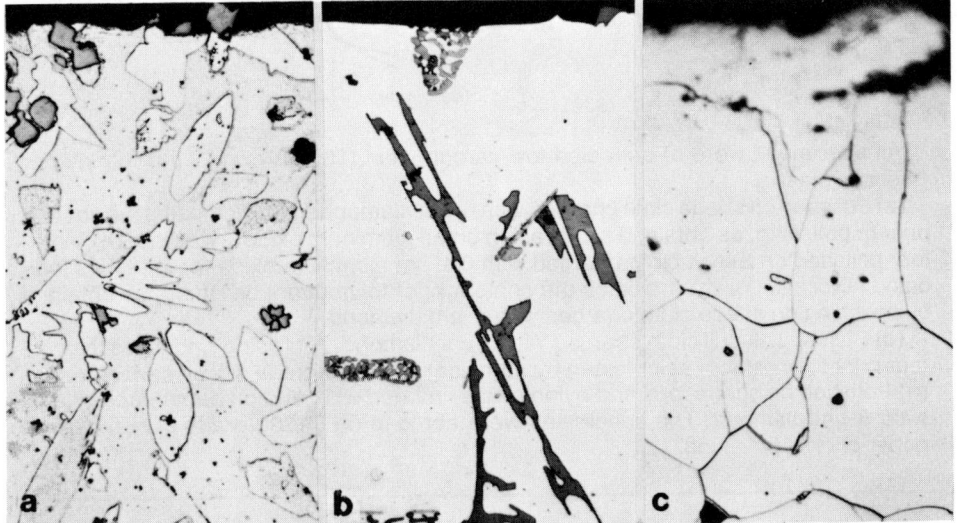

(a) Alpha-beta brass. The polishing rate of this metal is considerably greater than that of the plastic. (b) Aluminum alloy. The polishing rate of this metal is slightly greater than that of the plastic. (c) Steel. The polishing rate of this metal is lower than that of the plastic. Magnification (all), 500×.

FIG. 9.18. Contours at the plastic/metal interface following abrasion.

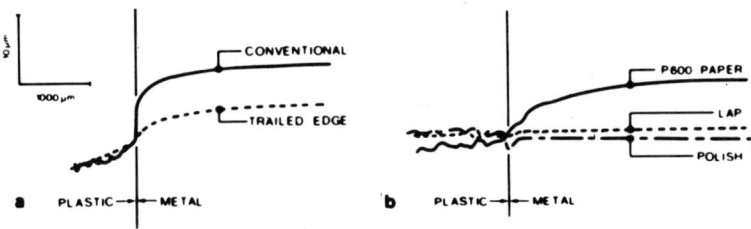

(a) Annealed low-carbon steel (130 HV) mounted in phenolic plastic and abraded on 600-grade silicon carbide paper. Results are compared for conventional and edge-trailing techniques. (b) Same steel as for (a), mounted in a filled epoxy molding plastic. "Lap" refers to a 15-μm aluminum oxide – wax lap; the "polish" was carried out on a suede cloth charged with 6-μm diamond abrasive.

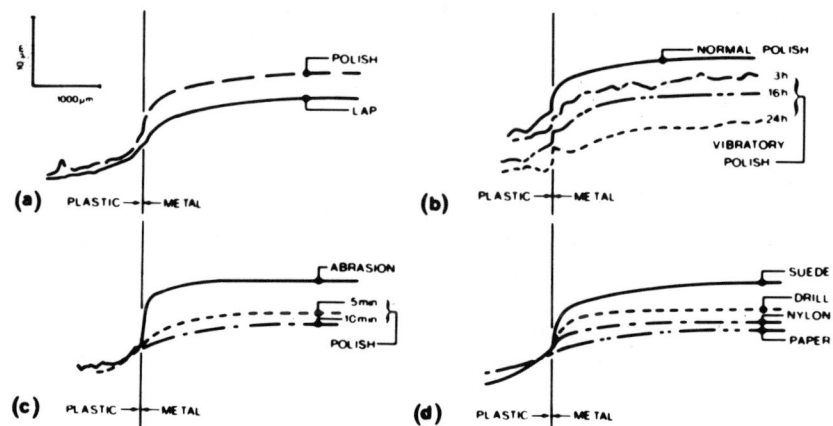

FIG. 9.19. Contours at the plastic/metal interface after polishing.

All specimens were of annealed low-carbon steel (130 HV) and were mounted in phenolic plastic.

(a) Polished on suede cloth charged with 6-μm diamond. "Lap" indicates the contour prior to polishing, as obtained by abrasion on an aluminum oxide – wax lap. (b) Vibratory polished on Selvyt cloth charged with 0.1-μm aluminum oxide for the times indicated. "Normal polish" indicates the contour prior to vibratory polishing, as obtained by polishing on suede cloth charged with 1-μm diamond.

(c) Polished on drill cloth charged with 6-μm diamond. "Abrasion" indicates the contour prior to polishing, as obtained by abrasion on P1200-grade silicon carbide paper.

(d) Polished on suede, drill and nylon cloths charged with, and on paper impregnated with, 6-μm diamond. The specimens were abraded on P1200-grade silicon carbide paper prior to polishing.

Effects of the Polishing Stages

We have seen that the specimen will almost invariably be standing above the plastic at the end of the abrasion stages of preparation.* This situation may either improve or deteriorate during the following polishing stages, depending on four factors. They are: (a) the relative polishing rates of the specimen and the mounting plastic, (b) the polishing time, (c) the nature of the polishing cloth and (d) the manner in which the specimen is handled.

The polishing rate of a plastic can be less than or greater than that of the metal specimen which it contains (cf. Tables 5.2 and 9.6). The specimen will tend to polish down to the same level as the plastic when its polishing rate is greater than that of the plastic, the edge retention then being improved. For such specimens, little edge rounding usually occurs during abrasion anyway. It is consequently found that good edge retention is obtained by quite standard preparation sequences with metals which have comparatively high polishing rates (Fig. 9.17a and b). The

---

*Exceptions may be found only when some soft metals that have comparatively high abrasion rates are mounted in plastics that have comparatively low abrasion rates (cf. Tables 3.3 and 9.6).

plastic used for mounting under these circumstances is of little consequence.

On the other hand, edge retention will tend to deteriorate during polishing when the polishing rate of the specimen is lower than that of the mount (Fig. 9.19a). Such a specimen generally will be of a metal which developed comparatively severe edge rounding during abrasion. Comparatively poor edge retention will therefore be obtained after polishing by standard techniques using napped cloths (Fig. 9.17c). The severity of edge rounding will now depend significantly on the plastic used for mounting, because this determines both the degree of edge rounding present after abrasion and the relative polishing rates of specimen and mount.

One possible corrective measure under these circumstances is to increase the relative polishing rate of the specimen, such as by employing an etch-attack technique (Fig. 9.19b and 9.20). This technique has to be watched, however, because peculiar contours develop after lengthy polishing times and edge retention is then likely to deteriorate (Fig. 9.19b and 9.20). Another possible corrective measure is to reduce the polishing

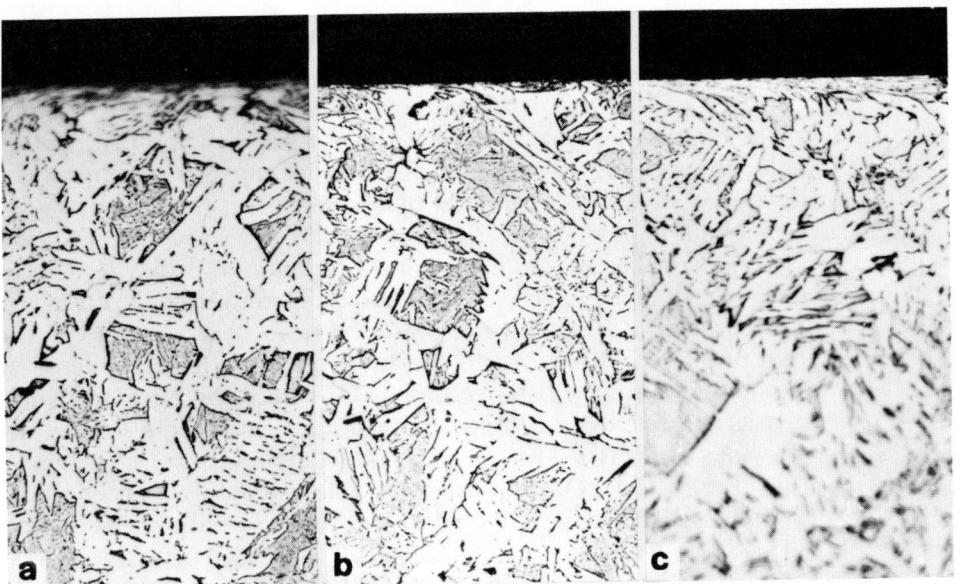

FIG. 9.20. Effect of etch attack during vibratory polishing on edge retention when the basic polishing rate of the specimen is lower than that of the plastic.

Steel specimens mounted in phenolic resin and finish polished using fine aluminum oxide abrasive. **(a)** Finished by standard polishing on Selvyt cloth; comparison standard. **(b)** Finished by vibratory polishing for 3 h; improved edge retention. **(c)** Finished by vibratory polishing for 24 h; deteriorated edge retention. Magnification (all), 500×. Compare these micrographs with the surface-contour traces in Fig. 9.19(b).

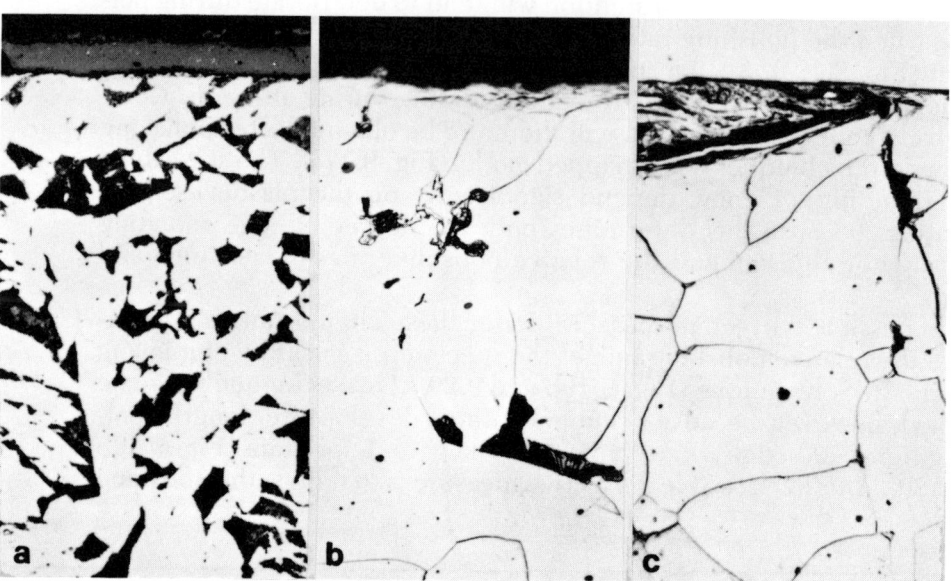

FIG. 9.21. Effects of special techniques for improving edge retention.

Steel specimens mounted in an epoxy resin and finish polished using diamond abrasive. **(a)** Polished on a drill cloth. **(b)** Steel shot incorporated in the mount; polished on a napped cloth. **(c)** Edge protected by an electrodeposited coating of nickel; polished on a napped cloth. Magnification (all), 500×.

rate of the plastic by incorporating in it an abrasion-resistant filler. This approach will be discussed in more detail soon.

The next important variable is the nature of the polishing cloth, a variable which becomes of increasing importance as the difference in level developed between metal and plastic increases during abrasion. A cloth with a long nap will tend to increase the edge rounding whenever there is any difference in level between specimen and plastic. A completely nonresilient cloth, on the other hand, will contact preferentially the high spots produced in the metal-plastic system by abrasion; the metal will then tend to polish down to the level of the mount (or vice versa), and so edge rounding will be reduced. However, practical polishing cloths have at least some resilience. Consequently, an equilibrium difference in level is established between plastic and metal after a certain polishing time (Fig. 9.19c). This equilibrium difference in level is characteristic of the polishing cloth, and is smaller the less the resilience of the cloth (Fig. 9.19d).

Thus, considering all of the cloths discussed in Chapter 5, the least edge rounding is obtained by polishing on the impregnated paper, followed by the nylon cloth, followed by the cotton drill (Fig. 9.19d), other things being equal. The less resilient cloths produce, as we discussed earlier, more severely scratched surfaces and lower polishing rates. A compromise in the choice of polishing cloth consequently may be necessary. The cotton drill is usually a satisfactory compromise for specimens of medium hardness; it can produce both acceptable edge retention and an acceptable finish (Fig. 9.21a). The less resilient nylon cloth, and then the even less resilient impregnated paper, can if necessary be used with harder specimen materials. It may be possible to follow polishing on the harder cloths with a brief finishing treatment on a softer napless cloth, or even on a napped cloth. The finishing treatment must then, however, be of limited duration if edge retention is not to be degraded to an unacceptable degree.

The method by which the specimen is handled during polishing has, as with abrasion, only a minor influence on edge rounding. It certainly has a much smaller influence than the other factors we have mentioned. The exception again is the technique of trailing the edge. The degree of rounding produced by a polishing cloth which has any resilience can well be expected to be more marked at the edge at which the specimen enters the cloth than at the edge at which it leaves. The trailing technique can thus also be successfully applied to polishing. But it is exceedingly laborious and tedious, even more so than for abrasion. It is to be considered only when other techniques are not applicable. An example of the standard of edge retention that can be achieved by edge trailing through the entire preparation procedure is given in Fig. 9.22. The edges of two per-

FIG. 9.22. Edges of two mutually perpendicular sections of a partly transformed steel. *(McDougall and Kennon, Ref 44)*

Both edges are excellently retained by the technique of "trailing the edges". Retention of both edges could not have been attained by any other technique. Etched in picral. Magnification, 1000×.

pendicular sections had to be retained simultaneously in this specimen. Moreover, the edge retention had to be held to a level which permitted examination of both edges at a high magnification. The edges could not simultaneously be protected from rounding by any of the other methods discussed here.

Special Mounting Techniques

Much attention has already been given to the effects of the mounting plastic on edge retention. We have seen that, as a method of protecting edges, the use of simple plastics has limitations. We shall now be concerned with additional precautions that can be taken at the mounting stage. The general aim is to surround the specimen by a material which more closely matches its abrasion and polishing characteristics than does a normal mounting plastic.

The simplest method is to incorporate in the working face of the mount a layer of chips[45] (Fig. 9.23a) or flakes[46] of a metal whose abrasion and polishing rates are similar to those of the specimen. Quite substantial improvements in edge retention can thereby be achieved (cf. Fig. 9.17c and 9.21b). Note again that similar polishing and abrasion rates are the critical factors, not similar hardnesses. For example, chilled iron shot is satisfactory for steel specimens of all types. The chips or flakes must be small and must be packed closely along the edge of interest. Chips can be packed in with the specimen in a mounting die set before the plastic is added. Alternatively, fine flaked powder, which must be free from stearic acid, may be premixed with the plastic molding powder in the

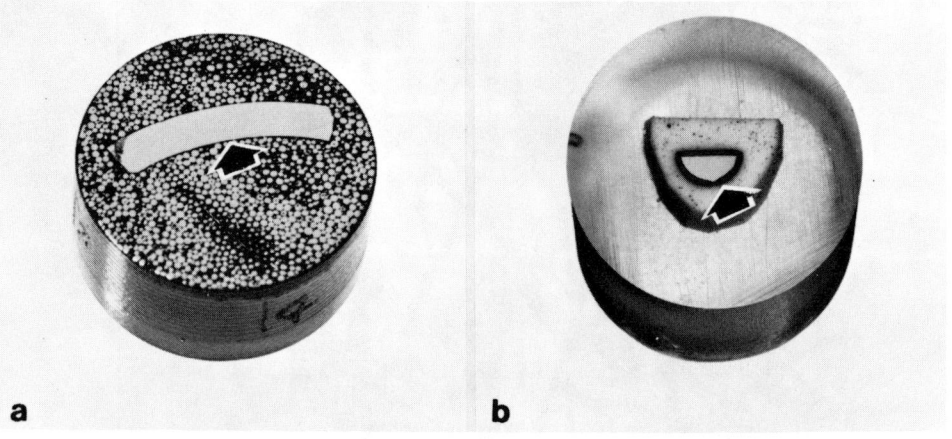

FIG. 9.23. Epoxy mounts of which the working surfaces have been filled to improve specimen-edge retention.

(a) Filled with chilled-iron shot. (b) Filled with pellets of sintered aluminum oxide. Filling has been confined to regions adjacent to the edges of interest by a two-stage mounting process. In each case, the specimen is indicated by an arrow.

proportion of 3 to 4 parts of flake to 10 parts of plastic.[46] A variant of this technique suitable for epoxy casting plastics is to coat the specimen with a low-viscosity resin, using vacuum impregnation if necessary, and then to apply the metallic flake to this coating while the resin is still tacky. After the coating has hardened, the specimen can be cast in a plastic to which the maximum possible amount of flake has been added consistent with reasonable pouring characteristics.

A range of molding plastics is available in which a comparatively large volume fraction of mineral or glass particles has been incorporated by the manufacturer. Two examples are the epoxy molding plastic and the allyl plastic listed in Table 9.6. A number of filled plastics of this nature are sold specifically as metallographic mounting media.

Filled plastics have abrasion characteristics similar to those of group 3 metals (p. 56). That is, they cause rapid deterioration of abrasive papers with the result that the abrasion rate is reduced effectively to zero after several thousand traverses. Thus two values are given in Table 9.6 to characterize filled plastics. The first is the maximum thickness that can be removed on a track of abrasive paper; these values are larger than those for metals. The second is the average abrasion rate over the first 500 to 1000 traverses; these values generally are an order of magnitude smaller than those for unfilled plastics, and are comparable to those for metals (cf. Tables 9.6 and 3.3). The differences in level developed between plastic and metal during treatments on abrasive papers consequently are considerably less than for unfilled plastics (Fig. 9.18b), at least where the abrasive paper is comparatively fresh. The improvement is not so marked after abrasion on harder laps.

Moreover, the polishing rates of filled plastics are much lower than those of corresponding unfilled plastics. This is not much of an advantage when the polishing rate of the metal is greater than that of the unfilled plastic. It is a considerable advantage, however, when the polishing rate of the metal is lower than that of the unfilled plastic. Now, instead of the difference in level tending to increase (as we saw above), the metal can be expected to polish down to the level of the filled plastic even with napped cloths. Considerable improvement in edge retention is achieved by using filled plastics with standard polishing techniques (Fig. 9.18b).

The filled plastic listed in Table 9.6 as a molding epoxy is of special interest in this context. This is a compression molding material that can be handled in normal molding apparatus for thermosetting plastics. The plastic adheres well to metal surfaces, and the adhesion is enhanced by the high filler content (see p. 15). Good adhesion is an advantage in many situations where good edge retention is needed, because it helps to ensure that a gap does not develop between the plastic and the specimen (see p. 14). It requires, however, that precautions be taken to ensure that the plastic does not adhere too well to the molding die set. The die must be maintained with a good surface finish, and a mold-release compound must be applied at the commencement of each molding cycle.

Casting-type epoxy plastics can also be filled with abrasion-resistant material. The filler can be mixed in with the liquid components before casting, and it is possible to fill only the plastic that is immediately adjacent to the section edges of interest. Aluminum oxide abrasive (say 400 to 600 mesh) can be used as the filler. Commercially available pellets of sintered aluminum oxide are also suitable. These sintered pellets are manufactured in comparatively fine mesh sizes (80 and 150), and so can be packed closely against the specimen edges. They are also made in several hardness grades. The abrasion and polishing rates of the filled plastic can thus be varied by changing both the volume fraction and the abrasion resistance of the filler (Table 9.6).

However, the use of abrasion-resistant fillers in mounting plastics has some consequences that need to be kept in mind, because a heavy price may have to be paid for their use. The first is that the rate of removal of material from the specimen during both abrasion and polishing is reduced virtually to that of the plastic. This can be a serious disadvantage in specimen materials that are sensitive to preparation artifacts (Chapter 8). Moreover, the filled plastic causes rapid deterioration of abrasive papers with specimen materials for which this would not otherwise have occurred. For these reasons, filling should not be used unless it is really necessary, and then the type and amount of filler should be chosen so that the abrasion resistance of the mount is no greater than necessary to achieve the desired degree of edge retention. It is also desirable that the area of filled plastic be kept to the minimum; it should be confined if possible to the region of the edge of concern. It may be possible to do this by preferential charging of the mounting mold. Alternatively, a two-stage mounting process can be used. For example, a filled epoxy may be first cast around the specimen and allowed to harden. This mount may then be trimmed to remove the filled plastic from all but the regions where it is really needed. The trimmed mount may then be cast into a normal epoxy mount (Fig. 9.23b).

The most satisfactory method of protecting a specimen edge is, without doubt, to electrodeposit a thick layer of metal of appropriate characteristics on to the surface concerned before sectioning. Excellent edge retention is then obtained by standard preparation techniques (Fig. 9.21c). The electrodeposit can be a dissimilar metal, provided that its polishing and abrasion rates are reasonably similar to those of the specimen material and that it does not interfere with subsequent etching processes. For example, nickel deposits are quite satisfactory on steel. Another advantage of electroplating is that the resultant coatings are brightly reflecting. Poorly reflecting surface layers, such as oxide scales, then show up in better contrast under the microscope (Fig. 9.24). The electroplating process is, however, a rather troublesome one and is justified only where other procedures fail (for example, in an application for which use of a taper-sectioning technique is intended). Information on a number of electroplating baths likely to be of use is set out in Appendix 9-B (p. 333). "Electroless" coating methods, which are very convenient, also can be useful. Some information on a method of electroless coating with nickel also is given in Appendix 9-B. Solutions for electroless plating are also available commercially.

A problem with many electroplating baths is that they have limited throwing power, which restricts penetration of the deposit into fine con-

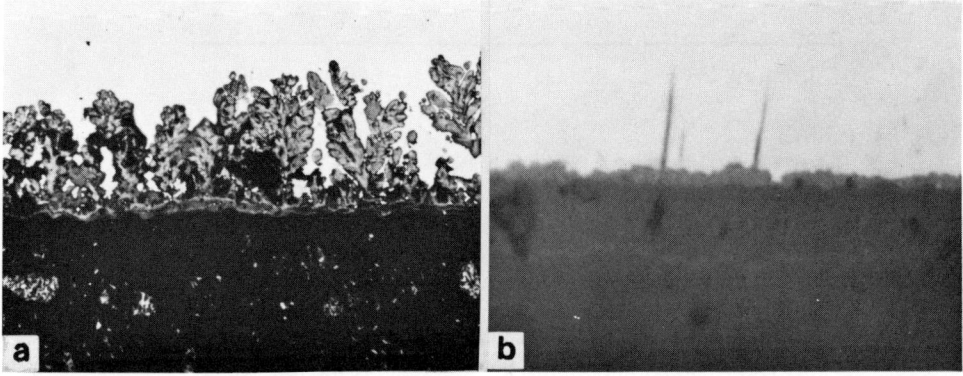

FIG. 9.24. Edges of specimens that are covered with a nonconducting layer.

The surfaces have been silvered by a chemical deposition process and then electroplated with nickel. **(a)** Lead oxide on a lead substrate. The electrodeposit has penetrated into the interstices between oxide dendrites. Magnification, 100×. *(Burley and Dale, Ref 47)*. **(b)** Outer layers of an oxide on iron. The spikes are sections of whiskers projecting from the outermost layer of hematite. Magnification, 1000×.

volutions in the surface of the specimen. A further difficulty is encountered when a nonconducting oxide or scale is present on the specimen surface. Both of these difficulties can be overcome by depositing a thin layer of silver on the surface prior to electrodeposition. The silver may be deposited either by vacuum evaporation or by chemical deposition using the standard Brashear process (Fig. 9.24).[47] Details of the Brashear process of silvering are set out in Appendix 9-C (p. 335). It may also be possible to deposit a thin electroless nickel layer before electrodeposition.

A major problem with the electroplating technique is that it is difficult to obtain adequate adhesion of the deposit. It is not possible to use the rather severe pre-etching treatments used in industrial practice because they might remove the very surface layers that are the subject of the examination. Cleaning in detergent solutions, preferably assisted by ultrasonic cleaning, should first be tried. If this is inadequate, a normal metallographic etch may be tried. An etch of this nature is not likely to produce detectable changes in surface topography.

Actually, only moderate adhesion is required. A useful precaution that can be taken to reduce the probability of separation when the adhesion is marginal is to ensure that all machining and abrasion is carried out so that the cut is made from the electrodeposit into the specimen.[49] Additional tensile stresses are not then imposed across the interface. Moreover, when the specimen is mounted in a plastic that adheres well to metals (e.g., an epoxy), it is desirable to adopt the precautions discussed on p. 14 that reduce to the minimum the tensile stresses developed at the plastic/metal interface.

TABLE 9.7. Comparative Values of the Internal Stress in Electrodeposited Metals (After Kuschner, Ref 48)

Metal	Stress in electrodeposit(a), lb/in.²
Cadmium	−500 to −300
Zinc	−1,000 to −2,000
Silver	+2,000
Gold	+1,500
Copper	+2,000
Nickel	+20,000
Iron	+40,000
Chromium	+60,000
Rhodium	+100,000

(a) The values are intended to indicate only the relative order of magnitude of the internal stress in deposits up to about 0.025 mm (0.001 in.) thick: − indicates compressive stresses; + indicates tensile stresses.

Difficulties with adhesion are aggravated greatly when internal stresses build up in the electrodeposit. The magnitude of these stresses tends to be a characteristic of the metal deposited, and the comparative figures listed in Table 9.7 serve as a guide to the type of deposits with which the greatest difficulties are to be expected. In some instances (e.g., nickel), special and complex plating baths have been developed to reduce the internal stresses in the deposit. But it is virtually impossible to maintain these solutions in the desired conditions in small laboratory-size plating baths. Electroplating solutions of this nature must normally be obtained from large industrial installations, and then have to be replaced frequently. It is also desirable from the present point of view to keep the thickness of the deposit to the minimum.

We mentioned earlier that retaining the junction region of a composite specimen is a special case of edge retention. An example of a composite specimen would be a metal with a thick surface coating of a dissimilar metal. In one sense this case is simpler than the one just discussed. The differences in abrasion and polishing rates between the two metals usually will be smaller than those between either metal and an unfilled plastic. In fact, a significant problem exists only in combinations where these differences are comparatively large (see Tables 3.3 and 5.2). Other difficulties are, however, introduced. Some of the corrective measures discussed above cannot be applied. Those that can be applied are the use of fine laps to finish abrasion, the use of polishing cloths of minimum resilience, and the use of etch-attack methods to enhance the polishing rate of the slower-polishing member of the pair. Another problem arises when it is desired to retain both the interface and an outer edge; it may not be possible to do both simultaneously. Finally, electrochemical differences between component metals may interfere with both polishing (see p. 247) and etching.

Summary

Summarized below are a number of steps that can be taken to improve edge retention when standard procedures prove to be inadequate. The steps are set out in the order in which they might be tried. They can also be tried in combination.

1. Finish abrasion on a fixed abrasive lap.
2. Choose a mounting plastic with a low abrasion rate and, if possible, with a lower polishing rate than that of the specimen material.
3. Incorporate metallic chips or ceramic particles in the plastic mount, or use a mineral-filled plastic, but preferably only to the extent that is really necessary.
4. Adopt an etch-attack technique of polishing, particularly in the case of specimens which have polishing rates lower than that of the mounting plastic.
5. Use a hard napless cloth during polishing, but only as hard a cloth as is really necessary.
6. Use manipulative procedures which reduce rocking of the specimen. If all else fails, try edge trailing.
7. Protect the surface concerned with a deposited coating.

Note, finally, that this entire discussion of edge retention has been based on the assumption that diamond abrasives are used for the rough and intermediate polishing stages. The same principles apply to other polishing abrasives, but the problems that arise at the polishing stages are considerably greater (see, for example, Fig. 5.15, p. 163).

CORRECT REPRESENTATION OF CAVITIES

We saw earlier that the shapes of cavities may be distorted by collapse of the cavities during abrasion. Collapse occurs, however, only for cavities that have entered the comparatively thin surface fragmented layer. Consequently, the first precaution that must be taken to ensure correct representation of cavities is to adopt preparation procedures which remove this layer easily, and we have seen in Chapter 8 that this can be done readily enough. There is every assurance, therefore, that large cavities will be correctly represented after specimen preparation by any reasonably well-designed procedure.

The correct representation of small cavities is, however, more difficult to ensure. This applies to any cavity whose dimensions in section are on the order of a few micrometres. First, small cavities may appear to be narrower than they actually are if polishing is not taken to a fine enough finishing stage.[50] For example, the specimen of brass illustrated in Fig. 9.25 contains chains of cavities at the grain boundaries — cavities which developed during creep straining. The cavities visible after polishing on 1-μm diamond (Fig. 9.25a) are narrower than those visible after further polishing by skidding on magnesium oxide (Fig. 9.25b); this is because the mouths of the cavities were partly filled with material during polishing on the diamond abrasive (Fig. 9.25c). The filling material seemed to

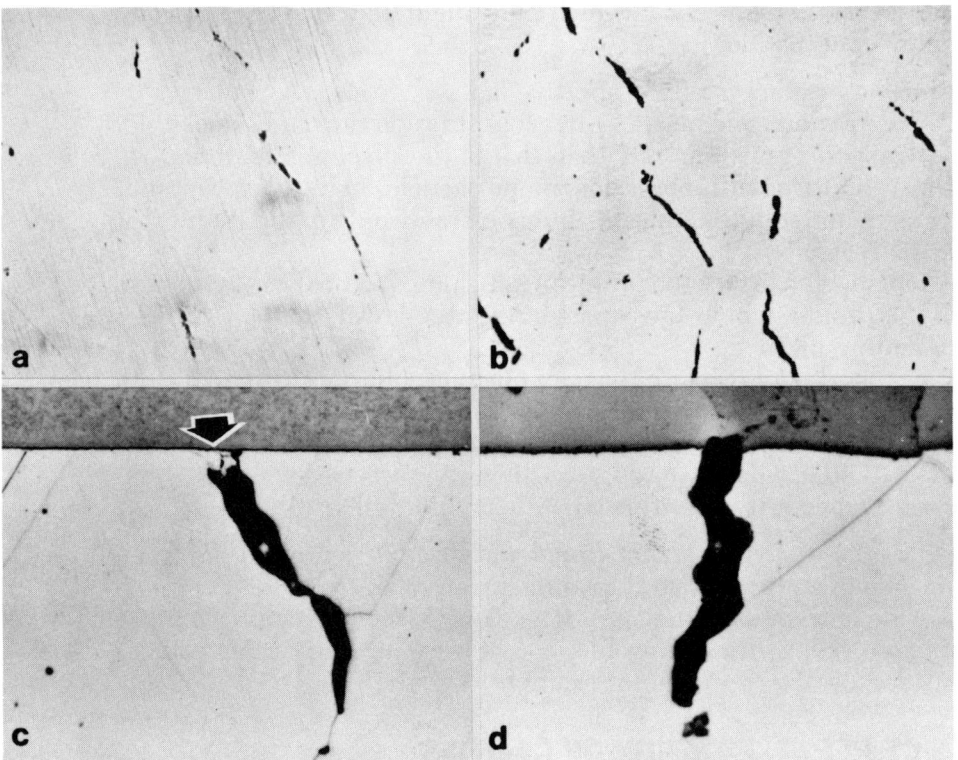

FIG. 9.25. Grain-boundary cavities in a 70:30 brass which has been subjected to creep strain.

(a) After polishing on 1-μm diamond. Magnification, 200×. (b) After further polishing by skidding on magnesium oxide. It is now apparent that the cavities in (a) are not shown at their true widths. Magnification, 200×. (c) Taper section of the surface shown in (a). The arrow indicates apparent polishing debris partly closing the mouth of a cavity. Taper ratio, 9.0. Magnification, 1000×. (d) Taper section of the surface shown in (b). The mouth of the cavity is clear, but note the slight rounding of the edges of the mouth. Taper ratio, 9.7. Magnification, 1000×.

be polishing debris, because it could be removed by ultrasonic cleaning. Lehtinen and Melander[51] have shown that the filling material can also be removed by an ion-polishing technique, but this requires special equipment. On the other hand, it is always possible that small cavities could collapse in the fragmented layer during polishing, as large cavities do during abrasion.

The apparent dimensions of the cavities visible after skid polishing on magnesium oxide are very close to the actual values, as confirmed by the taper section in Fig. 9.25(d). It is not possible to be so sure, however, that the fine details of the contours are correctly represented. The section surface produced by this polishing method is not absolutely flat (Fig. 9.25d; cf. Fig. 9.25c) as a result of the chemical component of the polishing mechanism. The rounding of the edges of cavities that occurs with this type of polishing process could distort slightly the details of the outline of a cavity as seen on the section surface.

These limitations of mechanical polishing are, nevertheless, minor compared with those for electrolytic polishing. Gross enlargement of the mouths of the cavities occurs with this method of polishing (Fig. 9.26), and this makes the cavities appear to be much larger than they really are (Fig. 9.27). It also seriously distorts their contour as seen in section.

Thus, although it is easy enough to reveal small cavities at their correct dimensions, the correct representation of their shape is one of the most difficult problems in metallography. Mechanical preparation methods can be relied upon in all but the most critical cases, but only if good procedures are followed. These remarks also apply to cracks and similar narrow discontinuities.

RETENTION OF NONMETALLIC INCLUSIONS

Industrial metals invariably contain inclusions consisting of particles of nonmetallic materials that were introduced or formed during manufacture. These inclusions vary widely in nature, size, and volume fraction. Their presence often affects significantly some important property of the metal; consequently, the identification and assessment of nonmetallic inclusions constitutes an important facet of microscopical metallography. The inclusions must be fully retained in their enclosing cavities, and their surfaces must be well polished, before examinations of this nature can be carried out effectively.

FIG. 9.26. Taper section of an electrolytically polished surface of the specimen shown in Fig. 9.25.

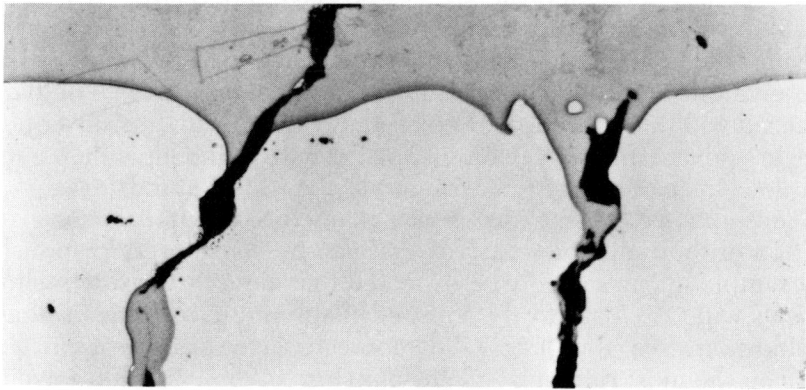

Marked enlargement of the mouths of the cavities has occurred during electrolytic polishing. Taper ratio, 9.6. Magnification, 250×.

306 / Metallographic Polishing by Mechanical Methods

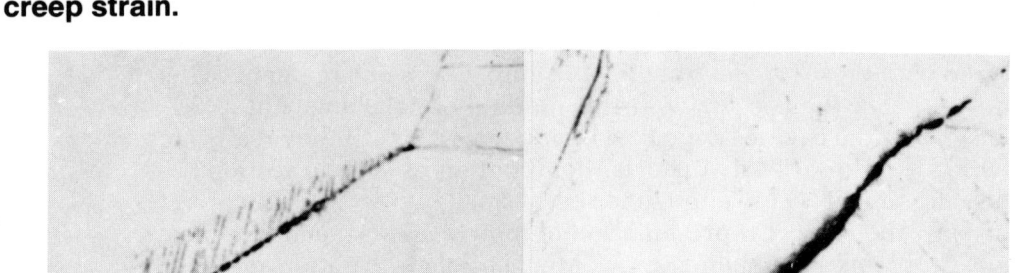

FIG. 9.27. Very small cavities in 70:30 brass which has been subjected to creep strain.

(a) Mechanically polished. **(b)** Electrolytically polished. Considerable enlargement of the cavities has occurred during electrolytic polishing; details of their arrangement have consequently been lost. Both etched in ferric chloride reagent. Magnification, 2000×.

The chemical and electrochemical characteristics of nonmetallic inclusions are usually so different from those of metals that they cannot be retained, let alone be polished, by chemical or electrochemical polishing methods. Consequently, the use of mechanical polishing methods of preparation usually is obligatory. The mechanical abrasion and polishing characteristics of inclusions usually are also considerably different from those of the metal in which they are contained. They often can be classified as brittle materials in the sense that was covered in Chapter 7. They are often hard, and could themselves be classified as abrasives.

As a consequence of their brittleness, inclusions tend to fracture during abrasion, particularly during coarser abrasion stages. Examples of this were given in Fig. 3.30(c) (p. 78) and Fig. 4.25(c) (p. 119). The damage so caused will then have to be repaired during polishing. Consequently, there are advantages in finishing abrasion with a fine lap which cuts the inclusions instead of fracturing them (cf. Fig. 3.30c and d).

A consequence of the abrasive-like characteristics of inclusions is a restriction of the range of abrasives that can be considered for polishing. For example, it can scarcely be expected that aluminum oxide inclusions in steels will be satisfactorily polished if aluminum oxide is used as the polishing abrasive (Fig. 9.28a). Diamond abrasives produce a satisfactory polish on the most refractory inclusions (Fig. 9.28b). Conventional abrasives such as aluminum oxide tend also to scour out softer inclusions whereas diamond abrasives do not (cf. Fig. 5.15e and f). The use of diamond abrasives is almost obligatory when examination of inclusions is important. They are, indeed, to be recommended whenever significant numbers of inclusions are present, even if the inclusions themselves are not of interest. This is because poor retention of inclusions greatly degrades quality of polish.

In practice, excellent retention of nonmetallic inclusions is obtained when diamond abrasives are used for polishing with any of the preparation procedures listed in Table 8.3 (p. 242). A procedure in which a fine lap is used as the final abrasion stage is to be preferred.

RETENTION OF GRAPHITE IN FERROUS ALLOYS

Several industrially important groups of ferrous materials contain free graphite in various shapes and forms, and this graphite may be present in comparatively large volume fractions. Gray cast irons and malleable cast irons are examples. Metallographic preparation procedures are needed which will reliably retain this graphite sufficiently well for its morphology and structure to be studied. For these groups of materials, chemical and electrochemical methods of polishing are not applicable.

We saw in Chapter 4 that the abrasion stages of preparation can remove graphite from its containing cavities in the surface layers, and can even

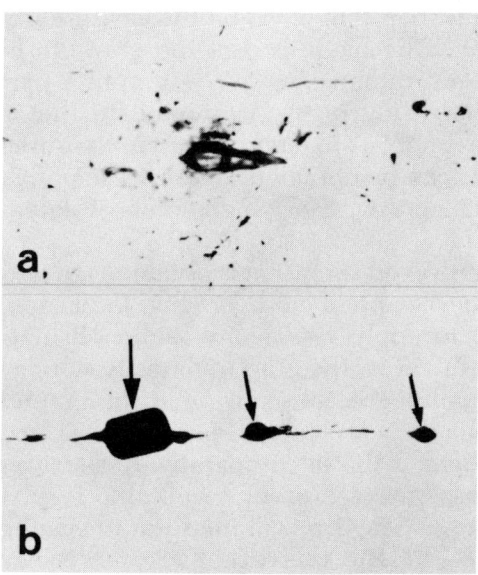

FIG. 9.28. Aluminum-killed steel containing aluminum oxide and silicate inclusions.

(a) Polished on a napped cloth charged with aluminum oxide abrasive. The aluminum oxide inclusion in the center of the field has not been polished, and much irregular scouring has occurred around this inclusion. Magnification, 1000×. (b) Polished on a suede cloth charged with diamond abrasive. The aluminum oxide inclusions (arrows) are now well retained and are polished. Magnification, 1000×.

cause the resultant empty cavities to collapse (Fig. 4.26, p. 120). However, we also saw that sufficiently fine abrasion stages do not do this; the graphite is then shown in its correct shape and dimensions, and may even stand slightly above the matrix surface (Fig. 4.26c and d). The continuation of abrasion to a stage, such as a fine abrasive lap, which achieves this desirable condition is obviously desirable. It is then not necessary to worry about repairing the abrasion damage to the graphite during polishing. It is difficult enough to ensure that further damage is not introduced during polishing without having to worry about repairing existing damage. The larger the graphite particles the more pertinent are these remarks.

The choice of abrasive is crucial in the polishing of these materials. A satisfactory standard of retention is much more likely to be achieved with diamond abrasives, used in the manner discussed in Chapter 5, than with any other abrasive. Consider first a napped, but short-napped, polishing cloth such as a synthetic suede. When a conventional abrasive, such as aluminum oxide, is used on this cloth, there is a strong tendency for the graphite particles to be removed from their containing cavities and for the cavities to be enlarged. The graphite particles will then appear to be wider than they actually are, their outlines will be distorted, and the structure of the graphite itself will not be able to be examined (Fig. 9.29a). Actually, it appears that the abrasive first removes preferentially the matrix material adjacent to the matrix-graphite interface (Fig. 9.29c); the graphite itself is then removed and the cavities are subsequently enlarged by scouring. The likelihood of the graphite being removed increases as the lateral dimensions of the graphite particles increase. It decreases with increasing pressure applied to the specimen, but the tendency for large scratches to be produced in the surface then increases. Satisfactory results can be obtained under these polishing conditions by use of conventional abrasives, but high levels of skill and experience are required.[26]

Considerably improved retention is obtained when diamond abrasives are used with a short-napped cloth (cf. Fig. 9.29a and b); it is again desirable, however, to apply reasonably high polishing pressures. After being polished in this way, the graphite mostly stands slightly above the surface, with a small degree of scouring at the graphite/metal interface which tends to damage unfavorably oriented flakes (arrow in Fig. 9.29d). Nevertheless, provided that a comparatively short-napped cloth and a heavy polishing pressure are used, acceptable results can be obtained with many cast irons. Gray irons of medium to small graphite flake size would be examples.[29,30] But the results are rarely perfect (Fig. 9.29b); flakes which happen to have been sectioned acutely may not be well preserved even in moderately fine-grain irons. The proportion of poorly retained graphite increases with increasing flake width, until eventually the problem becomes a serious one. Even greater difficulties are experienced with spheroidal-graphite and malleable cast irons because the lateral dimensions of the individual graphite particles in these irons tend to be large.

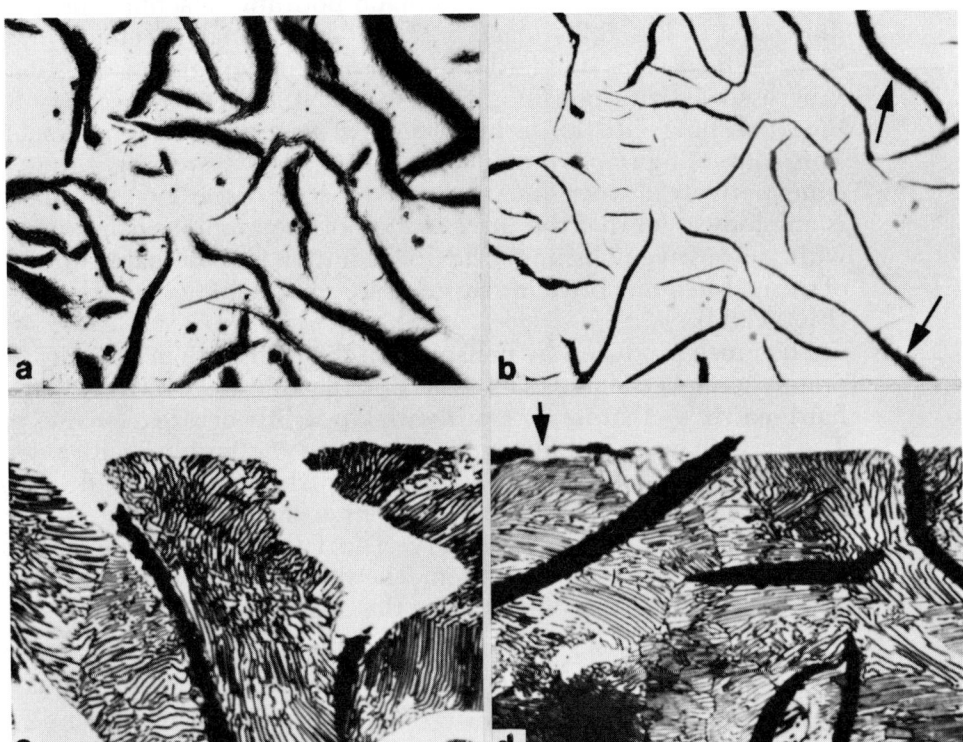

FIG. 9.29. Pearlitic gray cast iron.

(a) Polished on a suede cloth charged with 15-μm aluminum oxide. Most of the graphite flakes have been damaged. Magnification, 100×. (b) Polished on a suede cloth charged with 6-μm diamond. Most of the graphite flakes have been well retained, but a few (arrows) have not. Magnification, 100×.

(c) Taper section of the surface shown in (a). The matrix has been eroded adjacent to the graphite flakes, the flakes have then been removed, and the resultant cavity has been enlarged. Taper ratio, approx. 10. Magnification, 1000×. (d) Taper section of the surface shown in (b). Most of the graphite flakes now stand above the level of the matrix, but some scouring (arrowed) has occurred at the interface. Taper ratio, approx. 10. Magnification, 1000×.

Retention of graphite is improved greatly by the use of napless polishing cloths (see p. 167), although a compromise is necessary because of the concurrent tendency of such cloths to produce severe polishing scratches. A cloth similar to a cotton drill is a good compromise. Harder cloths (e.g., nylon or the impregnated paper) are rarely needed to achieve satisfactory graphite retention anyway.

The drill cloth preferably should be used on a medium-speed mechanical wheel, a circumferential track being charged with a 1-μm-grade diamond carrier paste in the usual way and lubricated with kerosine so that it is kept just moist. A heavier load than normal should be applied to the specimen. A convenient way of doing this manually is to hold the specimen stationary against the polishing wheel while oscillating it slowly about its own vertical axis. Automatic polishing machines of the types illustrated in Fig. 8.26(b) and 8.27 also enable high specimen pressures to be applied. The standard of preservation of graphite should increase progressively with polishing time so that it is merely necessary to continue polishing until the desired degree of preservation has been achieved. This continuing improvement is achieved, however, only if the cloth is comparatively unworn, and the cloth must be discarded as soon as there is any indication that the preservation of the graphite is not improving with progressive polishing. The deterioration is presumably due to some of the cotton fibers breaking during use, so becoming equivalent to a nap (Fig. 5.25d).

The finish produced by polishing on the drill cloth in this way is moderately scratched, but usually is acceptable for irons with a comparatively hard matrix — that is, for irons with a pearlitic or a martensitic matrix. The finish may not be fully acceptable, however, for irons with essentially ferritic or austenitic matrices even when the finest available grades of diamond abrasive are used. In this event, a final polish using a softer (i.e., napped) cloth becomes necessary. The following finishing treatments might then be tried in succession. The finish obtained will improve with each step but the risk of damaging the graphite will also increase.

1. Synthetic suede charged with 1-μm diamond in a carrier paste, using kerosine as the polishing fluid. This is likely to produce a result in which scratches are not detectable in the ferritic areas at low magnifications (Fig. 9.30a), but one in which they are detectable at high magnifications (Fig. 9.30b).
2. As for step 1, but using 0.1-μm-grade diamond. An acceptable polish usually is produced in ferrite provided that the abrasive is true to grade (Fig. 9.30c and d).
3. A soft napped cloth (e.g., Selvyt) charged either with a fine aluminum oxide or with magnesium oxide (Fig. 9.30e). Aluminum oxide is the simpler of the two to use, but magnesium oxide gives slightly better results. The retention of graphite will eventually deteriorate, however, with either of these finishing techniques (Fig. 9.30e and f). The question is whether an adequate polish can be achieved on the matrix before noticeable deterioration occurs in the graphite. Certainly, the duration of the polishing treatment has to be controlled with this in mind.

Note, moreover, that the graphite is not always brightly reflecting even when it is fully retained and its surface well polished (Fig. 9.30e; cf. Fig. 9.30c). It may then not be seen in good contrast, particularly against a dark-etching matrix. Shorter-napped cloths tend to produce more brightly reflecting surfaces on the graphite (cf. Fig. 9.30c and e). Poor reflectivity often can be improved, when the graphite is well retained, by rubbing the polished surface briefly on a fairly dry section of the polishing cloth without abrasive. Etching times also must be kept to the minimum. In

FIG. 9.30. Ferritic gray cast iron.

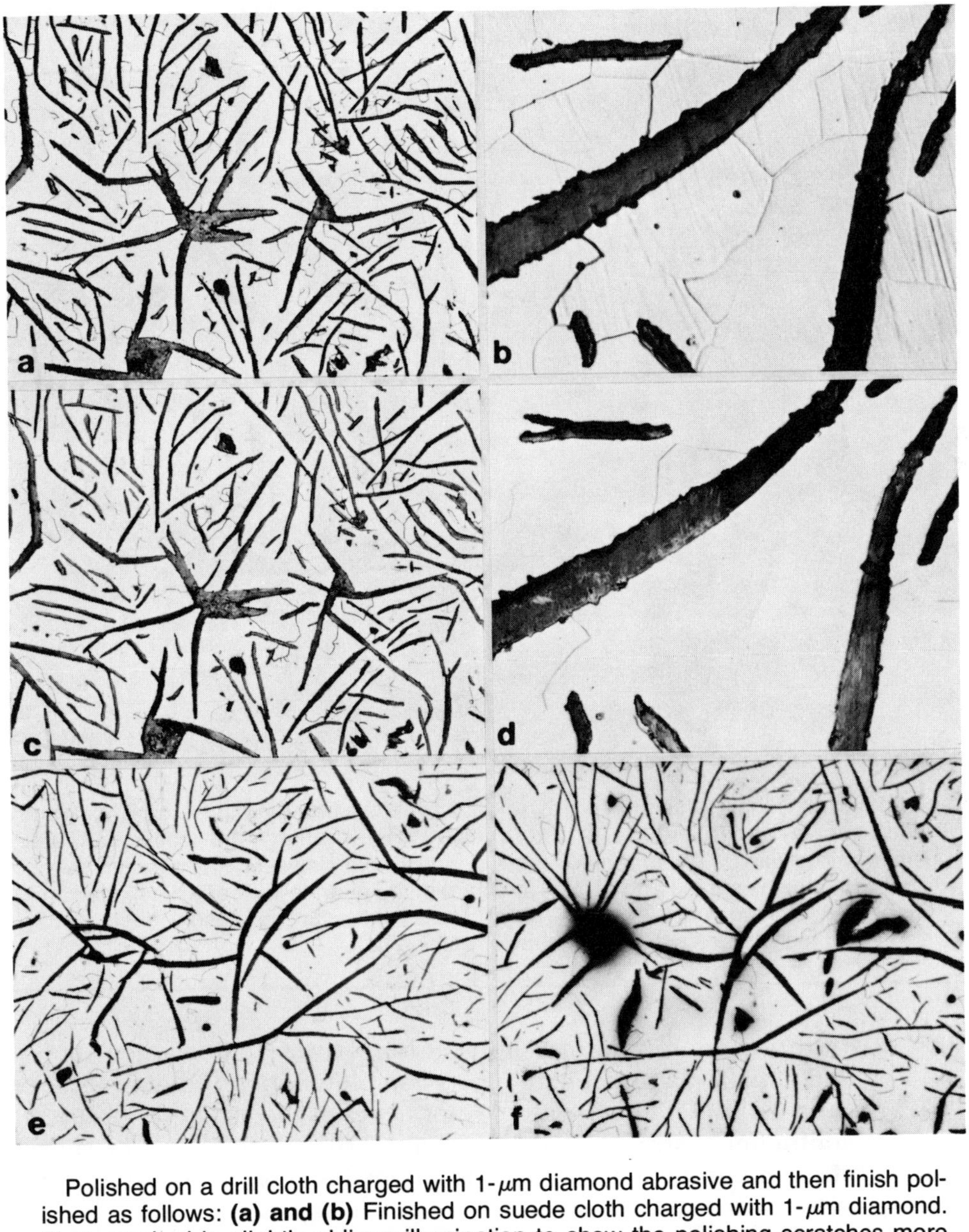

Polished on a drill cloth charged with 1-μm diamond abrasive and then finish polished as follows: **(a) and (b)** Finished on suede cloth charged with 1-μm diamond. Photographed in slightly oblique illumination to show the polishing scratches more clearly.
 (c) and (d) Finished on suede cloth charged with 0.1-μm diamond. Photographed in slightly oblique illumination. **(e) and (f)** Finished on Selvyt cloth charged with magnesium oxide. The minimum polishing time necessary to remove the rough-polishing scratches was used in **(e)**; a slightly longer polishing time was used in **(f)**. Magnifications: **(a), (c), (e) and (f)**, 100×; **(b) and (d)**, 500×.

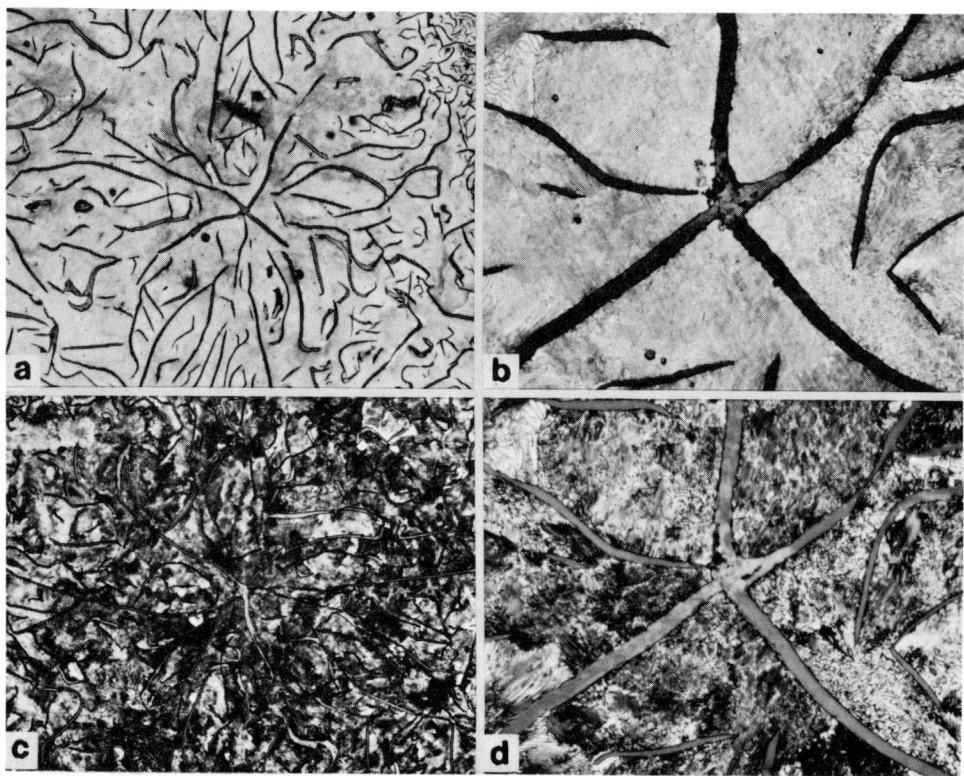

FIG. 9.31. Pearlitic gray cast iron.

(a) and (b) Lightly etched in picral. **(c) and (d)** More heavily etched in picral. Magnifications: **(a) and (c),** 100×; **(b) and (d),** 500×. Retention of graphite is equally satisfactory in all cases. The graphite is more clearly visible, however, at low magnification after lighter etching (cf. **a and c**). The structure of the pearlite is clearly developed only by heavier etching (cf. **b and d**). The graphite flakes can be distinguished readily enough at high magnification after heavier etching **(d)**.

fact, a compromise may be necessary with etching when the matrix is potentially dark-etching (e.g., a completely pearlite matrix). A comparatively light etch which does not develop the structure of the matrix in optimum contrast may be necessary if the graphite is to be shown in optimum contrast at low magnifications (cf. Fig. 9.31a and c). A heavier etch is necessary if the structure of the matrix is to be revealed (cf. Fig. 9.31b and d); the graphite then can be seen reasonably well at high magnifications (Fig. 9.31d), but not at low magnifications (Fig. 9.31c). It is up to the metallographer to decide which has to be illustrated most clearly. Correct choice of exposure and tone contrast range during photography is also necessary to achieve optimum appearance in the final photomicrograph.[52]

Graphite-containing cast irons are susceptible to staining if washed in hot water. Staining may occur around the sulfide inclusions that are invariably present, in a manner similar to that for the steel illustrated in Fig. 8.34(a) (p. 262). It sometimes, but not always, occurs around the graphite particles themselves (Fig. 8.34c). The risk of such staining can be reduced by washing in cold water only (Fig. 8.34b and d). It is always good practice to ensure that water remains in contact with the surface for the minimum period of time.

Summarizing, acceptable results may be obtained with cast irons in which the graphite particles are comparatively small by using standard polishing procedures (Table 8.3, p. 242). Preferably, however, the following modifications should be incorporated.

Stage I: Abrade on aluminum oxide papers to 400 grade.
Stage II: Abrade on 15-μm aluminum oxide – wax lap.
Stage III: Rough polish with 1-μm diamond charged on a cotton drill cloth. Use a fresh cloth and polish until an acceptable degree of graphite retention has been achieved.
Stage IV: Brief final polish, if necessary, on a napped cloth charged with fine diamond, magnesium oxide or fine aluminum oxide. Discontinue if retention of graphite begins to deteriorate.

The standards of graphite retention that can be expected for these procedures are illustrated in Fig. 9.30 and 9.32. The irons illustrated in Fig. 9.32 are representative of the most difficult cases that are encountered.

VERY SOFT MATERIALS

The group now to be considered includes metals and alloys which have hardnesses of less than about 20 HV. This includes, for example, lead, indium, tin and high-purity aluminum; it may also include some dilute alloys of these metals. Specific difficulties that arise are as follows:

1. Specimens could easily be distorted during handling. Moreover, the microstructure may be altered drastically if the material is heated to temperatures even slightly above ambient. Thermally induced structural changes are particularly likely to occur if the material has been distorted during handling.
2. A deep layer of significant deformation is produced during machining and abrasion. The deformation is likely to induce a recrystallized layer and, in the case of noncubic metals, massive deformation twins (see p. 123).
3. Large numbers of abrasive particles are likely to become embedded in the section surface during abrasion (see p. 131).
4. It is difficult to obtain an adequately scratch-free final polish by purely mechanical methods.

If they are to be mounted at all, these materials must be cast in plastics which set at the lowest possible temperatures. The mounting mold should be cooled if there is any doubt on this point; cooling may be effected by a stream of cool air or by immersion in crushed ice.[53] The possibility of abrasive fragments being embedded during abrasion is a major

FIG. 9.32. Graphite excellently retained by simple procedures in cast irons for which such retention is comparatively difficult.

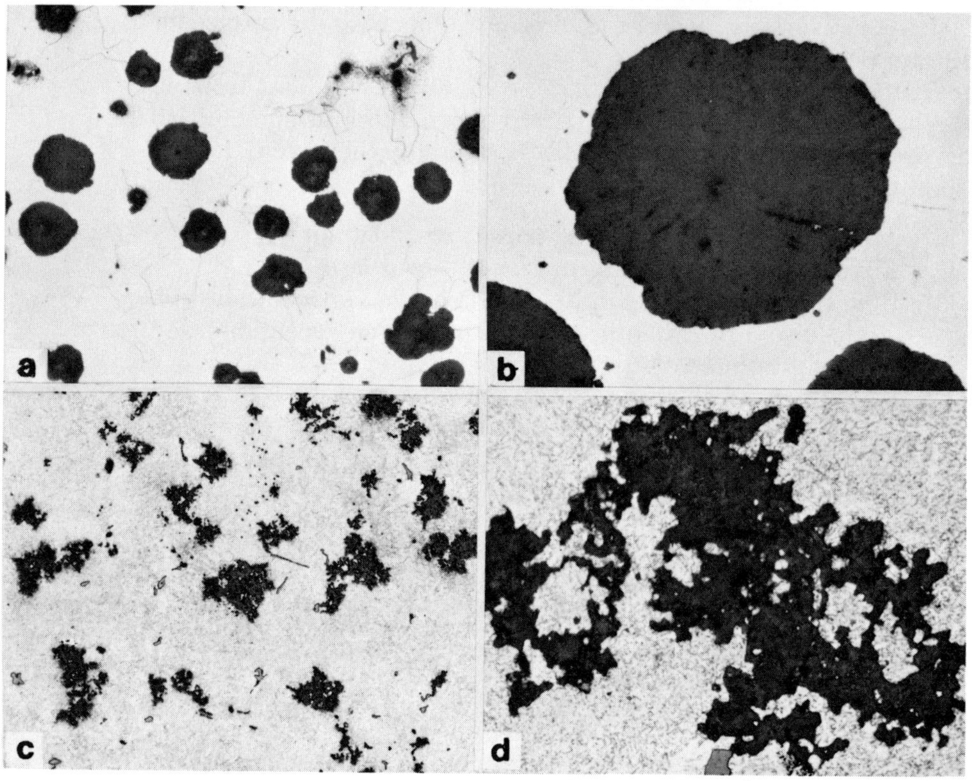

(a) and (b) A spheroidal-graphite iron with a ferritic matrix. Etched in nital. Magnifications, 100× and 500×. **(c) and (d)** A malleable cast iron with a spheroidized matrix. Etched in nital. Magnifications, 100× and 500×. The critical feature of these procedures is the use of a drill cloth charged with diamond abrasive.

problem, if these materials are abraded by standard techniques (Fig. 9.33a). A very rough surface containing many abrasive fragments inevitably is produced. This applies, for example, to annealed pure metals such as lead (4 HV), tin (6 HV), super-purity aluminum (25 HV), gold (25 HV) and silver (25 HV). The embedded particles become quite obvious after a short period of polishing (Fig. 9.33b). It is possible to remove the embedded particles by polishing, but longer polishing times than usual are required because the fragments are deeply embedded and because the presence of the particles reduces the polishing rate. Moreover, and of greater consequence, the released particles contaminate the polishing pads. It is clearly much better practice to avoid the embedding in the first place. One simple way of doing this effectively enough for most purposes is to load the working surface of the abrasive paper with wax and then to abrade dry (cf. Fig. 9.33a and c; see also p. 131). Low abrasion pressures are desirable.

An alternative measure is to substitute microtome cutting for abrasion. Lucas[54] originally developed this alternative technique for soft alloys, and he suggested that such a good finish could be obtained that only a brief final polish was necessary after microtoming. Worner and Worner[37] subsequently pointed out, however, that much skill and effort is required in sharpening the microtome knife if such a standard is to be reached, and that this is scarcely worthwhile. It is more effective to microtome to only a moderate finish (Fig. 9.33d) and then to follow microtoming by standard rough polishing. Microtoming, which will be discussed further in Chapter 10, also has some practical limitations. Special equipment is required. It is difficult to clamp many types of specimens in the holding vice of the microtome without distorting them. If the specimens are mounted in a plastic, the plastic is likely to degrade rapidly the performance of the microtome knife.

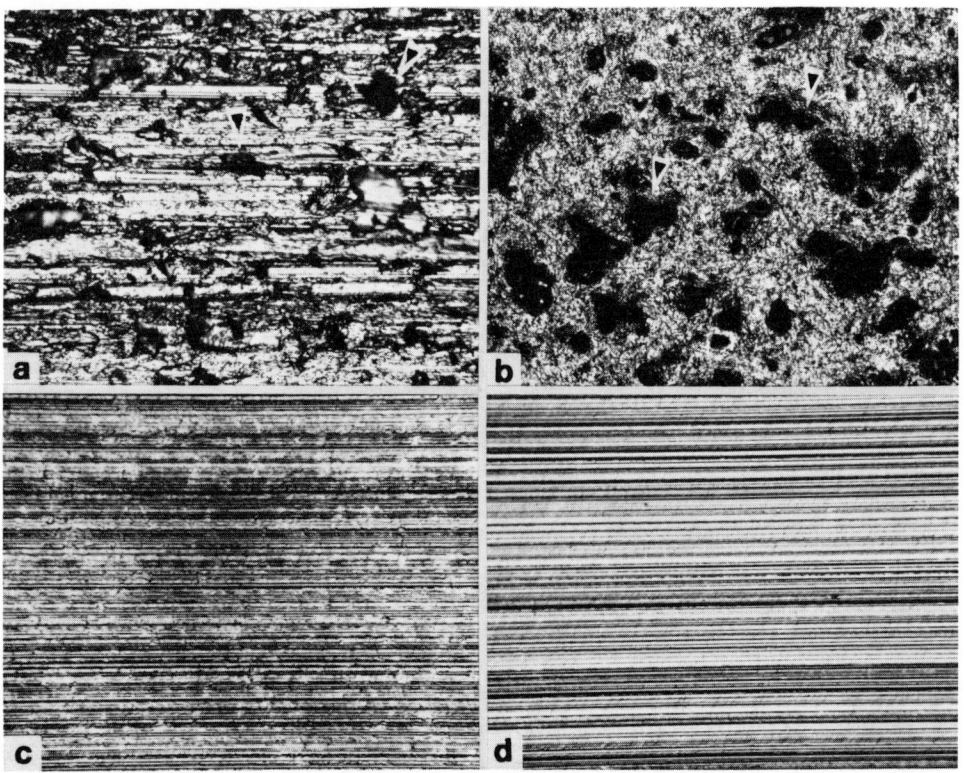

FIG. 9.33. Commercial lead.

(a) Abraded on P1200-grade silicon carbide paper, using water as the abrasion fluid. The dark spots (arrows) are embedded fragments of abrasive. Magnification, 250×. (b) As for (a), but polished briefly. The dark embedded abrasive fragments are now more apparent. Magnification, 100×. (c) Abraded on P1200-grade silicon carbide paper loaded with a soft wax. Magnification, 250×. (d) Cut with a microtome. Magnification, 250×.

Conventional rough-polishing treatments using diamond abrasives are applicable directly to most of the very soft metals, including lead. Unusual problems arise only during final polishing. Good results can be obtained with the advanced finishing techniques discussed earlier in this chapter when the specimen hardness is, say, 10 HV or higher. For example, results obtained by the skidding technique using magnesium oxide abrasive are illustrated in Fig. 9.2 and 9.3 for high-purity aluminum and in Fig. 8.20(c) for zinc.

More serious difficulties arise with conventional polishing methods, however, when the specimen hardness is less than about 5 HV. Metals as soft as this cannot be polished on magnesium oxide at all; a series of roughly spherical indentations merely is produced in the surface. Fine aluminum oxide abrasive can be used in a conventional way, but this produces a rather severely scratched surface (Fig. 9.34a). Vibratory polishing methods cannot be used with conventional abrasives because polishing abrasive becomes embedded in the surface being polished (Fig. 9.13). However, an excellent bright finish can be obtained by vibratory polishing on a synthetic suede cloth flooded with a proprietary solution of colloidal silica* (Fig. 9.34b).[55] This solution probably operates by a chemical-mechanical mechanism. Relief is developed between the individual grains of polycrystalline single-phase material, and the polishing time must be controlled to keep this relief within acceptable limits. The technique is reported also to produce a good polish on many soft metals of this group.[55]

Brief chemical polishing is a possible alternative method of finishing soft metals, and comparatively simple polishing solutions are available for many of them (Table 9.4). Most of the advantages of mechanical polishing can be retained if the duration of the chemical treatment is kept short, which is practicable when a reasonably high-quality finish is first obtained by the mechanical methods that have been described. The general quality of the polish obtained by chemical methods is good, but less satisfactory than that obtained by vibratory polishing using the colloidal silica (cf. Fig. 9.34b and c).

VERY HARD MATERIALS

We saw in Chapter 3 that an abrasive cannot be expected to cut a material that is harder than itself. Indeed, for effective cutting the hardness of the abrasive needs to be several times that of the specimen material. It follows that conventional abrasives are not appropriate for preparing materials whose hardness is much greater than 1000 HV. Diamond abrasives then have to be used exclusively. Very few strictly metallic materials have hardnesses higher than 1000 HV, but compounds of metals, such as carbides and nitrides, do, and metallographers sometimes have to deal with these materials. Perhaps the commonest examples are the sintered tung-

---

*SYTON HT-40, a product and trademark of the Monsanto Company. This solution has been developed specifically for the industrial polishing of semiconductor wafers.

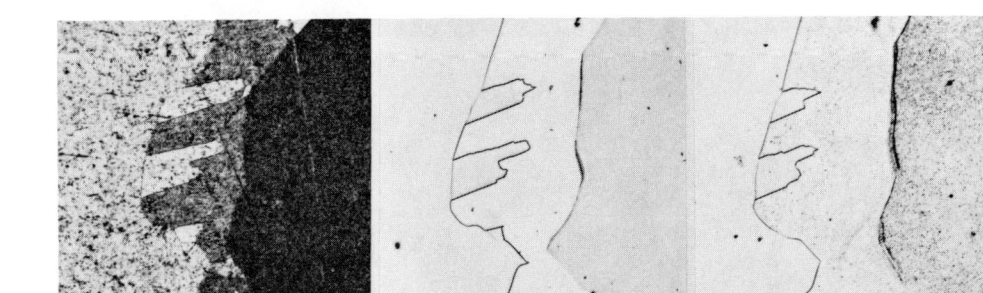

FIG. 9.34. Commercially pure lead.

Prepolished to a 1-μm diamond stage by conventional procedures and then finish polished as follows: **(a)** Finished on a suede cloth charged with 0.1-μm gamma aluminum oxide. **(b)** Finished by vibratory polishing using a colloidal silica solution. **(c)** Finished by chemical polishing. All specimens etched in a 10% ammonium molybdate – 10% citric acid solution. Magnification, 100×.

sten carbides used for tools and dies. Other examples are ceramics and cermets. Most of these materials must be classified as brittle materials in the context used in Chapter 7.

The original section will have been cut by a diamond saw, which usually produces a comparatively good surface finish. It will be assumed that specimens are received for metallographic preparation in this condition. Preliminary abrasion must then be carried out on diamond-impregnated laps, for which purpose suitable laps are available commercially. The characteristics of one of these have been described in Chapter 3 (p. 71). These commercial laps are available in comparatively fine mesh grades, but finer laps can be made in the laboratory (see Appendix 3-C, p. 82), and the use of laps down to about a 6-μm grade is desirable. Metal laps of the type used for lapping diamond gems or industrial stones might also be considered (see below).

Comparatively high abrasion rates are obtained with diamond laps, considering the hardness of the specimens, and this is because fracture-chipping mechanisms are operating at least partly. A comparatively good finish is obtained on the finer laps, although fracture pits may be present in the surfaces of the most brittle materials (Fig. 9.35a). These pits, when

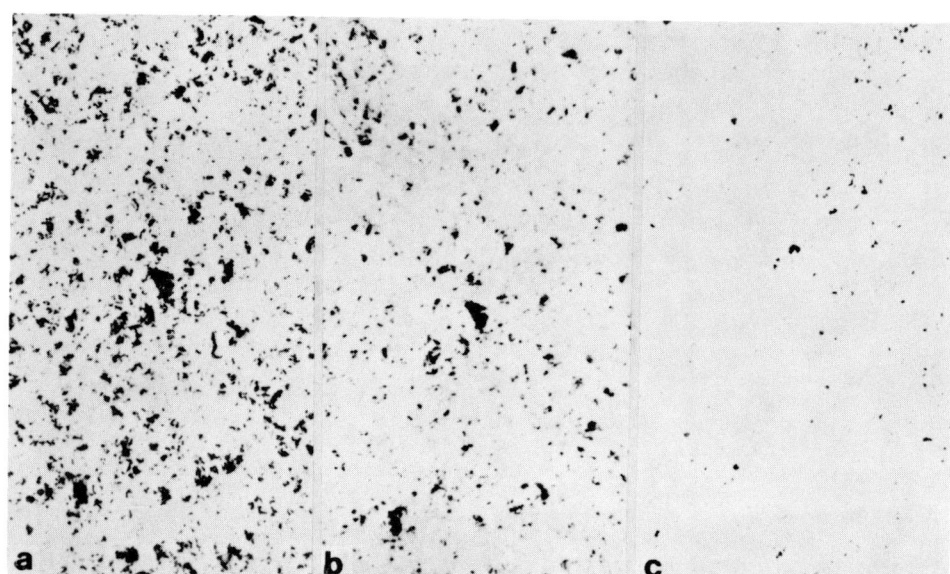

FIG. 9.35. Sintered tungsten carbide containing 15% titanium.

(a) As abraded on a 6-μm diamond-plastic lap. (b) Same area as in (a), but subsequently polished for a short period. (c) Same area, but subsequently polished for a longer period. Magnification (all), 500×. Most of the dark areas in (a) and (b) are artifacts resulting from chipping during abrasion; they might be mistaken for sintering pores. The true distribution of porosity is shown in (c).

produced, constitute the significant abrasion damage that has to be removed during rough polishing. This is particularly so when the material has been produced by sintering, which is a common method of manufacture for this class of material. Sintered materials inevitably contain some porosity, and one of the aims of a metallographic examination may be to assess the nature and extent of this porosity. The abrasion chips could easily be mistaken for sintering pores.

The polishing rates obtained with these materials are low, but not as low as might be expected on the basis of their hardness (Table 5.2, p. 172). Maximum polishing rate is obtained with about a 6-μm grade of diamond abrasive, but there is not much change in polishing rate with either an increase or a decrease in abrasive size.[56] Thus, there is still no point in using abrasives that are coarser than about a 6-μm grade. Hard napless polishing cloths need to be used for polishing, because softer cloths tend to wear rapidly. The differences in polishing rate among the different types of cloth are less than those for polishing softer materials.

A polishing treatment of a few minutes' duration on a drill cloth charged with 6-μm diamond, used as discussed in Chapter 5, will suffice to remove the scratches produced on a fine abrasion lap. However, much longer polishing times will be required when the abrasion has caused fracture chipping. In this event, the rough-polishing time needed to produce an artifact-free surface has to be established for each type of specimen material. This can be done in a trial of the type illustrated in Fig. 9.35. A particular area on the section surface is identified and re-examined after successive polishing periods. Artifacts gradually disappear (cf. Fig. 9.35a, b and c), but new features which appear can be taken to be genuine. The genuine structure for the material illustrated in Fig. 9.35 is that in Fig. 9.35(c).

The polish produced on a napless cloth will be adequate for many purposes, although some scratches may be visible after etching. Many of the etchants used for carbides and nitrides, for example, darken some of the phases present, and this makes any scratches more readily visible. If the degree of scratching is unacceptable, a finish polish on a synthetic suede cloth charged with 1-μm diamond may be tried. Further improvement will be achieved, if need be, by using finer grades of diamond (0.5 μm and 0.1 μm), provided they have been well graded by the manufacturer.

An acceptable general procedure for this class of material is as follows:

Stage I: Abrade on 300-mesh industrial diamond lap. Preferably follow by abrasion on finer diamond-plastic laps, down to about a 6-μm grade.

Stage II: Rough polish on a hard napless cloth (impregnated paper, nylon, or drill) charged with 6-μm diamond abrasive and used in the conventional way.

Stage III: Finish polish on a synthetic suede cloth charged with 1-μm diamond and used in the conventional way. If necessary, polish on finer grades of diamond abrasive.

Examples are given in Fig. 9.36 of the standard of results that can be expected when sintered tungsten carbides are prepared by these procedures.

The procedure described has its limits, however. For example, it cannot be expected to handle materials as hard as diamond or boron nitride. Specialized procedures have been developed for these extremely hard materials, procedures based on lapidary rather than metallographic techniques.[57] A cast iron lap, the surface of which has been suitably conditioned, is charged by spreading abrasive over its working surface; the abrasive is then forced into the lap surface by means of a hardened-steel roller. The lap is rotated at a comparatively high speed and the specimen held against it in a loaded holder. The holder is made to traverse radially across the surface of the lap. Very long polishing times, a day or more at each of a range of successively finer abrasives, are required.

BRITTLE MATERIALS

Some metallic or semimetallic materials are brittle but not particularly hard (hardness, <1000 HV). This group includes the semiconductors and compounds such as oxides, sulfides, silicates and glasses. It might also be taken to include many minerals and ores. However, preparation of

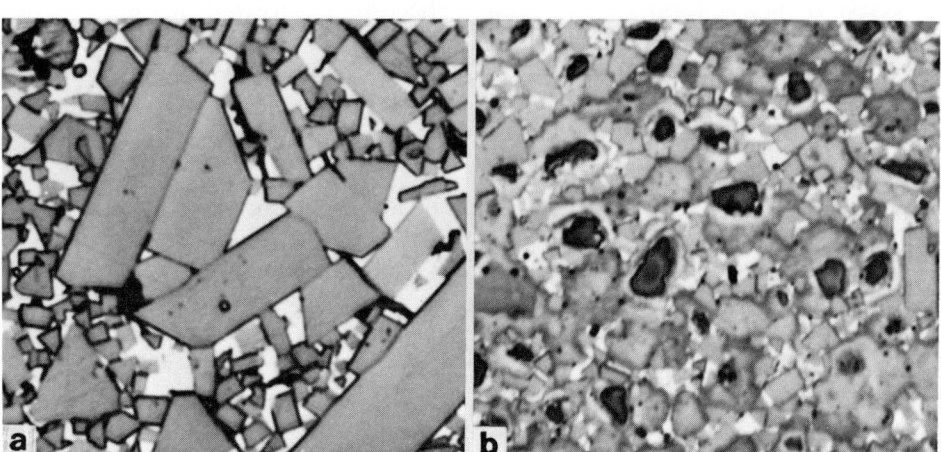

FIG. 9.36. Sintered tungsten carbides finish polished on 1-μm diamond.

(a) A standard 12% cobalt grade, but grossly oversintered, producing very large carbide particles. Only a few polishing scratches are visible, mostly in the large carbide particles. (b) A 15% titanium grade. No polishing scratches are visible. Both specimens etched in a boiling potassium hydroxide – potassium ferricyanide solution. Magnification, 2000×.

specimens for mineragraphic examination is a separate subject, and the present discussion should be taken only as a guide to how metallographic procedures might be applied.

The mechanisms involved in the abrasion and polishing of these types of materials were outlined in Chapter 7, where we saw that the mechanism of abrasion for brittle materials is likely to be different from that for ductile metals but that the mechanism of polishing is likely to be the same. The main difference is that abrasion can occur by a fracture-chipping mechanism instead of a ductile-cutting mechanism. Rolling abrasive particles are as effective at removing material as fixed particles in the chipping mode; indeed, they usually are more effective. Consequently, it is reasonable to base the choice between rolling and fixed abrasives in this instance on other criteria, such as the depth of the surface damage produced. The fracture-chipping mode of material removal tends to be replaced by a chip-cutting mode with increasing fineness of the abrasion process. The transition may commence with, and may even be completed by, the finest abrasion stages. It will invariably be complete by the time the diamond polishing stages have been reached. All the principles described earlier for the polishing of metals apply when the transition has become well advanced.

Standard metallographic procedures (Table 8.3, p. 242) consequently can be applied to many of these materials with a few modifications. The first problem arises in rough polishing, because the surface has to be reduced to below the base of the abrasion-chip cavities. This tends to be more difficult than removal of ductile cutting grooves, because the pits tend to be deeper. The problem arises particularly in materials which have planes of easy cleavage — materials in which the chip cavities are particularly deep. When this problem becomes severe, it is desirable to explore alternative abrasion procedures. For example, abrasion with rolling abrasives (e.g., a slurry of loose abrasive on a glass plate) may in some materials produce shallower cavities than abrasion on coated papers. The continuation of abrasion to the finest possible stage, such as to the abrasive-wax lap described in Appendix 3-D, also becomes desirable. The chip cavities present at the end of abrasion will then be shallower, and there is a possibility that a chipping mechanism of material removal will have been replaced completely by a ductile-cutting mechanism.

The next problem is to remove the crack-containing damaged layer which lies beneath the fracture chips. If this is not done, a series of artifact cracks will be left in the surface and will become particularly obvious after etching (see Fig. 8.11a, p. 229). The removal of this crack-containing layer is, however, no more than a standard problem of removal of abrasion damage.

The scratches produced during polishing will, even in these materials, have plastically deformed zones associated with them similar to those present in ductile metals. Consequently, some etching treatments, particularly those which develop dislocation etch pits, will enlarge the scratches (see Fig. 8.11b, p. 229) just as they do for metals. Polishing may be continued to the finest-available diamond grades, but this still may not be completely successful in producing a satisfactorily scratch-free finish. A finish polishing treatment using vibratory methods on the colloidal silica solution (SYTON HT-40) is successful for silicon[58] and may be successful for other semiconductors. A final chemical polish is also a possibility for several of these materials (see Table 9.3).

There is a need in some applications to produce optically flat surfaces on materials of this group, but metallographic preparation methods do not do this nearly well enough. Quite different procedures have had to be developed for this purpose[59] — procedures which are based on classical optical-working techniques. On the other hand, these optical techniques do not necessarily produce surfaces that are ideal for microscopical examination, and certainly do not necessarily do so as simply as the metallographic procedures discussed above. The choice of the more appropriate of these two types of procedures depends on the requirement of the final surface that is of dominating importance.

Mineragraphic specimens that are to be examined in reflected light can also be prepared by these basic procedures, the following modifications then being desirable.[60]

1. Rolling abrasives, rather than coated abrasive papers, are desirable for abrasion when minerals are present which cleave easily (e.g., galena).
2. A fine fixed-abrasive lap is desirable as the final abrasive stage. This ensures that surface chipping is reduced to a minimum, and perhaps is even eliminated.

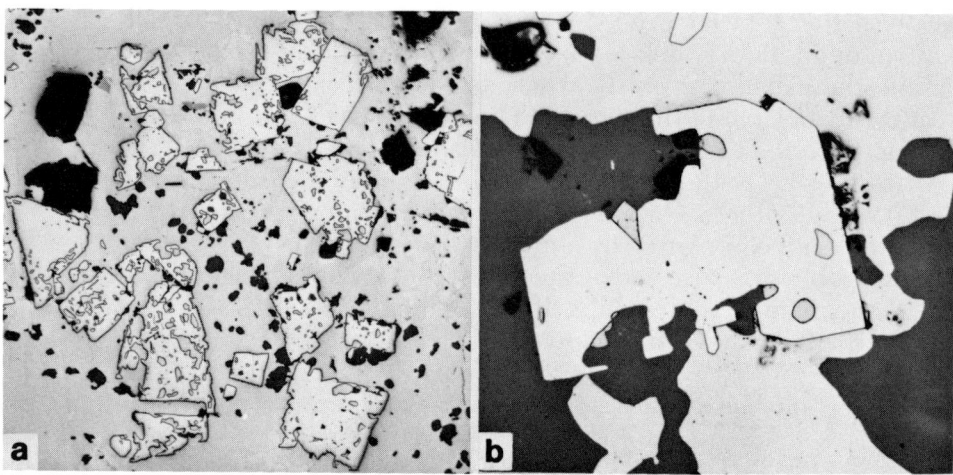

FIG. 9.37. Mineragraphic specimen prepared by a modified metallographic procedure.

Pyrite (off-white) with extensive replacement by galena (white). Sphalerite (gray) is present also. Magnifications: **(a)** 50×; **(b)** 250×. Galena is a difficult mineral to prepare because it has a strong tendency to develop cleavage chips during abrasion.

3. A napless cloth (e.g., an impregnated paper or a cotton drill) is essential for rough polishing of any specimen that contains a number of different minerals. The polishing rates of minerals differ considerably, and excessive relief develops between them if napped cloths are used. Moreover, rough polishing needs to be continued until the abrasion-chip cavities have been removed adequately — that is, until any cavities that were produced by abrasion have been removed.
4. Final polishing is best carried out with diamond abrasives also. The finest-available grade should be used on a napped, but preferably a short-napped, cloth (e.g., a synthetic suede). Final polishing must be discontinued when excessive relief begins to develop between the different minerals. Many minerals are comparatively hard, and specimens normally are examined unetched. It is comparatively easy, therefore, to obtain a finish that appears to be acceptably free from scratches when examined in vertical bright-field illumination.

The micrographs in Fig. 9.37 are examples of the standard of results that can be expected when mineragraphic specimens are prepared by these methods.

SURFACE OXIDES AND SCALES

The examination of nonmetallic surface layers (*scales*), such as those resulting from oxidation or corrosion, presents several special problems in specimen preparation. First, it may be necessary to retain well both the metal/scale interface and the outer edge of the scale; both of these regions may be of critical interest. Secondly, the scale may be friable and thus susceptible to chipping and cracking during preparation. The detection of porosity or cracking in the scale layer is frequently a significant feature of the examination. It is therefore necessary to avoid reliably the development of preparation artifacts which might be mistaken for pores or cracks. Thirdly, the polishing characteristics of the metal and the oxide layer are usually so different that it may be difficult to attain an adequately scratch-free finish in both while simultaneously satisfying the other criteria just discussed.

The retention of the outer edge of the scale is a special case of the problem of edge retention, the general problem having already been discussed in detail earlier in this chapter. Some special method of edge protection usually needs to be used in the case of surface scales, and the ideal one is that of electrodepositing a protective layer over the scale. To do this satisfactorily it may be necessary first to silver the surface (see p. 301). The presence of an electrodeposited layer has the further advantage that it enables the detailed topography of the oxide surface to be distinguished in strong contrast (e.g., see Fig. 9.24 and 9.38a; cf. Fig.

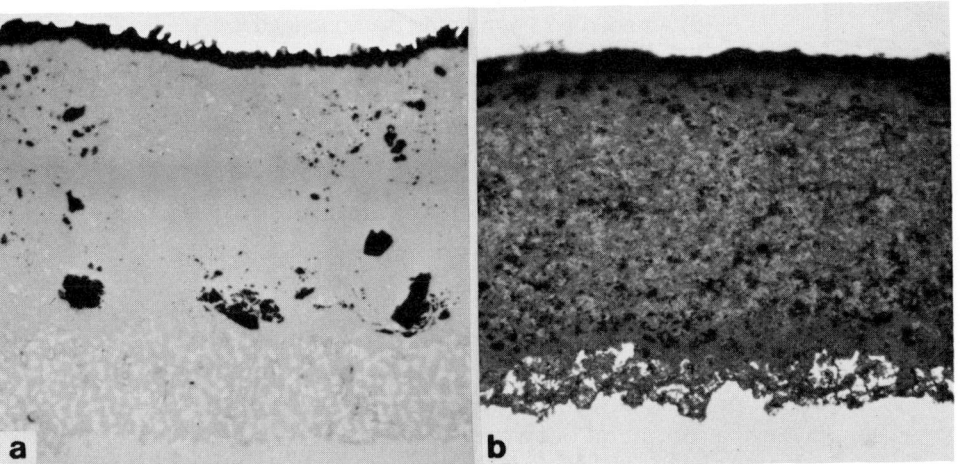

FIG. 9.38. Surface layers of nickel alloy blades from a marine gas turbine engine.

Protective aluminide layers were formed on the surfaces of the blades during manufacture. The outer surfaces (top) have been protected by electrodeposits of nickel. The sections were then prepared by standard methods.

(a) The aluminide layer has continued to be protective. The dark outermost layer is a protective layer of aluminum oxide formed in service. Most of the remainder of the field is occupied by the original aluminide layer; the dark patches in this layer are particles of aluminum oxide formed during aluminizing. **(b)** The aluminide layer has been consumed completely in service. It has been replaced by a layer of corrosion products. Magnification (both), 500×.

324 / Metallographic Polishing by Mechanical Methods

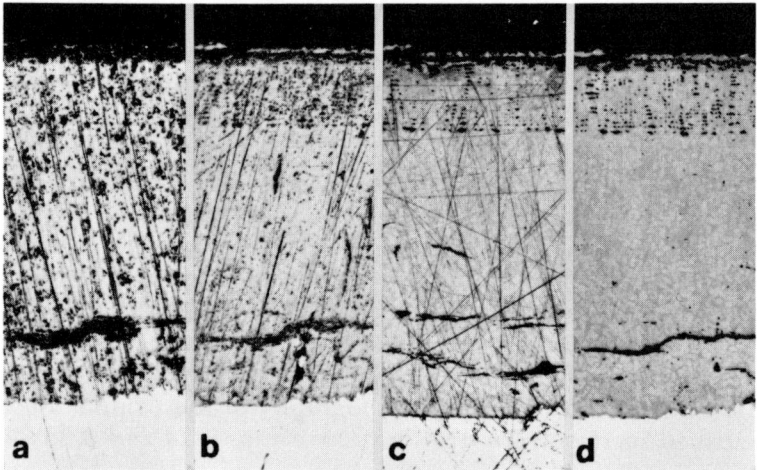

FIG. 9.39. Section of the oxide layer formed on a specimen of high-purity iron heated at 950°C for 20 min and slowly cooled. Abraded on 220-grade silicon carbide paper, and then abraded as noted below.

(a) Abraded on 400-grade silicon carbide paper. Numerous chipping artifacts are present in the oxide. (b) Abraded on 15-μm-grade aluminum oxide – wax lap. Minor chipping artifacts only. (c) Abraded on 1-μm-grade diamond-lead lap. Oxide is free of artifacts, but both oxide and metal are badly scratched.

(d) Abraded on 1-μm-grade diamond-lead lap, then polished on cotton drill cloth charged with 1-μm-grade diamond. Oxide is free of artifacts, and finish is adequately free of scratches. The same area is shown in all cases. Magnification, 100×.

9.39). A difficulty is, however, that internal stresses in the deposit may cause cracking to develop in the scale layer, particularly at or adjacent to the scale/metal interface. The layer may even become detached from the base metal.

Considerable experimentation may be needed to avoid this difficulty. It may be necessary, for example, to establish deposition conditions which produce low-stress deposits (see p. 302), or to deposit a buffering layer of a soft metal between the oxide and the main electrodeposit. The solution to the problem will be specific to the type of specimen involved, and its development usually is worthwhile only when a large number of specimens is involved. A simpler, and often acceptable, method of edge protection is to fill the mounting plastic with an abrasion-resistant material or to use a commercial plastic that is so filled (see p. 298). This was the precaution adopted in the preparation of the specimen illustrated in Fig. 9.39.

The magnitude of the problem involved in reducing to an acceptable figure the difference in level at the interface between the scale and the base metal depends on the relative abrasion and polishing rates of the two regions. The modifications of standard procedures that should be adopted when the problem becomes significant are as follows. First, abrasion should be continued to a fine abrasive lap, such as the aluminum oxide – wax lap described in Appendix 3-D (p. 83). Secondly, napless cloths should be used for polishing. The cloth chosen should, however, be no harder than is really necessary. Otherwise, unnecessarily severe scratching will be produced in the base metal. Thirdly, final polishing on napped cloths should be used only if absolutely required to produce an acceptably scratch-free surface. If it is used, the polishing treatment should be of minimum duration. It will usually be easy enough to obtain a satisfactory finish on the scale, but often it will not be possible to obtain a high-quality finish on the base metal while at the same time retaining the interface and polishing the scale well. The metallographer has to choose which is the more important in the particular examination.

Many scales are not particularly friable. Such scales can be satisfactorily polished by conventional techniques, selected to achieve good edge and interface retention as just discussed. The micrographs in Fig. 9.38 are examples of the standard of results that can be expected.

However, some scales are friable enough to behave as brittle materials in the context discussed in Chapter 7. The oxide layers that form on iron are examples. Extensive chipping occurs in scale layers of this nature when they are abraded on coated abrasive papers (Fig. 9.39a), and this damage is difficult to repair during polishing. Fine abrasive laps cause more cutting and less chipping than coated papers and consequently produce much more acceptable results. The use of such a lap is desirable in any event to improve edge and interface retention. The aluminum oxide – wax lap described in Appendix 3-D produces a reasonable, but not fully satisfactory, result (Fig. 9.39b). A diamond-lead lap of the type recommended by Jay[61] produces a more artifact-free result (Fig. 9.39c); the finish is a little more severely scratched than for the abrasive-wax lap, but this is acceptable. Although Jay recommended lead sheet for the lap, annealed commercially pure aluminum is more satisfactory. A portion of the standard diamond-carrying paste is spread over the intended working surface of the aluminum, and a blank specimen of hardened steel is rubbed over the track to embed the abrasive particles into the track surface. A few drops of kerosine are added, and the specimen is abraded by hand in the usual way. Mechanized versions of this technique probably could be developed.

Rough polishing is usually best carried out on a napless cotton drill cloth charged with 1-μm diamond abrasive and operated in the usual way. The result obtained is free from artifacts, the retention of both the outer surface and the interface is good, and the finish is adequately scratch-free when the specimen is observed in the unetched condition (Fig. 9.39d). A final polish on a napped cloth charged with fine alumina may be desirable, however, if the specimen is to be etched. The final polish must be kept brief to ensure that excessive edge rounding and relief do not develop.

The important thing is to obtain, if possible, an artifact-free finish at the lap stage. If there is any doubt on this point, the polishing treatment on the napless cloth should be continued and the same field viewed at intervals. New features which appear can be taken to be real with a high degree of certainty. There are, for example, distinct differences between Fig. 9.39(c) and (d) in this respect.

Summarizing, an acceptable procedure for scale layers which are adherent and not friable is as follows:

Stage I: Abrade on coated papers to P600 grade.
Stage II: Abrade on 15-μm aluminum oxide – wax lap.
Stage III: Rough polish on cotton drill cloth charged with 1-μm diamond abrasive.
Stage IV: Final polish on short-napped cloth charged with 0.1-μm gamma alumina. This stage must be kept brief.

However, the abrasive-wax lap of stage II should be replaced with a soft-aluminum lap charged with 1-μm diamond abrasive if problems arise because the scale is friable or poorly adherent. Treatment on this lap should then be made the key stage of the procedure and should be continued until the scale is free from chiplike artifacts and until a satisfactory standard of edge and interface retention has been attained.

PRECIOUS METALS AND REFRACTORY METALS

The term *precious metal* is applied to silver, gold, platinum, and the platinum group of elements (palladium, rhodium, iridium, ruthenium and osmium). A basic characteristic of these metals is that the surface oxide film which forms at ambient temperature is extremely thin, but is protective. These metals are rare and have aesthetically pleasing colors. The term *refractory metal* is applied to metals which have high melting points. Examples are chromium, molybdenum and tungsten. With the exceptions of silver and gold, the precious metals also can be classified as refractory metals.

The important characteristic of these two groups of metals from the present point of view is that they, and their alloys, have low abrasion rates and particularly low polishing rates (Table 9.8). Silver is the only exception. It has polishing and abrasion rates comparable to those of the common metals (Tables 3.3 and 5.2) and exhibits group 1 abrasion characteristics (see p. 56). Pure silver and silver alloys can be prepared satisfactorily by standard procedures. Methods of achieving a satisfactory final polish will be discussed later.

Gold and chromium exhibit group 2 abrasion characteristics, their stable abrasion rates being very low for their hardnesses. Platinum and tungsten, the only remaining metals of the group that have been investigated so far, have group 3 abrasion characteristics (Table 9.8). Only very limited amounts of material can be removed from platinum or tungsten specimens by abrasion on one track of a coated abrasive paper. These are circumstances where the use of diamond laps might well be considered.

TABLE 9.8. Characteristics of Some Precious and Refractory Metals

Metal	Crystal structure	Melting point, °C	Hardness, HV	Abrasion(a)		Polishing rate(b), μm/100 m
				Rate, μm/m	Maximum thickness removable, μm	
Silver	fcc	961	25	...	...	3.0
Gold	fcc	1063	25	0.26	...	0.10
Platinum	fcc	1769	40	...	2.5	...
Chromium	bcc	1875	200	0.25	...	0.42
Tungsten	bcc	3410	225	...	20	0.15

(a) Abrasion on P240-grade silicon carbide paper, determined under the conditions outlined in Chapter 3. An abrasion rate is quoted for metals that have group 1 or group 2 abrasion characteristics. The maximum thickness that can be removed on a track of abrasive paper is quoted for metals with group 3 abrasion characteristics. (b) Polishing on a suede cloth charged with 6-μm diamond abrasive. Used under optimum conditions, as outlined in Chapter 5.

The polishing rates of most precious and refractory metals are an order of magnitude smaller than those of common metals (Tables 5.2 and 9.8), but the reason for these low polishing rates is not known. Because the depth of significant damage will be comparable to that for common metals of comparable hardnesses, it follows that the abrasion and polishing times used must be much longer than for common metals under otherwise comparable conditions. This is the basic difference between the preparation procedures used for precious and refractory metals and those used for common metals. Automated abrasion and polishing machines and high specimen pressures are desirable, but the limited effective lives of silicon carbide and aluminum oxide coated papers need to be kept in mind for metals with group 3 abrasion characteristics.

Most precious and refractory metals have cubic crystal structures and are not particularly susceptible to abrasion artifacts other than those originating in the fragmented layer (p. 216). The type of artifact to be expected is similar to that illustrated in Fig. 8.3 for iron, an example for high-purity chromium being given in Fig. 9.40(a) and (b). Two metals of the platinum group (ruthenium and osmium), however, have hexagonal close-packed crystal structures. These metals are susceptible to twin-type abrasion artifacts similar to those illustrated for zinc in Fig. 8.8 (p. 226). An example for ruthenium is given in Fig. 9.40(c) and (d). Much longer rough-polishing times are required to remove abrasion artifacts of this type than to remove those likely to be found in cubic metals, and this presents considerable practical difficulties. Ruthenium is also very susceptible to twin artifacts produced by accidental deformation of the specimen, including that which can occur during specimen mounting.[62]

The problems that arise in polishing of precious and refractory metals as a result of their low polishing rates are alleviated to some extent at final polishing. Apparently scratch-free finishes can be produced by comparatively coarse abrasives. Moreover, these materials generally are not very sensitive to polishing artifacts. Finish polishing can first be attempted on

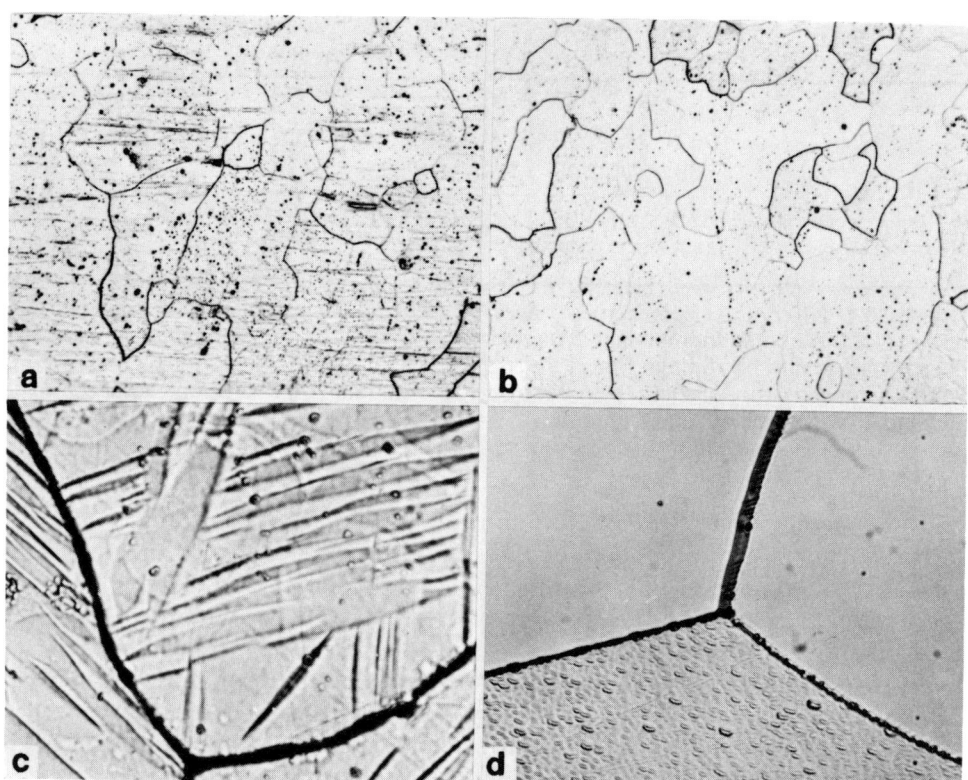

FIG. 9.40. Abrasion artifacts in refractory metals.

(a) High-purity chromium rough polished for too short a time. The longitudinal markings are abrasion artifacts from the final abrasion stage. Etched electrolytically in oxalic acid. Magnification, 250×. (b) As for (a), but rough polished for a longer time. The surface is free of artifacts. Etched electrolytically in oxalic acid. Magnification, 250×.

(c) Ruthenium finish polished by conventional mechanical methods, but for a time insufficient to remove massive twin abrasion artifacts. Magnification, 500×. *(Piotrowski and Accinno, Ref 62.)* (d) As for (c), but finish polished by an electromechanical method. The abrasion artifacts have been removed in an acceptable polishing time. Magnification, 500×. *(Piotrowski and Accinno, Ref 62.)*

suede cloths charged with a 1-μm diamond abrasive; if necessary, finer grades of diamond abrasive may be tried. For many of these metals (e.g., platinum, palladium and tungsten), a satisfactory result is obtained with 1-μm. abrasive, even though the material is quite soft (Fig. 9.41). For others, even though harder (e.g., chromium), a further treatment on Selvyt cloth charged with a fine aluminum oxide is necessary (Fig. 9.40b). The use of an automatic polishing machine is desirable.

The use of etch-attack or electromechanical polishing methods should be considered for this group of metals, suitable electromechanical techniques having been developed for a number of them (Table 9.2). Electromechanical methods in particular achieve much higher polishing rates than straight mechanical methods, and so are more likely to remove abrasion artifacts in a reasonable period of time (Fig. 9.40d). They also produce a scratch-free finish (Fig. 9.5b). Electromechanical methods are comparatively simple to use if the appropriate equipment is available, and usually produce better results than etch-attack techniques (Fig. 9.5). However, they are not always applicable. For example, they could not have been used to prepare satisfactorily the liquid-phase sintered tungsten illustrated in Fig. 9.41(c) and (d). The nickel-iron binder would not have been satisfactorily polished, and the oxide particles in the binder would not have been detected (Fig. 9.41c). The detection of these particles was an important feature of the examination. The fact that the frac-

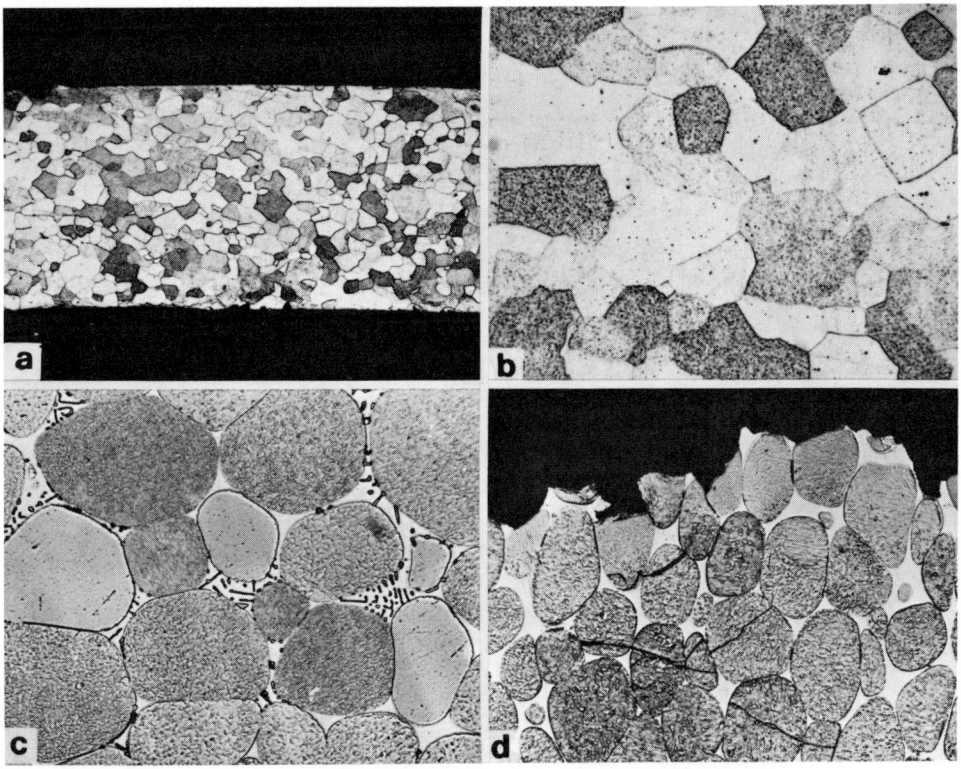

FIG. 9.41. Refractory metals finish polished on suede cloths charged with 1-μm diamond.

(a) and (b) Pure platinum wire. Etched electrolytically in a solution of sodium chloride, hydrochloric acid and water. Magnifications: **(a)** 100×; **(b)** 500×.

(c) Tungsten alloy compacted from powder by liquid-phase vacuum sintering. The binder is a nickel-iron alloy and contains eutectic oxide particles. These particles indicate that the vacuum used for sintering was inadequate. Consequently, the compact fractured in tension at the binder/tungsten interfaces. Magnification, 250×.

(d) Edge of a fracture in an alloy similar to that in **(c)** but satisfactorily sintered. The fracture runs through the tungsten particles. It can be seen that cleavage cracks have initiated in the particles and propagated from one to another at their points of contact. Magnification, 250×.

ture cracks visible in Fig. 9.41(d) had propagated at the contacting points of the tungsten particles would not have been apparent either. Note also that it was possible to retain well the fracture edge shown in Fig. 9.41(d) by the fully mechanical procedure.

Thus it is necessary to be able to prepare specimens by fully mechanical methods when the need arises, even if reliance is placed on electromechanical or etch-attack techniques for more routine applications.

The two metals of the group under consideration with which the most difficulty is experienced in obtaining a satisfactory final polish are silver and gold. Skidding techniques with magnesium oxide can be used with marginal success for silver, but not for gold. An acceptable but not particularly good finish can be obtained on silver by polishing with a fine aluminum oxide. Piotrowski and Accinno[62] recommend the use of 1-μm alpha alumina for finish polishing of gold. Etch-attack techniques are also available for both silver and gold (Table 9.1, p. 274).

POLARIZATION CONTRAST

Noncubic metals may need to be examined in polarized light while they are in the unetched condition. The aim usually is to distinguish between regions with differing crystal orientations, and to do this with maximum discrimination. Occasionally, the aim may also be to determine quantitatively the degree of polarization rotation introduced by the crystals exposed at the polished surface. There is, consequently, a need to obtain maximum polarization contrast and to know the cause of this contrast.

The polarization contrast observed can be due to any combination of the following three effects:

1. The intrinsic anisotropy of the crystal exposed at the section surface.
2. The presence of regular facets on the surface. The facets may be developed either inadvertently during specimen preparation or deliberately by etching. Inclined reflecting surfaces introduce elliptical polarization into the reflected beam.
3. The presence of an epitaxial film of oxide or any other reaction product. Polarization contrast will then be introduced if the oxide is optically active or has a faceted surface.

The effects of these factors are illustrated in the series of photomicrographs in Fig. 9.42. The specimen is a section of a plastically deformed single crystal of zinc and the area illustrated contains a massive twin which has induced a triangular kink band; the misorientation between the kink band and the matrix crystal is only 0.5°. The massive twin was clearly visible under all circumstances but the kink band was only just discernible when the section had been polished by good mechanical procedures (Fig. 9.42a). The kink band became visible in better contrast, however, when the mechanically polished surface was subsequently etched very lightly (Fig. 9.42c) or allowed to oxidize in air at ambient temperature (Fig. 9.42d).

On the one hand, this experiment illustrates procedures that can be tried in an attempt to increase the polarizing contrast. On the other hand,

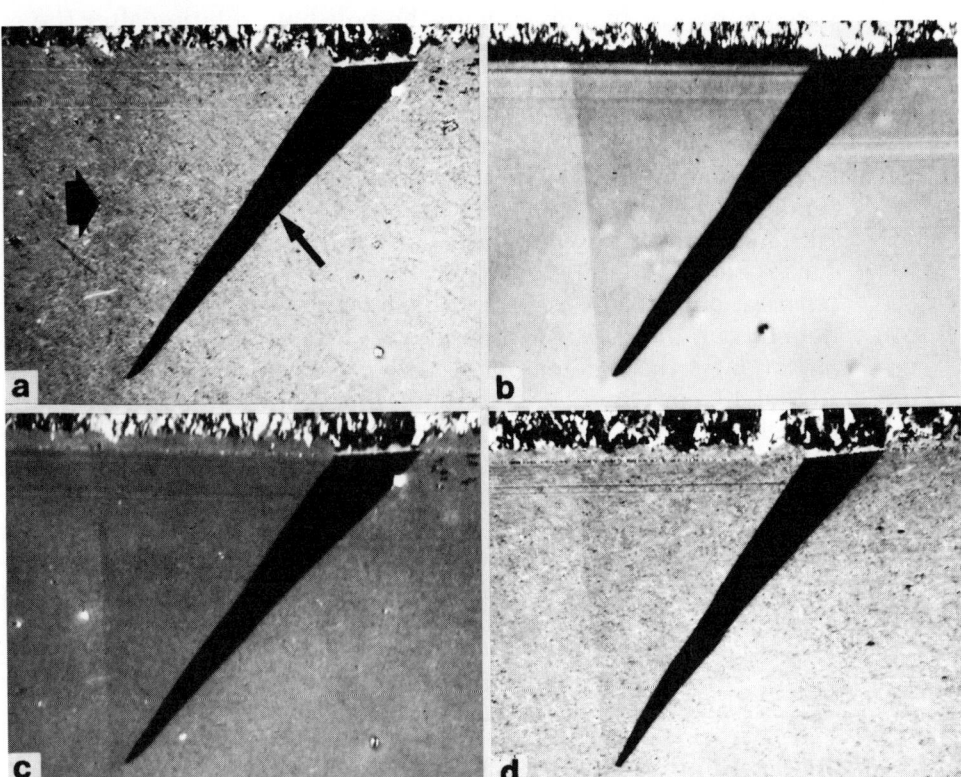

FIG. 9.42. Compressed single crystal of zinc.

(a) Mechanically polished. Finish polished by skidding on magnesium oxide. (b) Electrolytically polished. (c) As for (a), but immersed for 1 s in the electropolishing solution without applying an emf. (d) As for (a), but oxidized in air for 1 day. The area shown in these four micrographs contains a massive twin, indicated by the small arrow in (a), which has induced a triangular kink band, indicated by the large arrow in (a). The misorientation between the kink band and the matrix is only 0.5°. Viewed in polarized light. Magnification, 250×.

it also illustrates how difficult it is to be sure of the extent to which the contrast observed is directly attributable to the basic characteristic of the crystals that have been sectioned. If we accept that there is no Beilby-like layer on the surface illustrated, then we must conclude that the contrast due to basic crystal properties is probably close to that illustrated in Fig. 9.42(a). But even this contrast may be slightly exaggerated due to very minor etch faceting which could have developed during final polishing. The degree to which the presence of faceting has increased the polarization contrast can be established by depositing a thin layer of silver on the surface. Any polarization contrast then observed is attributable to faceting.

These problems are no less severe after polishing by chemical or electrochemical methods. Both are likely to produce a surface with minor but variable etch faceting. The degree of this faceting will depend on the exact polishing conditions and, in some instances, on the time that the specimen is left in contact with the polishing solution after the emf has been disconnected from the polishing cell. The result obtained by electropolishing the specimen illustrated in Fig. 9.42 is shown in Fig. 9.42(b). The polarization contrast is good, slightly better than for the as-mechanically-polished condition, but some of this contrast is due to etch faceting.

The above remarks apply to surfaces which are free from polishing scratches within the limits of resolution. Scratch grooves are revealed with more sensitivity in polarized light than in bright-field illumination. This is because the inclined surfaces of the scratches introduce elliptical polarization into the reflected beam. The degree to which this occurs depends on the relative orientations of the scratch and the plane of polarization of the incident light. For example, the surface illustrated in Fig. 9.43 contained a system of fine unidirectional scratches. The scratches are not visible directly when aligned either parallel to (Fig. 9.43a) or perpendicular to (Fig. 9.43c) the plane of polarization. They are, however, so apparent in the 45° position (Fig. 9.43b) that the polarization contrast due to the underlying crystal structure has been completely obscured. The relative prominence of the scratch and of the crystal polarization contrast varies progressively between these two limits. Lines of twinned material are also detected at the more severe scratches even when the scratches themselves are not interfering with the polar-

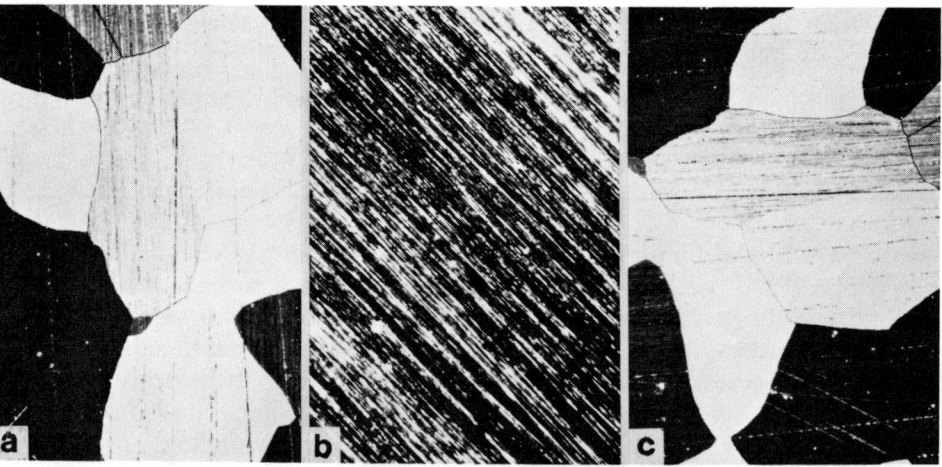

FIG. 9.43. Polycrystalline zinc finish polished on 0.1-μm aluminum oxide to produce a system of unidirectional scratches.

(a) Scratches aligned parallel to plane of polarization. (b) Scratches aligned at 45° to plane of polarization. (c) Scratches aligned perpendicular to plane of polarization. Viewed in polarized light, with the plane of polarization vertical in all cases. Magnification, 250×.

ization effects (Fig. 9.43a and c). The reasons for this were discussed earlier (p. 238).

It is more common for scratches in polished surfaces, where present, to be randomly arrayed. In this event, a proportion of the scratches will be so oriented as to cause significant degradation of true polarization contrast at all rotations. Another proportion will always be appropriately oriented for the line of artifact twins to be visible. Consequently, the use of advanced final-polishing techniques which produce essentially scratch-free surfaces is necessary when mechanically polished specimens are to be examined in polarized light.

APPENDIX 9-A
A Water – Propylene Glycol Polishing Fluid

Constituents

Water (distilled)	100 ml
Propylene glycol	250 ml

Method of Manufacture

Saturate the water with methaldehyde (as a preservative), filtering off any undissolved residue. Add the propylene glycol and adjust the pH of the mixture to about 7.2 with additions of 1N sodium hydroxide. Finally, adjust the pH to 7.4 using N/5 sodium borate (pH 9.2 buffer solution). Make up to 500 ml with distilled water.

APPENDIX 9-B
Electroplating Methods for Edge Protection

1. COPPER : ACID BATH

Solution:	$CuSO_4 \cdot 5H_2O$	170 g/l
	H_2SO_4 (conc.)	60 g/l
Temperature:	15-50°C	
Voltage:	1-4 V	
Current density:	10-20 mA/cm^2	
Agitation:	Mild stirring preferable	
Anode:	Copper	
Anode bags:	Desirable but not essential	
Uses:	General use for copper alloys	

2. COPPER : CYANIDE BATH

Solution:	CuCN	20 g/l
	NaCN	30 g/l
	NaOH	1.5-3 g/l
Temperature:	45-60°C	
Voltage:	4-6V	
Current density:	0.5-1 mA/cm^2	
Agitation:	Mild stirring	

Anode:	Copper
Anode bags:	Desirable but not essential
Uses:	As a preliminary to an acid deposit to provide better contrast at the section line, particularly in the case of taper sections.

3. IRON (Jenkinson[63])

Solution:	$FeCl_2 \cdot 4H_2O$ 288 g
	NaCl 57 g
	Water (distilled) 1 l
	Filtered for use.
Temperature:	70-100°C
	(This necessitates a constant-level device to make up for evaporation losses with distilled water.)
Current density:	0.5-4 A/dm^2
Agitation:	Specimen (cathode) suspended from a spindle and rotated at 50 rpm
Anode:	Ingot iron plate.
Uses:	Excellent for all ferrous specimens, but usefulness restricted by the difficulties in operating and maintaining the bath.

4. NICKEL

Solution:	$NiSO_4 \cdot 7H_2O$ 300 g/l
	$NiCl_2 \cdot 6H_2O$ 60 g/l
	Boric acid 40 g/l
	pH 4
	(Note: Add 1 part of 30% H_2O_2 per 200 parts of solution once a day.)
Temperature:	40-70°C
Voltage:	1-3 V
Current density:	30 mA/cm^2
Agitation:	Vigorous stirring
Anode:	Anode nickel
Anode bags:	Essential
Uses:	Ferrous alloys; nickel alloys; convenient for most metals of moderately high melting point.

NICKEL : Electroless (quoted by Miley and Calabra[64])

Solution:	$NiCl_2 \cdot 6H_2O$ 30 g/l
	$NaHPO_2 \cdot H_2O$ 10 g/l
	$Na_3C_6H_2O_7 \cdot H_2O$ 100 g/l
	NH_4Cl 50 g/l
	NH_4OH to adjust pH to 8-10
	Dissolve in order in warm distilled water.
Temperature:	Bring solution to low boil
Operation:	Immerse specimen for 1-2 h

Uses:	Nickel, iron, and other metals of high melting point. Nonconducting surfaces.	

5. ZINC

Solution:	$Zn(CN)_2$	60 g/l
	NaCN	23 g/l
	NaOH	53 g/l
Temperature:	Room temperature	
Voltage:	1-4 V	
Current density:	10-15 mA/dm^2	
Anode:	Zinc	
Anode bags	Not necessary	
Uses:	Zinc alloys	

APPENDIX 9-C
Brashear Process for Silvering Prior to Electroplating

STOCK SOLUTIONS

A. Silver nitrate: 20 g
   Water to 300 ml
B. Potassium hydroxide: 14 g
   Water to 100 ml
   (Note: This solution must either be freshly prepared or be stored in polythene bottles.)
C. Ammonium hydroxide (s.g. 0.880)
D. Dextrose: 6.5 g
   Water to 100 ml
   (Note: This solution should preferably be freshly prepared but will keep for a few days. It should be discarded when cloudy.)

Pure chemicals should be used, particular attention being paid to the potassium hydroxide, many grades of which contain excessive amounts of chloride. The dextrose should preferably be the grade used for intravenous injections.

MIXING THE SILVERING SOLUTION

Take 3 volumes of A and add C from a burette until the precipitate which forms is just redissolved (avoid excess). Add A, a few drops at a time, until the solution is a straw color. Add 1 volume of B gradually, while stirring. Add solution C as before until the solution clears (disregard specks and avoid excess). Add solution A as before until a permanent precipitate forms. Filter, and use the same day.

Although the filtered solution is quite safe to use when fresh, it should be disposed of on the day that it is made up; otherwise, an insoluble explosive compound may form. Before using, filter off any black scum that has formed.

SILVERING

Three volumes of this silvering solution are mixed with one volume of reducer (D), and the specimen is immediately immersed in the mixture. Silvering is complete in approximately three minutes. The specimen is then rinsed under running distilled water and another coat is applied, care being taken to keep the specimen wet between coats. The specimen is then rinsed as before and transferred immediately to the plating bath.

REFERENCES

1. R. T. Pepper, *Trans. Amer. Inst. Met. Eng.*, 1960, *218*, 374.
2. R. L. Anderson, "Symposium on Methods of Metallographic Specimen Preparation", ASTM Special Technical Publication No. 285, 1960, p. 58.
3. L. E. Samuels, *J. Inst. Metals*, 1952-53, *81*, 471.
4. M. C. Udy, G. K. Manning and L. W. Eastwood, *Trans. Amer. Inst. Min. Met. Eng.*, 1949, *185*, 770.
5. D. Boyd Metz and H. W. Wood, U.S. Atomic Energy Commission Publication, 1950 (SEP-42).
6. H. H. Nausner and N. P. Pinto, *Trans. Amer. Soc. Metals*, 1951, *43*, 1052.
7. V. J. Haddrell, E. C. Sykes and B. W. Mott, *J. Inst. Metals*, 1955-56, *84*, 112.
8. N. Ogleby, private communication.
9. R. D. Buchheit, private communication.
10. V. J. Haddrell, *J. Inst. Metals*, 1963, *92*, 121.
11. C. B. Gilpin and F. J. Warzala, *Rev. Sci. Instruments*, 1964, *35*, 225.
12. R. Osadchuk, W. P. Koster and J. F. Kahles, *Metal Progress*, 1953, *64* (4), 129.
13. H. W. Wood, *Metal Progress*, 1947, *51* (2), 261.
14. U. E. Wolff and L. B. Fradette, *Metal Progress*, 1959, *76* (2), 111.
15. J. F. R. Ambler and G. F. Slattery, *J. Nuclear Materials*, 1961, *4*, 90.
16. K. Nagaiu and K. Mano, *Sci. Rep. RITU*, 1951, *1*, 391; 1953, *4*, 389.
17. G. Reinacher, *Metall*, 1957, *11*, 1.
18. G. Reinacher, *Z. fur Metallkunde*, 1957, *48*, 162.
19. J. M. Dickinson, *Metal Progress*, 1958, *74* (4), 142.
20. J. Just and W. Altgeld, *Z. fur Metallkunde*, 1961, *52*, 410.
21. C. W. Price, *Metallography*, 1968, *1*, 5.
22. H. E. N. Stone, *Metallography*, 1978, *11*, 105; 1979, *12*, 117.
23. P. Rothstein and F. R. Turner, "Symposium on Methods of Metallographic Specimen Preparation," ASTM Special Publication No. 285, 1960, p. 90.
24. T. Piotrowski and D. J. Accinno, *Metallography*, 1971, *4*, 521.
25. T. M. Kegley, Jr., D. M. Hewette and B. C. Leslie, *Metallography*, 1968, *1*, 237.
26. H. Morrogh, *J. Iron Steel Inst.*, 1941, *143*, 195.
27. H. L. Grange, *Metal Progress*, 1940, *38*, 679.
28. H. W. K. Honeycombe, *Proc. Australian Inst. Min. Met.*, 1944, No. 133, 29.
29. R. W. Turner, *BCIRA Journal*, 1960, *8*, 247.
30. L. E. Samuels, *J. Iron Steel Inst.*, 1955, *180*, 23.
31. J. Herenguel and R. Segond, *Rev. Mét.*, 1951, *48*, 262.
32. J. Herenguel, *Rev. Aluminium*, 1953, *30*, 261.
33. J. McAfee, "Symposium on Recent Advances in Physical Metallurgy", Aust. Inst. Metals, 1944.

34. J. J. de Jong, *Metalen*, 1954, 9, 2.
35. S. C. Carapella and E. A. Peretti, *Metal Progress*, 1949, 56, 666.
36. M. R. Plichta, H. I. Aaronson and W. F. Lange, *Metallography*, 1976, 9, 455.
37. H. W. Worner and H. K. Worner, *J. Inst. Metals*, 1940, 66, 45.
38. C. G. Rhodes and R. A. Spurling, *Trans. Met. Soc. AIME*, 1965, 233, 1193.
39. H. J. Levinstein and W. H. Robinson, *Trans. Met. Soc. AIME*, 1962, 224, 1292.
40. W. Vinaver and P. Dreulle, *Rev. Mét.*, 1955, 52, 612.
41. F. M. Cain, "Symposium on Methods of Metallographic Specimen Preparation", ASTM Special Publication No. 285, p. 37.
42. R. C. Gifkins, *Metallurgia*, 1961, 57, 209.
43. W. A. Roman, *J. Less-Common Metals*, 1966, 10, 150.
44. P. G. McDougall and N. F. Kennon, *J. Aust. Inst. Metals*, 1960, 5, 200.
45. A. L. Schaeffler, *Metal Progress*, 1946, 50, 659.
46. J. T. N. Atkinson and G. A. Gooden, *Metallurgia*, 1961, 63, 151.
47. N. A. Burley and J. J. Dale, *Metallurgia*, 1962, 65, 203.
48. J. H. Kruschner, *Metal Progress*, 1962, 81 (2), 88.
49. K. L. Hsu, T. M. Ahn and D. A. Rigney, "Wear of Materials 1979", Amer. Soc. Mech. Eng., 1979, p. 12.
50. G. J. Cocks and D. M. R. Taplin, *Metallurgia*, 1967, 75, 229.
51. B. Lehtinen and A. Melander, *Metallography*, 1980, 13, 283.
52. L. E. Samuels, "Interpretive Techniques for Microstructural Analyses", edited by J. L. McCall and P. M. French, Plenum, New York, 1977, p. 17.
53. M. D. Allen, "Microstructural Science", Vol. 6, edited by J. E. Bennett, L. R. Cornwall, and J. L. McCall, Elsevier, New York, 1978, p. 31.
54. F. Lucas, *Trans. Amer. Inst. Min. Met. Eng.*, 1928, 78, 481.
55. R. M. Slepian and G. A. Blann, *Metallography*, 1979, 12, 195.
56. H. E. Exner and K. Kuhn, *Practical Metallography*, 1971, 8, 453.
57. A. S. Holik, R. R. Russell and D. G. Fink, "Microstructural Science", Vol. 5, edited by J. D. Braun, H. W. Arrowsmith and J. L. McCall, Elsevier, New York, 1977, p. 247.
58. R. M. Slepian, J. M. Driggers and M. W. Spisiak, *Metallography*, 1981, 14, 213.
59. G. W. Fynn and W. J. A. Powell, "The Cutting and Polishing of Electro-optic Materials", Halsted Press, New York, 1979.
60. R. L. Stanton, *Canadian Mineralogist*, 1957, 6, 87.
61. G. T. F. Jay, *Metallurgia*, 1962, 66, 47.
62. T. Piotrowski and D. J. Accinno, *Metallography*, 1977, 10, 243.
63. J. E. Jenkinson, *J. Iron Steel Inst.*, 1940, 142, 89.
64. D. V. Miley and A. E. Calabra, "Metallographic Specimen Preparation", edited by J. L. McCall and W. M. Mueller, Plenum, New York, 1973, p. 1.

CHAPTER 10

Nonabrasive Techniques

THE GREAT BULK of surface preparation is carried out by the processes which rely on the action of abrasives and which we have already discussed. However, several techniques which rely on entirely different principles have come into reasonably common use. Several of the more important of these will now be discussed.

ETCH CUTTING AND MACHINING

The principle of these techniques is illustrated in Fig. 10.1. A wire (or thread) is traversed under slight tension across the surface to be cut. The wire also traverses through a solution of a reagent which will dissolve the specimen material, the solution being carried onto the specimen by surface tension. An etching solution chosen from those used for micrographic or macrographic etching can be used. The aim is to achieve maximum cutting rate consistent with an acceptable finish on the cut surface. The direction of motion of the wire usually is reversed periodically, although it may be continuous. The wire must be of a material that is resistant to the solution concerned; fabric-insulated metal wires may also be used, in which event the dissolution may be made electrolytic by applying a potential between the wire core and the specimen. A number of designs have been described,[1-4] and several are available commercially.

This is the only method of cutting that potentially can produce an absolutely strain-free surface, and it will do this if a film of liquid is maintained between the wire and the slot being cut so that the wire does not physically contact the surface. This method has two disadvantages. First, it is very slow; cutting rates of only about 0.1 mm/h are achieved. Secondly, a rather grooved surface is produced at best. For both of these reasons the technique is not well suited to normal metallographic specimen preparation. It has found application mostly for cutting single crystals of soft metals for investigation by x-ray diffraction and like techniques.

Apparatus based on the same principle can be used for facing flat surfaces or for turning cylindrical surfaces. A disk or wheel of the required shape is covered with an etchant-resistant napped cloth. The disk or wheel dips into a bath of etchant and rotates so that the cloth sweeps past the surface being machined.

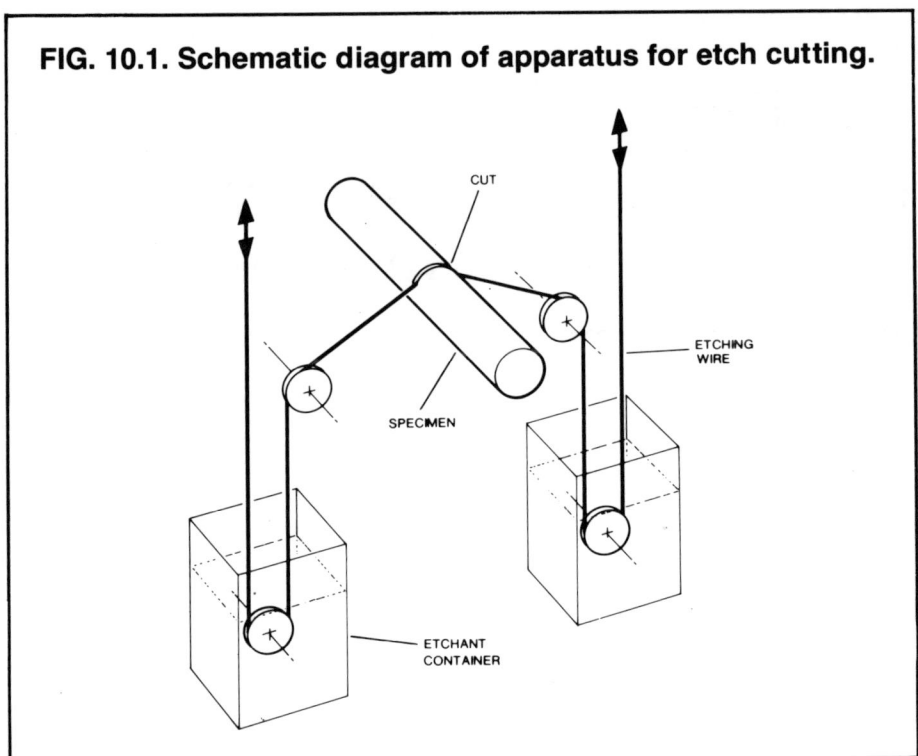

FIG. 10.1. Schematic diagram of apparatus for etch cutting.

SPARK CUTTING AND MACHINING

In spark machining, also called electric-discharge machining, pulsed sparks are generated between an electrode and a conducting workpiece while the spark region is immersed in a dielectric. The electrode is advanced into the workpiece as both are consumed. This is a standard production machining process, but can be adapted to the needs of metallographic surface preparation.[5] The spark is made controllable so that material may be removed from the workpiece at a variable rate. The electrode is advanced by a servomechanism which maintains a constant spark gap, so ensuring that the electrode never contacts the surface that is being cut. Slots or holes of any shape can be cut by shaping the electrode appropriately. Surfaces can likewise be machined as in conventional turning and milling.

The details of the mechanism by which material is removed during spark machining are in some dispute, but a coherent working model of the process has been developed.[6] The arcs cause localized impact of electrons on the anode and of ions on the cathode, and a large amount of heat energy is released. A molten pool of metal is formed by this localized release of energy and this pool flies apart as droplets when the arc current pulse is terminated. Not all of the pool is ejected, however, and a quantity of liquid is left in contact with the arc crater. The surface that is produced has the following characteristics:

1. The surface is covered by shallow overlapping craters which have lips of partly ejected droplets (Fig. 10.2). The diameter:depth ratio of the craters ranges from 5 to 50, but crater shape is independent of crystal orientation. Crater diameter is approximately constant for a particular set of arc conditions, being larger the more intense the arc (cf. Fig. 10.3a and b). Crater depth, and hence the volume of material ejected, varies with the material. The volume ejected is not simply related to the thermophysical properties of the material, but roughly is inversely related to its ultimate tensile strength.[7]
2. Material which has melted and then solidified epitaxially is present in a layer which contours the craters (Fig. 10.4a);[8] the thickness of this layer is somewhat less than the depth of the craters. The molten material may absorb extraneous alloying elements from the surroundings. In particular, it may absorb (a) carbon from pyrolysis products of the dielectric, (b) elements contained in the electrode or (c) oxygen or nitrogen from the atmosphere. For example, ferrous materials may absorb up to 3.5% carbon,[9] and titanium may absorb sufficient carbon to form a layer of titanium carbide,[6] both when kerosine is used as the dielectric. Steel may absorb up to 12% Cu when copper electrodes are used.[9] The absorbed elements may also diffuse into the layers beneath the molten layer. The molten layer and the immediately adjoining heated regions cool very rapidly, and appropriate structures may be expected.[6,9]

FIG. 10.2. Spark-cut surface of 70:30 brass.

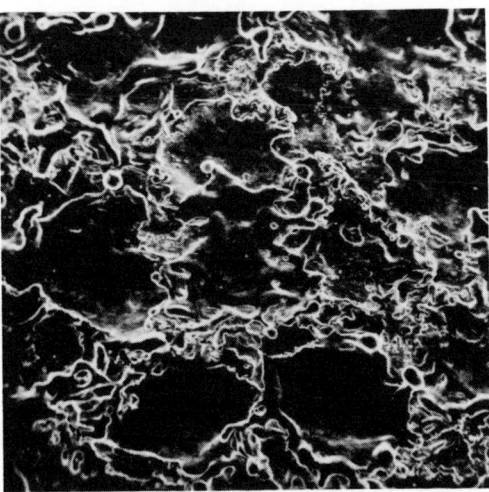

The surface is composed of contiguous craters which have lips of partly ejected droplets. Scanning electron micrograph. Magnification, 100×.

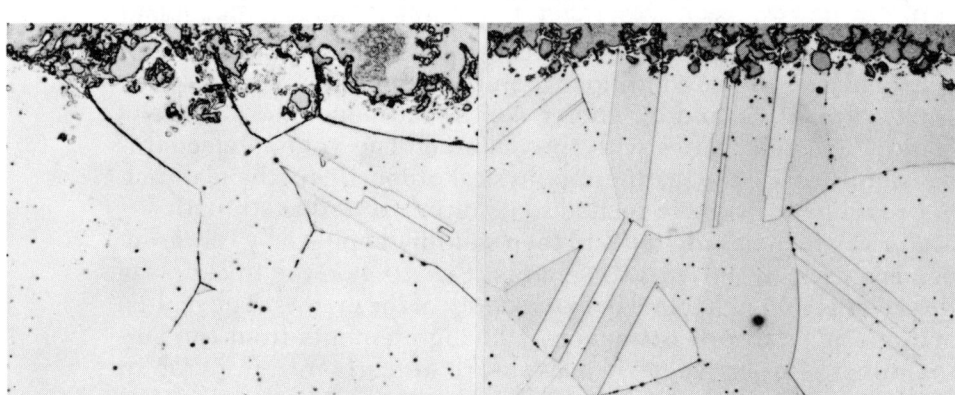

FIG. 10.3. Taper sections of spark-machined surfaces of 70:30 brass.

(a) Surface cut with high-energy sparks. The spherical contours of the surface cavities are apparent. Intercrystalline cracks are present. (b) Surface milled with low-energy sparks. The surface craters are shallower than those in (a). No cracks are present. Both sections etched in ferric chloride reagent. Taper ratio, approx. 10. Magnification, 50×.

3. A layer containing cracks may also be present immediately beneath the surface (Fig. 10.3a). Intercrystalline cracks have been found in brass,[8] austenitic steels,[6] some martensitic steels[5] and antimony.[10] Transgranular cleavage cracks have been found in chromium,[10,11] tungsten,[11,12] molybdenum[12] and silicon iron.[13] The depth of the crack-containing layer decreases with a decrease in the intensity of the spark, and this layer may not be produced at all when the energy is small enough (Fig. 10.3b; Table 10.1).[8,12] The cracks, when present, may be tens or even hundreds of micrometres deep. The layers containing cleavage cracks in brittle materials are likely to be particularly deep.

4. The thermal or acoustic effects of the arcs always induce plastic deformation in the surface layers. The strained layers are deep, but the maximum strains in them are small compared to those produced by mechanical cutting. For example, slip strain markings are observed in spark-cut brass surfaces after etching by suitable methods (Fig. 10.4b; cf. Fig. 4.10d), but not the twin strain markings found in mechanically cut surfaces and indicative of larger strains. The depth of the strained layer decreases with decreasing spark energy (Table 10.1) but is always comparable to that produced by conventional cutting, machining and abrasion processes (cf. Table 10.1 with Tables 4.1 and 4.2, p. 112). It has also been established that dislocations are introduced to depths of at least 500 μm when single crystals of copper,[14] silicon iron,[15] bismuth,[16] antimony[16] and zinc[16] are spark cut.

TABLE 10.1. Depth of Damage in Spark-Cut and Spark-Machined Surfaces of 70:30 Brass

Treatment	Craters	Depth, μm, of:		
		Molted layer(a)	Cracked layer(a)	Deformed layer(a)
Cut with high-energy sparks......	50	50	90	280
Cut with medium-energy sparks...	15	6	40	100
Planed with low-energy sparks....	7	2	Not present	50

(a) Maximum depth beneath bases of craters.

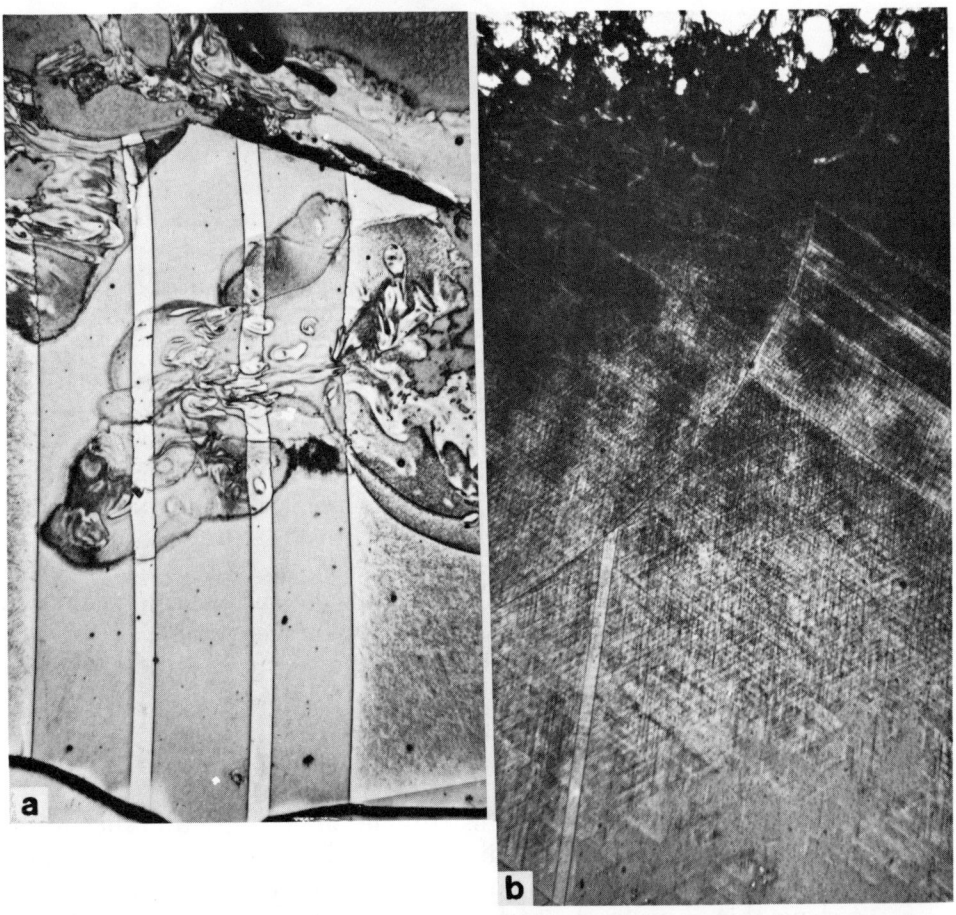

FIG. 10.4. Taper sections of 70:30 brass surfaces cut by spark machining.

(a) Cut with high-energy sparks. The regions containing swirl marks have been melted. They clearly have resolidified epitaxially. Etched in a ferric chloride reagent. Magnification, 500×. (b) Cut with low-energy sparks. A deep layer is present in which slip strain markings have been developed. Etched by the high-sensitivity sodium thiosulfate method. Magnification, 250×. Taper ratio (both), approx. 10.

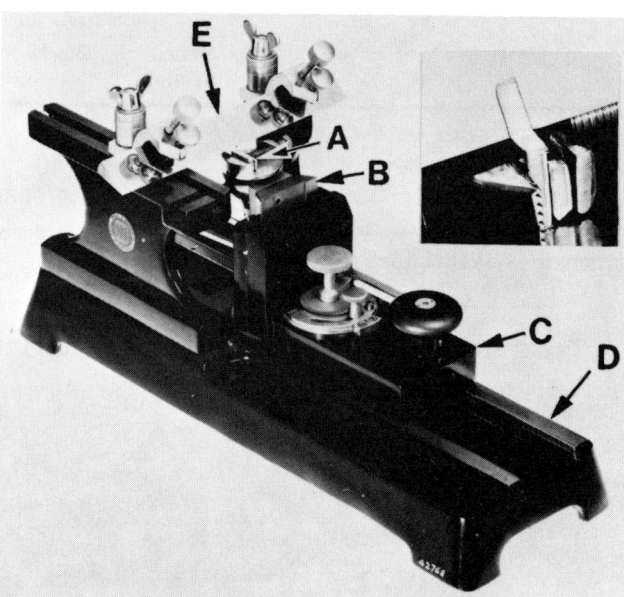

FIG. 10.5. A sledge microtome.

The specimen (A) is clamped in a vise (inset) which can be elevated in a slide (B) and in smaller increments by means of a lead-screw mechanism. The vise is part of a sledge (C) which slides in a base (D) past a knife blade (E).

In spite of these surface effects, spark-cut and spark-machined surfaces generally are good starting points for metallographic preparation. Only the approximate equivalents of the abrasion scratches and fragmented layers generally have to be removed by a rough-polishing stage. Exceptions are when cracks are developed in the surface layers and when the examination is sensitive to small strains; considerable thicknesses may then have to be removed by rough polishing to produce artifact-free surfaces.

MICROTOME CUTTING AND MACHINING

A *microtome* is basically an instrument for cutting a thin slice from a bulk specimen, the slice then being examined by transmission microscopy. In optical microscopy, the same general type of instrument can be used to shave thin chips from a surface, the machined surface instead of the chip then being examined. The advantage is that a tool of much more efficient shape than the point of an abrasive particle can be used to machine the surface.

There are two basic designs of instrument suitable for this purpose. In the first, a long knife is clamped rigidly in a holder and is used as the tool.

The specimen is clamped in a slide which can traverse past the knife (Fig. 10.5), and the surface of the specimen can be advanced into the knife in small increments. A machining chip is thereby separated from the full width of the specimen surface, as can be seen in the insert in Fig. 10.5. The device is made to be as rigid as possible to reduce chatter during the machining cut; the edge of the knife can be made of a variety of materials, including steel, tungsten carbide and diamond. Nevertheless, simple sledge microtomes of this type are effective in metallography only with comparatively soft metals (p. 313). The heaviest machines (heavier than the one illustrated in Fig. 10.5), which have diamond knives, can be used successfully for the preliminary preparation of materials with hardnesses only up to about 100 HV.[17]

The second type of microtome is basically a milling machine with a single-point fly cutter.[18,19] An attachment of this nature (Fig. 10.6) can be fitted to a sledge microtome as a replacement for the knife blade. Lathes can also be adapted for such purposes.[20,21] The machine preferably should be designed to run at high speed with a low level of vibration, and the depth of cut must be adjustable in small increments (typically, between 0.1 and 20 μm). A diamond tool must be used, and the cutting edge of the tool must be finished very well.

A surface which is very flat (or which has any other chosen geometry) and which is highly specularly reflecting can be produced (Fig. 10.7). The micromachining grooves present, which largely are reproductions of irregularities in the edge of the cutting tool, are comparable in dimensions to the scratches produced by conventional polishing with 1-μm diamond abrasives. Such a finish may be adequate to reveal the microstructure in systems that are not particularly sensitive to deforma-

FIG. 10.6. Schematic diagram of a micromilling head that can be substituted for the blade of a sledge microtome.

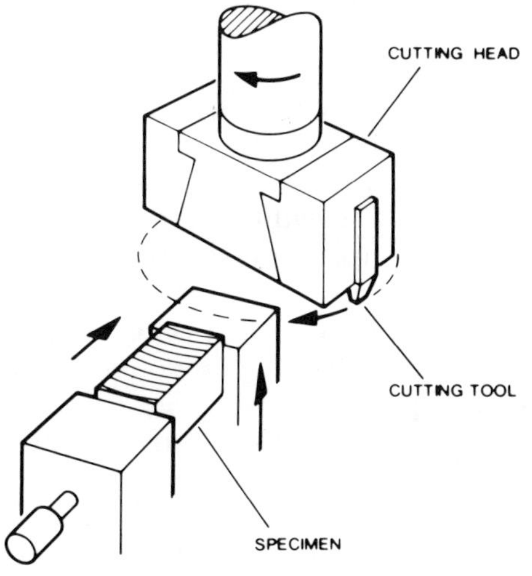

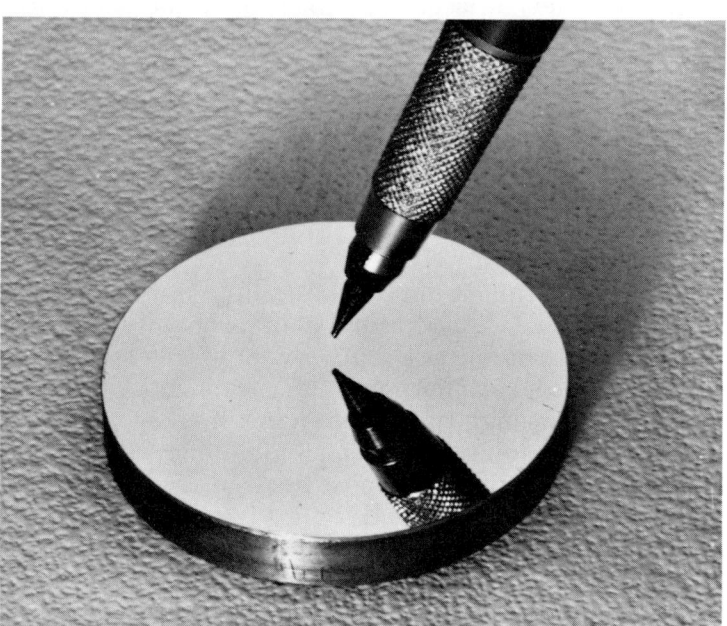

FIG. 10.7. A 4-cm-diam disk of silver finished by microtome turning.

The surface is very flat and is highly specularly reflecting.

tion;[18] it is certainly acceptable for macrographic examination.[19] Other advantages are as follows: (a) edges are well preserved, and can be preserved on mutually perpendicular surfaces if necessary;[18] (b) cavities and inclusions are correctly represented (Fig. 10.8a); and (c) serial sectioning is easy because layers of known thicknesses can be removed in succession[18] and because this thickness can be made to be much smaller than for conventional polishing.[20]

However, there are several type of specimens that cannot be satisfactorily machined by microtoming. The first type comprises ferrous materials, whether hard or soft. Diamond tools wear rapidly and so do not perform well when used to machine ferrous materials at high surface speeds. For example, the surface finish obtained on the gray cast iron illustrated in Fig. 10.8(b) is much poorer than those of the remaining materials illustrated in this group. Moreover, the widths of the graphite flakes exposed at the surface have been reduced considerably, as they were by coarse abrasion processes (see Fig. 4.26, p. 120). Both factors are due to rapid degradation of the edges of the cutting tools. Hard materials of any type (hardness > 500 HV) cannot be machined satisfactorily either. Many mounting plastics interfere with the machining process; either the specimen needs to be unmounted or the plastic must be below the level of the specimen. Finally, because the depth of cut is comparatively large, removal of material from brittle materials or phases consequently is likely to occur by the fracture-chipping mechanism. Silicon constituents in aluminum-silicon alloys, for example, are badly shattered except when quite small (Fig. 10.8c and d). Less brittle constituents, such as many of the other intermetallic phases present in aluminum alloys, may however be cut satisfactorily.[22]

On one count, the depth of the plastically deformed layer produced by microtoming might be expected to be small. The geometry of the cutting point is much more favorable than that of a typical abrasive point (see p. 40), and this would tend to reduce the depth of the surface de-

FIG. 10.8. Surfaces finished by microtome milling.

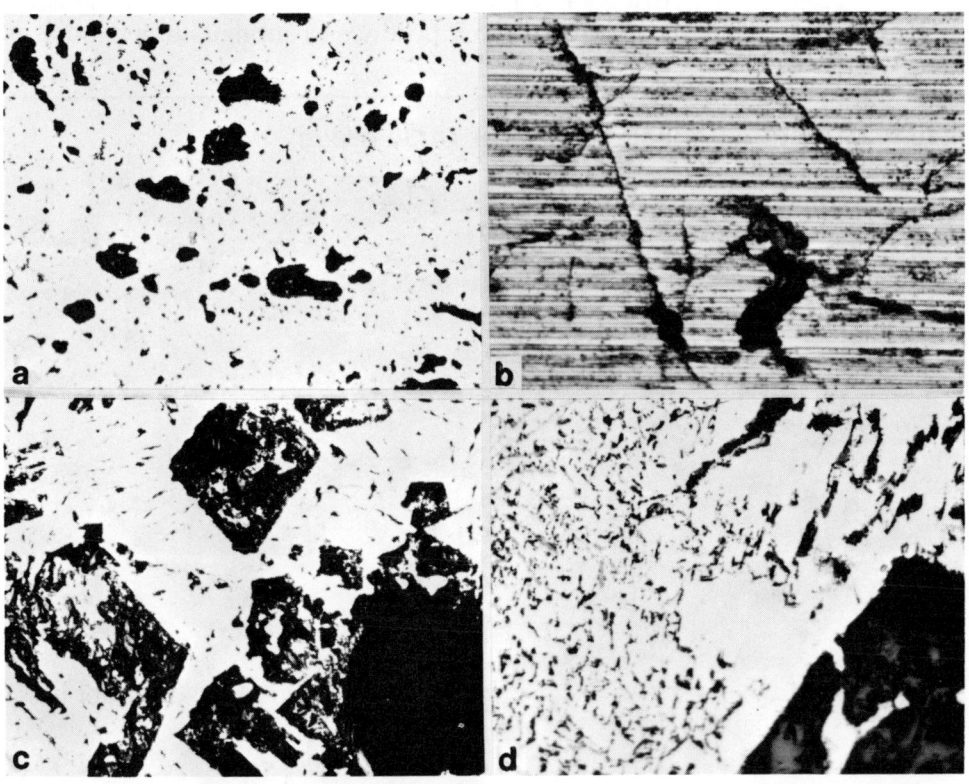

(a) A leaded tin bronze (16% Pb). The large particles of lead, although now darkened by oxidation, are excellently retained. The same specimen is illustrated after conventional abrasion and polishing in Fig. 8.13, p. 232. Magnification, 100×.

(b) A gray cast iron. The surface finish is not as good as those of the other specimens shown, and the graphite flakes have collapsed. This is due to deterioration of the cutting edge of the diamond tool during cutting of ferrous materials. The same specimen is shown after conventional abrasion in Fig. 3.30(d), p. 78; see also Fig. 4.26, p. 120. Magnification, 250×.

(c) An aluminum – 13% silicon alloy. The large silicon particles have shattered. The same specimen is shown after conventional abrasion in Fig. 3.30(b), p. 78. Magnification, 100×. (d) As for (c). Only the very smallest silicon particles have been well retained. Magnification, 250×.

formed layer.[23] On the other hand, the depth of cut is considerably greater than for individual abrasive points during abrasion or polishing, and this would tend to increase the depth of the surface deformed layer. The net result in practice is that the deformed layer is quite deep considering the fineness of the finish that is obtained.

For example, a surface of 70:30 brass finished by microtome milling with a depth of cut of 3 µm is shown in Fig. 10.9. The section has been etched in a ferric chloride reagent, which develops indications of plastic deformation with moderate sensitivity. Bands of artifact structures are visible at low magnification at the sites of the machining cuts (Fig. 10.9a), and these can be seen at higher magnification to be bands of twin strain markings (Fig. 10.9b). Note also that the strains have been sufficiently large and local to bow twin and grain boundaries intersected by the cuts (Fig. 10.9a). Equivalent photomicrographs for a surface polished with 1-µm diamond abrasive, which produces a similar surface finish as far as

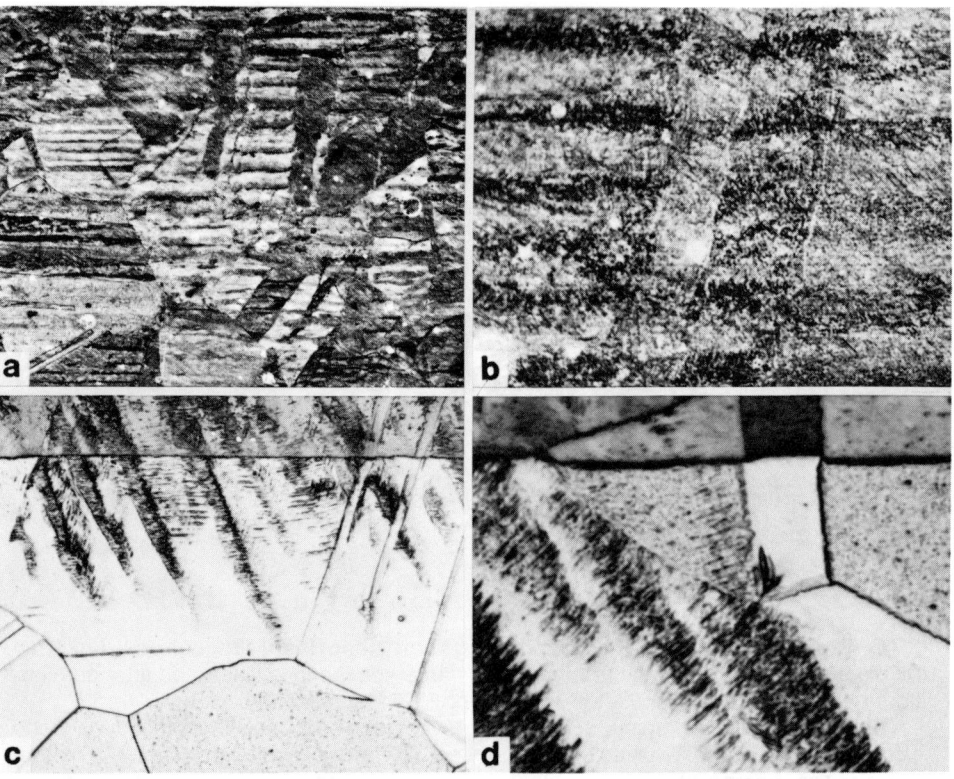

FIG. 10.9. Microtome-milled surface of 70:30 brass.

The depth of the finishing cut was 3 µm. Etched in a ferric chloride reagent. **(a)** Surface. Magnification, 100×. **(b)** Surface. Magnification, 500×. **(c)** Taper section of surface. Taper ratio, 10:1. Magnification, 250×. **(d)** Taper section of surface. Taper ratio, 10. Magnification, 1000×.

topography is concerned, are given in Fig. 8.17(c) and (d) (p. 236). The distortion of the structure produced by microtoming is in fact more equivalent to that found on an abraded surface from which the abrasion grooves have only just been removed by polishing (cf. Fig. 8.6a, p. 222).

Bands of massive twins are known also to form on zinc surfaces during microtoming,[21] and bowing of grain boundaries is known to occur in the soft metal indium.[21]

Rays of twin strain markings can be seen in sections of microtomed surfaces of 70:30 brass when the sections are etched in a ferric chloride reagent (Fig. 10.9c). These rays are similar to those that are developed in abraded surfaces (Fig. 4.14, p. 105). They clearly are located at the sites of the machining cuts. No fragmented layer can be seen by optical microscopy, but such a layer is present and can be detected by electron microscopy. Its thickness and its structure are similar to those of the fragmented layer produced by abrasive polishing (see p. 195).

Note, incidentally, that even these surfaces are not completely flat on a microscopic scale; small differences in level have developed between individual grains (Fig. 10.8d). These differences are probably due to local variations in elastic recovery after the passage of the machining tool.

The results of microtome milling thus are somewhat deceptive. The topography of the finish is at least the equal of, and in some respects is superior to, that obtained by conventional metallographic polishing processes. The deformation produced in the surface, however, is more the equivalent of, and in some respects is worse than, that resulting from conventional metallographic abrasion. Microtome milling has application where the latter is not significant. It has particular application where a high degree of surface flatness is desirable across large areas of a section. The process is rapid and requires little special skill, but it does require special equipment.

REFERENCES

1. T. R. McGuire and R. T. Webber, *Rev. Scientific Instruments*, 1949, *20*, 962.
2. R. Maddin and W. R. Asher, *Rev. Scientific Instruments*, 1950, *21*, 881.
3. U. Bonse, E. te Kaat and E. Kappler, *J. Scientific Instruments*, 1965, *42*, 631.
4. M. D. Hunt, J. A. Spittle and R. W. Smith, *J. Scientific Instruments*, 1967, *44*, 230.
5. M. Cole, I. A. Bucklow and G. W. B. Grigson, *Brit. J. Appl. Phys.*, 1961, *12*, 296.
6. H. K. Lloyd and R. H. Warren, *J. Iron Steel Inst.*, 1965, *203*, 238.
7. E. M. Williams and R. E. Smith, *Trans. Amer. Inst. Electrical Eng.*, 1957, *76*, 93.
8. L. E. Samuels, *J. Inst. Metals*, 1962-63, *91*, 191.
9. L. Massarelli and M. Marchionni, *Metals Tech.*, 1977, *4*, 100.
10. G. R. Wilms and J. B. Wade, *Metallurgia*, 1956, *54*, 263.
11. L. S. Palatnik and A. A. Levchenko, *Kristallografiya*, 1958, *3*, 612.
12. P. M. Beardmore and D. Hull, *J. Inst. Metals*, 1966, *94*, 14.
13. P. M. Brown and J. A. Robey, *J. Inst. Metals*, 1969, *97*, 63.

14. W. T. Brydges, *J. Inst. Metals*, 1967, *95*, 223.
15. B. Sestak and S. Libovicky, *Czech. J. Physics*, 1960, *10* (B), 759.
16. L. S. Palatnik, A. A. Levchenko and V. M. Kosevitch, *Soviet Physics — Crystallography*, 1962, *6*, 472.
17. G. Reinacher, *Metall*, 1957, *7* (11), 1.
18. G. Kiessler and G. Elssner, *Practical Metallography*, 1980, *17*, 536.
19. S. A. Levy, *Microstructural Science*, 1980, *8*, 375.
20. D. W. Stevens and R. N. Gillmeister, *Microstructural Science*, 1977, *5*, 277.
21. G. Elssner, G. Keissler and L. Gessner, *Practical Metallography*, 1977, *14*, 445.
22. R. Klockenkamper, A. Beyer and M. Mones, *Practical Metallography*, 1979, *16*, 53.
23. D. M. Turley, *J. Inst. Metals*, 1968, *96*, 82; 1971, *99*, 271.

CHAPTER 11

Procedures for Some Common Metals and Alloys

PRINCIPLES WHICH APPLY to the preparation of most types of specimens have been discussed throughout the preceding text. The information given in this chapter is intended mainly to draw these points together for some common metals and alloys, and to note any desirable variations on standard procedures.* Attention will be concentrated on two types of problems: those involving artifacts, particularly abrasion artifacts; and those that arise in obtaining satisfactory final polishing. Special problems that arise during mounting of some of these metals and alloys were discussed in Chapter 2; they will not be discussed further here.

The procedures that are recommended are internally consistent and conform to the principles developed throughout this book. In particular, they are centered about the use of diamond abrasives for polishing. Other procedures, such as those described in Ref 1, may be satisfactory in specific circumstances. Many published procedures need, however, to be reviewed in light of the principles that have been discussed earlier to determine if they are really adequate or if they are excessively complicated. Representative photomicrographs therefore have been included in this chapter as indications of the standards of results that should be attainable.

ALUMINUM

Three groups of aluminum metals need to be considered. The first group includes high-purity aluminum and dilute alloys; these materials are soft, particularly when in the annealed condition. The second group comprises more highly alloyed materials that are normally used in a heat treated condition; these alloys are moderately hard even when in the annealed condition. The third group consists of alloys containing significant volume fractions of intermetallic phases.

---

*The "standard" procedures and methods referred to in this chapter are those outlined in the section commencing on p. 241.

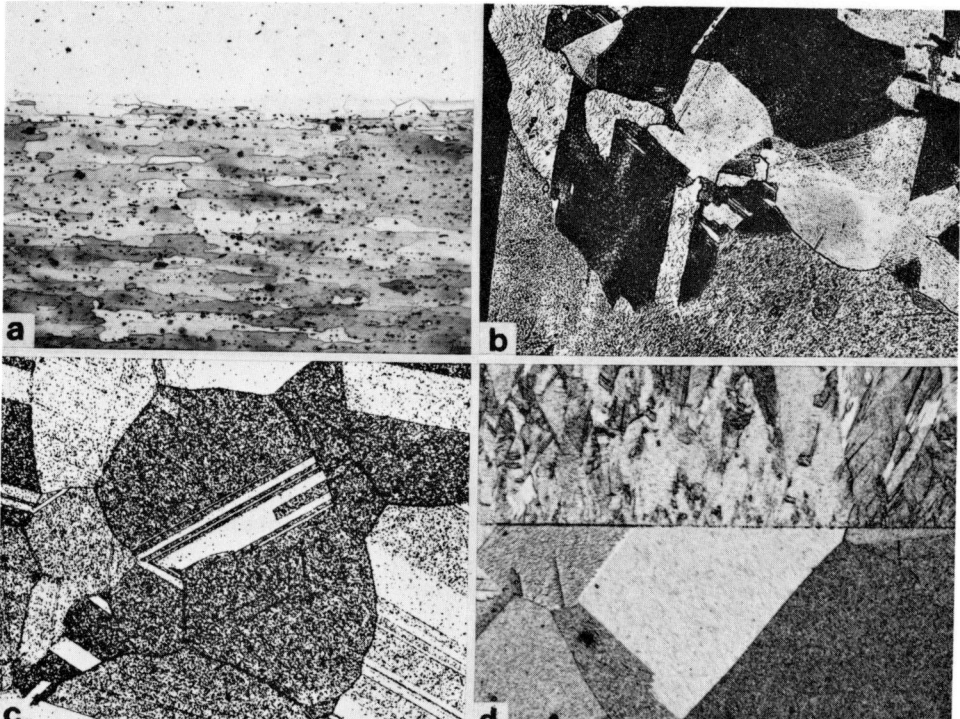

FIG. 11.1. Soft metals finish polished by standard methods.

(a) The cladding/core interface region in a clad aluminum alloy sheet. Finished with magnesium oxide abrasive. Etched in a mixed-acid reagent. The cladding, base metal and interdiffusion layer are all well polished. Magnification, 250×. (b) Gold alloy (24 carat), cast and annealed. Finished with 5-μm alpha aluminum oxide abrasive. Etched in a potassium cyanide – ammonium persulfate – hydrochloric acid solution. Magnification, 200×. (Piotrowski and Accinno, Ref 2).
(c) Silver – 9% cadmium oxide alloy. Finished with 0.5-μm diamond abrasive. Etched in an ammonium hydroxide – hydrogen peroxide solution. Magnification, 200×. (Piotrowski and Accinno, Ref 2). (d) Silver. Section of a silver electrodeposit on a silver substrate. Finished with magnesium oxide abrasive. Magnification, 250×.

Artifacts

Aluminum and its alloys generally are not susceptible to abrasion artifacts. Crack artifacts may be found in brittle intermetallic phases, particularly in the casting alloys (p. 230), but are easily eliminated if a reasonably efficient rough-polishing stage is used. Fragments of abrasive may embed in the surfaces of the softer aluminums during abrasion (p. 131).

Final Polishing

Soft Alloys. Only moderately satisfactory finishes can be obtained on the soft, high-purity materials by standard methods. Certainly, finishing at least on magnesium oxide or fine aluminum oxide is necessary for adequate results. A finish of increasingly high standard may be obtained by etch-attack vibratory polishing (Table 9.3, p. 279) and by skidding on magnesium oxide (Fig. 9.2b and 9.3b, p. 270). Vibratory polishing using conventional abrasives without etch attack is not satisfactory for the softest materials because abrasive particles embed in the surface (p. 284).

Alloys of Moderate Hardness. This group of alloys is comparatively easy to finish, standard final-polishing techniques giving satisfactory results; an example of the standard to be expected is presented in Fig. 11.1(a). Etch-attack techniques, such as those in which a dilute sodium hydroxide solution is added to the abrasive slurry, may often be used with profit. Etch-attack vibratory techniques are also successful (Fig. 9.11, p. 283).

Spurious etch pits may develop in aluminum alloys due to electrolytic effects between the specimen and other metals of the polishing apparatus; they are most likely to develop during the finest polishing processes when copper alloys are used in the polishing apparatus. The cure for such pitting was discussed on p. 271. Etch pits may also develop in the cladding layer on clad aluminum alloy sheet. The cure for this was discussed on p. 247.

Spurious etch pitting may also develop during etching of these alloys when deformation introduced during the rough-polishing stages is present. The development of this pitting is most likely with aggressive etchants and in alloys in which the matrix has some elements in solid solution. For example, etch pits developed in the matrix of the alloy illustrated in Fig. 11.2 when the surface was finished by a conventional fully mechanical method (Fig. 11.2c and d). The coarser the polish the more numerous the etch pits (cf. Fig. 11.2c and d). Etch pits are not developed, except at genuine structural features, when the surface is finished by a chemical-mechanical method (Fig. 11.2b) — i.e., when little or no polishing deformation is present. The etch pits illustrated in Fig. 11.2(c) and (d) are artifacts. They could easily be mistaken for effects associated with a real structure.

Alloys Containing Intermetallic Phases. These phases may be present in both wrought and cast alloys but are most significant in cast alloys. The special requirement is to ensure that excessive relief does not develop between the matrix and the intermetallic phases. The abrasion and polishing characteristics of these phases often differ from those of the matrix, but to a degree that varies considerably. The problem of avoiding excessive relief varies accordingly. Unmodified hypereutectic aluminum-silicon alloys, which contain large particles of primary silicon phase, constitute one of the more difficult cases. An alloy of this type was employed earlier to illustrate the advantages of using a fine lap for final abrasion (Fig. 3.30, p. 78) and of using diamond abrasives and short-napped cloths for polishing (Fig. 5.15, p. 163). The final result then to be expected from the standard procedure is illustrated in Fig. 9.14(b) (p. 285). A range of more typical intermetallic constituents is present in the specimen illustrated in Fig. 11.2(a). Very little relief has developed between the matrix and some of the phases present in this alloy. For

FIG. 11.2. A cast aluminum alloy (3% Cu, 5% Si, 0.6% Fe) finish polished by etch-attack and standard methods.

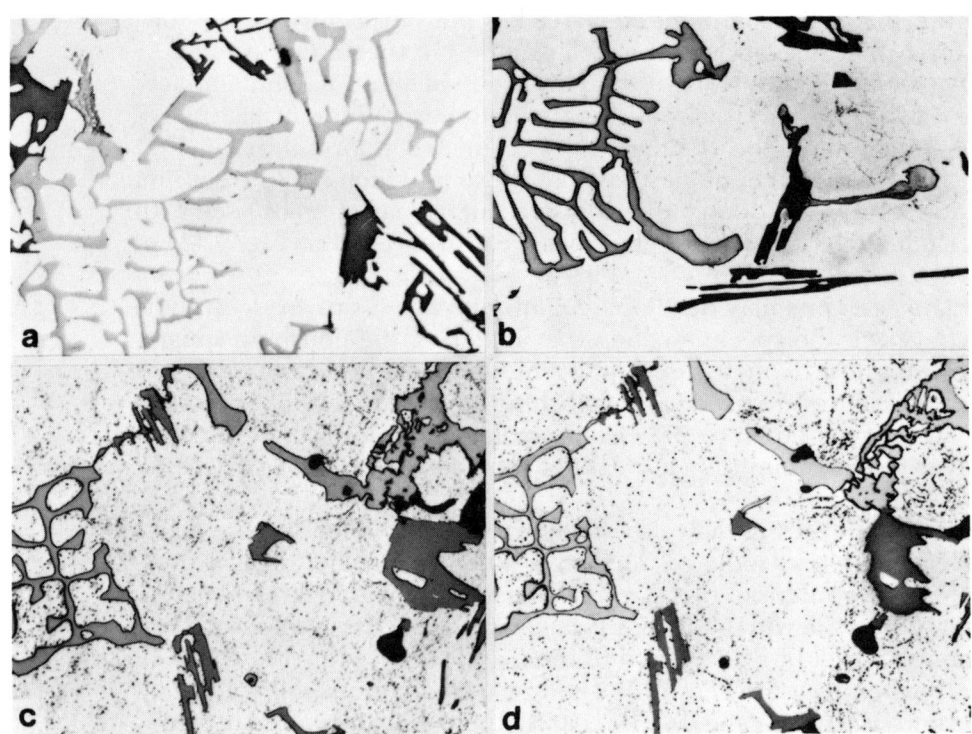

(a) Finished by an etch-attack vibratory method (see Table 9.3, p. 279). Unetched. (b) Finished as for (a). Etched in 10% sodium hydroxide at 70°C. (c) Finished by a standard method using 1-μm diamond abrasive. Etched in 10% sodium hydroxide at 70°C. (d) Finished by a standard method using 0.1-μm gamma aluminum oxide abrasive. Etched in 10% sodium hydroxide at 70°C. Magnification (all), 250×.

The etch pits in (c) and (d) are artifacts due to polishing deformation. The phases present are: AlMnSi compound (script), Si (dark), and $CuAl_2$ (light, rounded).

some, in fact, the relief is insufficient to delineate the constituents clearly. However, the outline of the constituents can then be readily sharpened by light etching (Fig. 11.2b). Considerable flexibility in examinational and etching methods becomes possible when relief between phases can be kept to a low level during polishing. An example of what is possible is given in Fig. 11.3.

BERYLLIUM

Artifacts

Artifacts of the massive twin type are produced to some depth during abrasion. The use of a rough-polishing stage having maximum cutting efficiency consequently is desirable. The use of electromechanical polishing methods (p. 273) is advantageous in this respect, although a less satisfactory finish is produced than when fully mechanical methods are used.

Cleavage cracks may be produced in the surface layers of beryllium, particularly during preliminary machining. Some beryllium alloys contain brittle constituents in which fracture cracks also may develop during abrasion. Reasonably efficient rough-polishing processes are capable of removing these crack-containing layers readily enough. The use of fresh

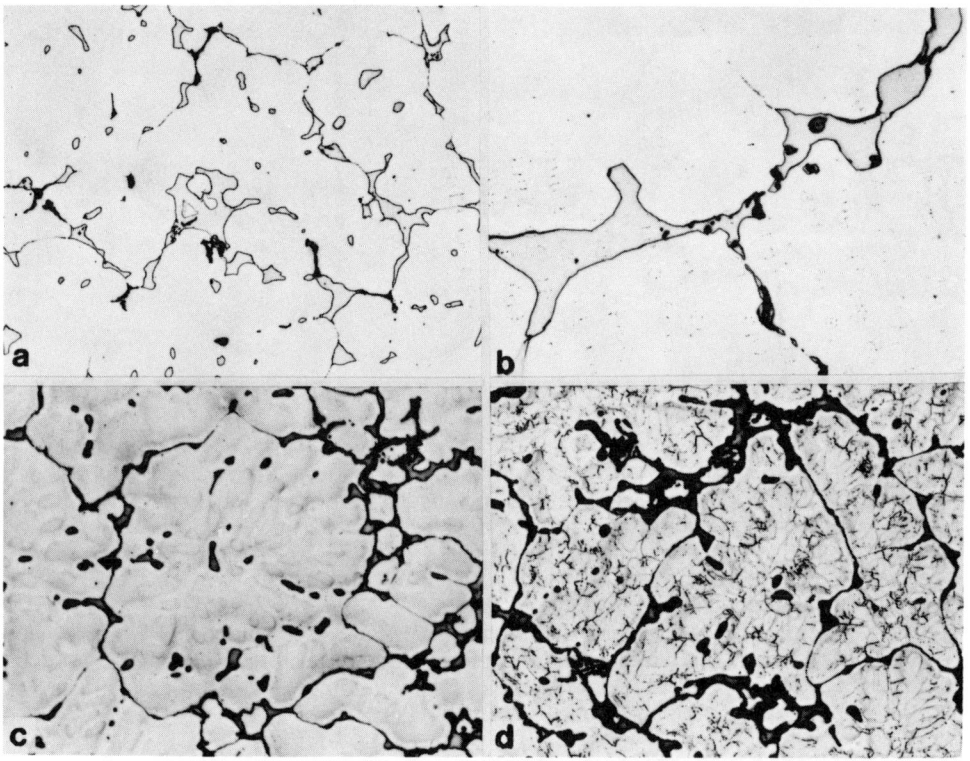

FIG. 11.3. A cast aluminum – 10% magnesium alloy finish polished by an etch-attack vibratory method.

(a) Lightly etched in 1% HF. The distribution of the Al-Mg phase is clearly visible. Magnification, 100×. (b) Lightly etched in 1% HF. Minor phases within the Al-Mg phase are now apparent. Magnification, 500×. (c) More heavily etched in 1% HF. Oblique illumination. The coring segregation in the matrix crystals is revealed. Magnification, 100×. (d) Etched in 0.5% HF – 1.5% HCl – 25% HNO_3 solution. Precipitation etch pitting has been developed at subgrain boundaries. Magnification, 100×.

This series illustrates the experimental flexibility that is possible when low relief between constituents as well as a scratch-free polish can be attained. The phases present are an Al-Mg compound (gray networks), an aluminum-rich solid solution (light matrix) and a MnFeAlSi compound (dark minor phase).

356 / Metallographic Polishing by Mechanical Methods

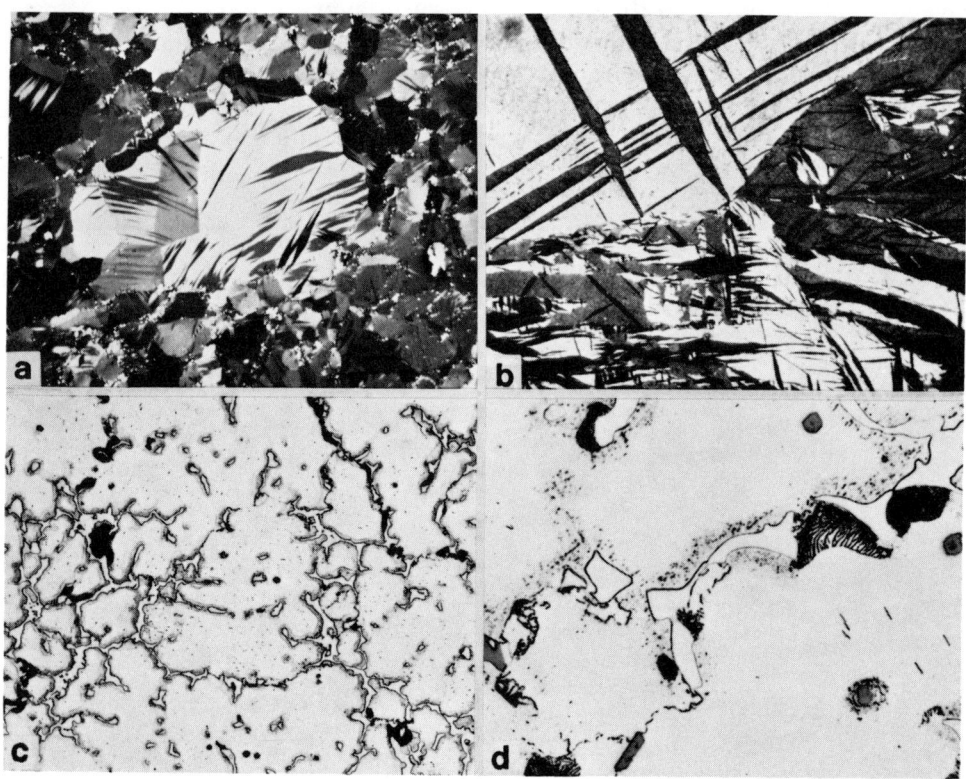

FIG. 11.4. Beryllium and magnesium alloys finish polished by advanced techniques.

(a) Beryllium finished by skidding on magnesium oxide. Polarized light. Magnification, 250×. (b) High-purity magnesium, plastically deformed. Finished by vibratory polishing with magnesium oxide suspended in propylene glycol. Polarized light. Magnification, 250×. (c) A magnesium casting alloy (8.8% Al, 0.8% Zn) finished by etch-attack vibratory polishing (see Table 9.3, p. 279). Etched in nital. Magnification, 100×. (d) As for (c). Magnification, 500×.

The structures in (a) and (b) are composed of equiaxed grains containing massive deformation twins. The structure in (c) and (d) is composed of a network of $Mg_{17}Al_{12}$ phase in a matrix of magnesium-rich solid solution; some areas of a lamellar transformation product are also present.

(sharp) abrasive papers and light specimen pressures is desirable during abrasion to reduce these problems to the minimum.

Final Polishing

Standard methods using either magnesium oxide or fine aluminum oxide abrasive are acceptable when the specimen is to be examined in brightfield illumination, such as when inclusions or cavities are to be studied. However, polarized light tends to be used to discern grain structure because it is not easy to develop this structure by etching. Vibratory polishing methods or etch-attack techniques employing 5 to 10% oxalic acid solutions then give reasonably satisfactory results, but more advanced techniques, such as skidding techniques, are necessary for good results (Fig. 11.4a). The use of magnesium oxide suspended in 30% hydrogen peroxide has been recommended specifically for skid polishing of this material. The hydrogen peroxide holds the magnesium oxide together, which facilitates skidding. It may also have a passivating oxidizing role. In addition, it reduces the attack by water on some reactive phases that might be present. An etch-attack technique for final polishing is also available (Table 9.1, p. 274). Electromechanical polishing methods are also applicable (Table 9.2, p. 277). Cavities and other discontinuities are then retained well, much more satisfactorily than for full electrolytic polishing. The factors that control the grain contrast observed in polarized light are discussed in Chapter 9, p. 330.

Note on Toxicity

Preparation of beryllium generates toxic dust. It should be carried out only in a glove box and only by persons trained in safe handling of this material.

CHROMIUM

See the section "Precious Metals and Refractory Metals" in Chapter 9, p. 326.

Artifacts

The only artifacts to be expected are those that would have originated in the outer fragmented layer produced during machining and abrasion (Fig. 9.40, p. 328). However, the polishing rates of chromium and chromium alloys are comparatively low. The elimination of this artifact-containing layer even by efficient rough-polishing methods consequently takes longer than usual. Hence abrasion practices which reduce to the minimum the final depth of significant abrasion damage should be used. Rough-polishing stages designed to achieve maximum cutting rates are also essential.

Final Polishing

A standard method using either magnesium oxide or fine aluminum oxide is adequate. An illustration of the standard of results that can be expected is given in Fig. 9.40(b) (p. 328). An etch-attack technique is also available (Table 9.1, p. 274).

COPPER

Three groups of copper alloys can be considered: first, unalloyed copper and dilute alloys of copper; second, more highly alloyed alpha solid solutions, particularly those with alloy contents which approach the limit of the alpha field; and third, even more highly alloyed beta-phase or complex alloys. This system of classification can also be applied to the individual phases in a multiphase alloy.

Artifacts

Few difficulties involving abrasion artifacts are experienced with the first and third groups of alloys, other than those associated with removal of the fragmented layer (see discussion of Fig. 8.9, p. 228). The fragmented layer is easy to remove by reasonably good procedures because compar-

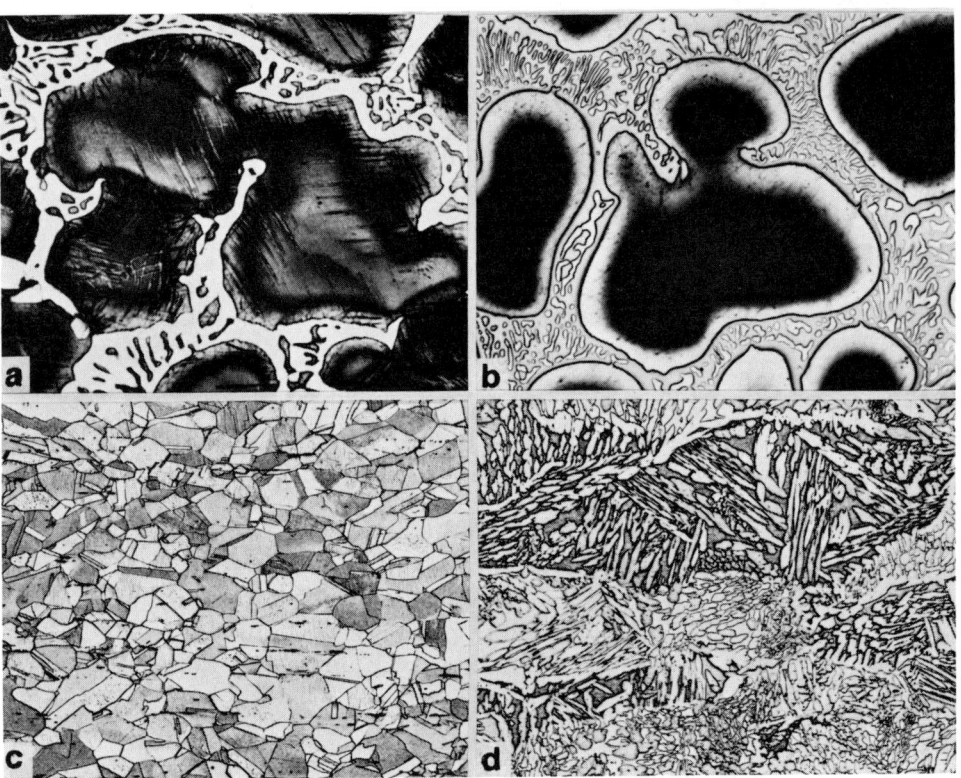

FIG. 11.5. Copper alloys finish polished by an etch-attack vibratory method.

(a) A cast copper – 4.5% phosphorus alloy that was rough polished for an insufficient period of time. The alpha dendrites contain abrasion artifacts (cf. Fig. 8.6, p. 222). Magnification, 500×. (b) As for (a), but rough polished for a longer time. The alpha dendrites are now free of abrasion artifacts. Magnification, 500×. (c) A tough-pitch copper, hot extruded. Magnification, 100×. (d) A copper – 40% zinc alloy, hot extruded. Magnification, 100×.

The structure in (a) and (b) is composed of primary dendrites of Cu-P solid solution and a Cu-Cu$_3$P eutectic. The structure in (c) is composed of equiaxed grains of copper. The structure in (d) is composed of plates of alpha phase in a matrix of beta phase. All specimens were etched in a ferric chloride reagent.

atively high abrasion and polishing rates are obtained. Medium to major difficulties involving both abrasion and polishing artifacts arise with the second group, the degree of difficulty depending on the closeness of the alloy to the limit of the alpha field and on the etchant used. The behavior of a specific alloy, 70:30 brass, has been discussed at some length throughout this book (see, for example, p. 220), and this discussion applies generally to many alloys of the second group. Remember also that it applies equally well to individual phases of appropriate composition in multiphase alloys (Fig. 11.5a and b; cf. Fig. 8.6b, p. 222).

Final Polishing

Final polishing is most difficult for the second group of alloys (alloy-rich alpha solid solutions) because of the sensitivity of these alloys to polishing artifacts. Examples of these polishing artifacts have been given earlier (p. 235, 236, 239 and 241) for 70:30 brass. Examples have also been given of the results obtained with a range of conventional and advanced polishing techniques (p. 268 and 278). These examples can also be taken as a guide for the whole group. Less difficulty is experienced with the less richly alloyed alpha solid solutions, but reasonably advanced methods are still required. The standard that can be attained with commercially pure copper is illustrated in Fig. 11.5(c).

Alloys which are composed substantially of alpha phase but which contain moderate volume fractions of beta phase are no more difficult to finish than fully alpha-phase alloys (Fig. 11.5d). It is more difficult, however, to obtain a high-quality finish with alloys composed substantially of beta phase (Fig. 9.12a, p. 283), and advanced techniques using controlled etch-attack are required (Fig. 9.12b). On the other hand, harder and more complex alloys, such as aluminum bronzes, can be satisfactorily finished by standard methods (Fig. 11.6a and b).

For alloys containing large volume fractions of lead (25 vol % or more), satisfactory retention and polishing of the lead phase requires special procedures. Etching at regular intervals in a reagent which attacks the copper-rich areas but not the lead areas is required (p. 285), intermediate polishing being continued only until the etch has been removed. Alloys containing up to about 15 vol % of lead phase can, however, usually be prepared satisfactorily by standard methods. Even when well retained and polished, the lead areas tend to oxidize and hence to appear blue or even dark and mottled (Fig. 8.13, p. 232).

GOLD

Artifacts

Only abrasion artifacts originating from the fragmented layer are to be expected in either pure gold or common gold alloys. Abrasion and polishing rates of these materials are, however, very low. Rough-polishing processes with maximum cutting rates and longer-than-usual polishing times consequently are required to produce artifact-free surfaces (p. 216). An electromechanical polishing process is available which alleviates

FIG. 11.6. Harder alloys of copper and of zinc finish polished by a standard method using magnesium oxide abrasive.

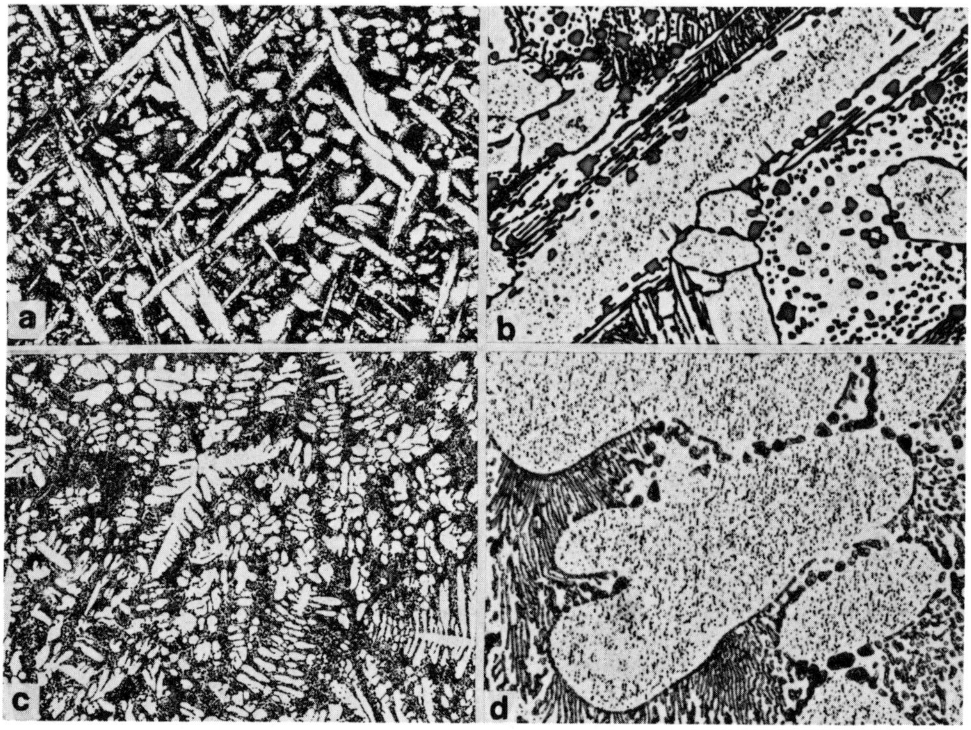

(a) An aluminum bronze (10% Al, 5% Ni, 5% Fe). Etched in an ammonium hydroxide – hydrogen peroxide solution. Magnification, 100×. (b) As for (a). Magnification, 1000×. (c) A zinc-base die-casting alloy (4% Al, 0.05% Mg). Etched in a chromic acid reagent. Magnification, 100×. (d) As for (c). Magnification, 1000×.

The structure in (a) and (b) is composed of particles of kappa phase in a matrix of alpha phase. The structure in (c) and (d) is composed of dendrites of zinc-rich solid solution, containing precipitates of aluminum (dark), in a zinc-aluminum eutectic.

this problem (Table 9.2, p. 277). Fragments of abrasive may embed in the surfaces of softer grades of gold during abrasion (p. 131).

Final Polishing

Gold and gold alloys are difficult to finish polish to an acceptable standard. Reasonable results are obtained by fully mechanical polishing using a soft long-napped cloth, a 5-μm grade of alpha alumina, light specimen pressure, and a wheel rotating at approximately 200 rpm (Fig. 11.1b).[2] An etch-attack technique is also available (Table 9.1, p. 274).

INDIUM

Treat as described in the section "Soft Metals" in Chapter 9, p. 313. A chemical polish is available (Table 9.4, p. 286).

IRON-BASE ALLOYS: AUSTENITIC

Artifacts

All austenitic iron-base alloys are sensitive to abrasion artifacts, particularly the nickel-rich alloys. These artifacts, which take the form of deformation-induced transformation products, have been illustrated earlier (p. 122 and 225). The use of an efficient rough-polishing stage is essential with these alloys.

Final Polishing

Conventional final-polishing methods can produce finishes that appear to be adequately scratch-free in the unetched condition. However, these alloys are sensitive to scratch enlargement during etching (see p. 234); the lower the carbon content of the alloy the greater its sensitivity to the effect. The finishes obtained by conventional final-polishing methods consequently will appear at best to be only moderately satisfactory after etching, unless special care is taken. An example of the standard of result that can nevertheless be obtained is given in Fig. 8.7(b). Skidding techniques of finish polishing are not applicable. No suitable etch-attack or electromechanical methods have yet been developed. An etch-attack vibratory method is available which produces good results (Table 9.3, p. 279). A typical result obtained by this method is illustrated in Fig. 11.7.

FIG. 11.7. A cast austenitic steel (10% Ni, 16% Cr) finish polished by an etch-attack vibratory method.

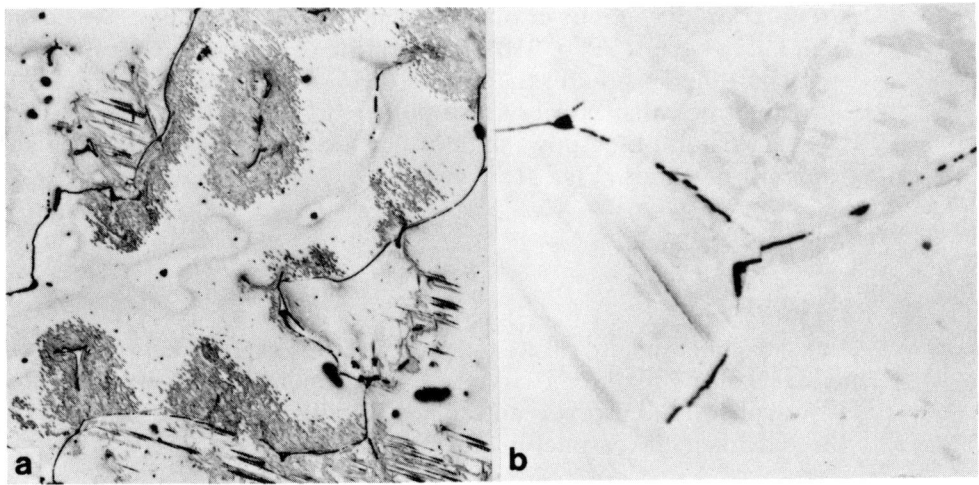

Etched electrolytically in an oxalic acid solution. Magnifications: **(a)** 100×; **(b)** 1000×. The structure is composed of a segregated solid solution with a large grain size; some carbides have been precipitated at the grain boundaries.

IRON-BASE ALLOYS: FERRITIC AND MARTENSITIC

Artifacts

Only abrasion artifacts due to the fragmented layer are likely to develop in ferritic and martensitic iron-base alloys (see p. 117, 217 and 218). Consequently, the use of a moderately efficient rough-polishing stage ensures an artifact-free result (see p. 217 and 218). Untempered or lightly tempered martensitic steels are sensitive to tempering artifacts caused by surface heating during preliminary machining or grinding (see p. 130 and 229). This problem can be avoided by taking suitable precautions at the preliminary machining stages (p. 126).

Account needs to be taken of the fact that these alloys cause rapid deterioration of coated abrasive papers during abrasion (see p. 62).

Final Polishing

It is easy to obtain acceptable results with most of these alloys by standard methods. Many examples of the standard to be expected have been published elsewhere.[3] Special care is necessary only with low-carbon alloys which are completely or almost completely ferritic. An etch-attack vibratory method is available for such alloys (Table 9.3, p. 279), a technique which produces highly satisfactory results (Fig. 9.8b).

IRON-BASE ALLOYS: GRAPHITE-CONTAINING

Preparation procedures for graphite-containing iron-base alloys were discussed at length in Chapter 9 (p. 307).

LEAD

Preparation of sections of pure lead and dilute lead alloys was discussed in detail in Chapter 9 (p. 313). Preparation of more concentrated alloys — with hardnesses greater than, say, 10 HV — is much less difficult. These alloys can be satisfactorily finish polished by slightly advanced methods (Fig. 11.8a and b). Alloys slightly harder than these can be finished by standard methods (Fig. 11.8c and d).

MAGNESIUM

Artifacts

Pure magnesium and dilute magnesium alloys are susceptible to abrasion artifacts of the massive twin type. Abrasion and polishing rates are comparatively high, however, and so no difficulty should be experienced with these artifacts if reasonably efficient procedures are used.

Final Polishing

Magnesium and most of its alloys are comparatively soft and so require advanced finishing techniques. Moreover, they are reactive to water and may require less-reactive polishing fluids. For example, propylene glycol

Procedures for Some Common Metals and Alloys / 363

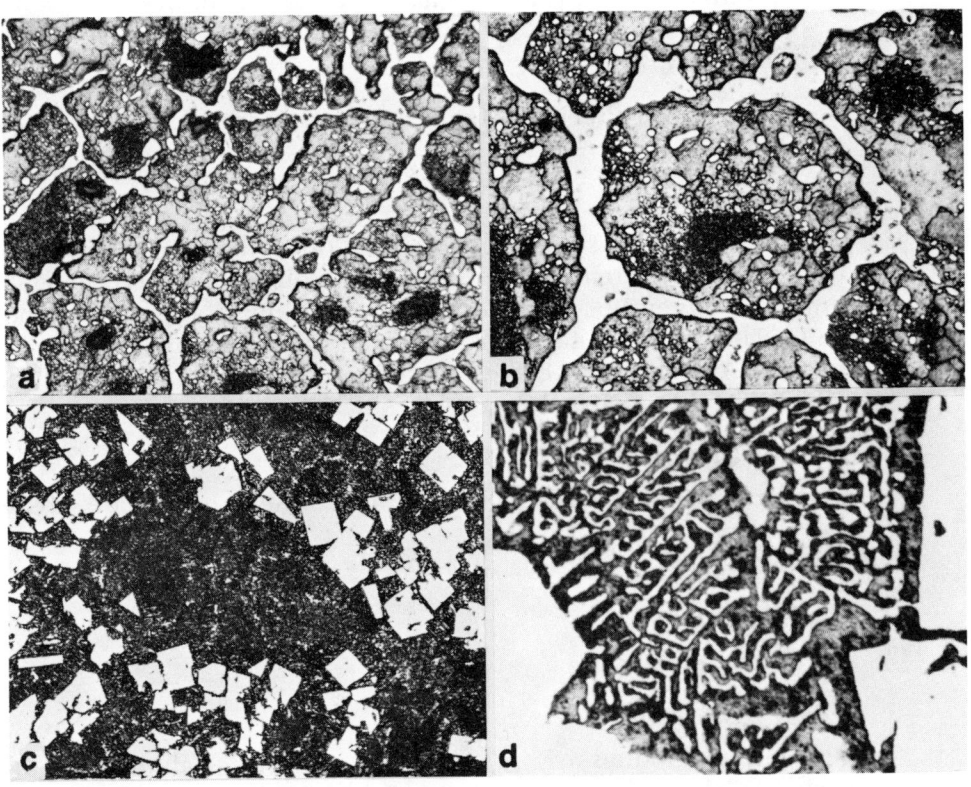

FIG. 11.8. Lead alloys finish polished by slightly advanced and standard methods.

(a) A lead – 20% tin alloy (hardness, 11 HV). Finished by vibratory polishing using magnesium oxide suspended in propylene glycol. Etched in nital. Magnification, 100×. (b) As for (a). Magnification, 500×. (c) A lead-base alloy used for type metal (10% Sn, 16% Sb; hardness, 19 HV). Finished by a standard method using 0.5-μm gamma aluminum oxide. Etched in nital. Magnification, 250×. (d) As for (c). Magnification, 100×.

The structure in (a) and (b) is composed of networks and spheres of tin in a matrix of lead-rich solid solution. The structure in (c) and (d) is composed of cubes of Sn-Sb compound in a matrix of ternary eutectic.

or propylene glycol – water mixtures are suitable. An example of the standard that can then be obtained with pure magnesium is given in Fig. 11.4(b). The specimen was finished by a vibratory method using magnesium oxide suspended in propylene glycol; but the surface is probably still slightly etched, so enhancing the polarization contrast. The skidding technique based on a diamond-impregnated paste described on p. 267 can also be applied successfully.

Harder alloys are somewhat easier to finish. A standard casting alloy, finished by the vibratory technique listed in Table 9.3 (p. 279), is illustrated in Fig. 11.4(c) and (d). The etch pitting visible in these photomicrographs was developed during final etching, but similar etch pitting would have developed during polishing if the conditions had been too aggressive chemically. An etch-attack technique, which effectively uses an inhibitor solution, is also available (Table 9.1, p. 274).

MOLYBDENUM

See the section "Precious Metals and Refractory Metals" in Chapter 9, p. 326.

Artifacts

No special problems are to be expected other than those that arise from the abrasion fragmented layer. However, polishing and abrasion rates are low. Consequently, an efficient rough-polishing stage needs to be used and each treatment stage has to be continued for longer than usual (see p. 216). An electromechanical technique is available (Table 9.2, p. 277) which greatly reduces this problem.

Final Polishing

Standard procedures using fine diamond abrasives usually are satisfactory.

NICKEL

Artifacts

No special problems.

Final Polishing

Standard techniques using magnesium oxide or fine alumina are the only ones available. They produce acceptable results in most instances, particularly in harder alloys.

PLATINUM

See the section "Precious Metals and Refractory Metals" in Chapter 9, p. 326.

Artifacts

No special problems except those that arise from the abrasion fragmented layer. However, abrasion and polishing rates are low, and abrasive papers are deteriorated rapidly. Consequently, frequent changes in abrasive papers are necessary and efficient rough-polishing stages are needed. Treatment times at each stage also need to be longer than usual. An electromechanical polishing method is available which alleviates this problem (Table 9.2, p. 277), but it uses a very dangerous solution.

Final Polishing

An acceptable finish usually can be obtained by standard methods using fine diamond abrasives. An example of the standard of results to be expected is given in Fig. 9.41(a) and (b) (p. 329).

SILVER

Artifacts

No special difficulties, except that fragments of abrasive may embed in the surface of the softest pure silver during abrasion (p. 131).

Final Polishing

It is difficult to obtain high-quality final polishes on silver and silver alloys. Adequate finishes can be obtained with fine diamond abrasives if it is acceptable for the surface to be etched fairly deeply (Fig. 11.1c). A surface that will accept a lighter etch can be obtained by standard methods using magnesium oxide (Fig. 11.1d), but with some difficulty.

A chemical polish is available (Table 9.4, p. 286) which might be considered for a brief final treatment after rough polishing by mechanical methods. This method was developed specifically for polishing of single crystals. For highly perfect single crystals, the use of apparatus of the type sketched in Fig. 11.9 is needed. The use of apparatus of this nature would be an advantage with all types of specimens, provided that the throughput of specimens justified its construction.

TIN

Artifacts

Tin and dilute tin alloys are susceptible to abrasion artifacts of both recrystallized grains and massive twins. Although abrasion and polishing rates are high, efficient rough polishing and reasonably long treatment times are necessary to produce artifact-free surfaces. Fragments of abrasive may embed in the surfaces of the softer tin alloys during abrasion (p. 131).

Final Polishing

Pure tin is not sensitive to polishing artifacts, and thus a moderate-quality finish produced by standard methods using the finest abrasives often is acceptable. Advanced techniques such as skidding on magnesium oxide are necessary, however, to produce highest-quality results. Surfaces can then be produced even on high-purity tin which can be examined either in polarized light without etching (Fig. 11.10a) or in bright-field illumination after etching (Fig. 11.10b). Harder alloys such as those used for bearing metals can be adequately polished by standard methods, particularly if magnesium oxide is used as the abrasive (Fig. 11.10c and d).

Tin is widely used to coat metals — particularly steels — for corrosion protection. These coatings tend to be very thin, and the interface between the tin layer and the base metal is often of interest. Examination of such coatings consequently presents a special case of edge preservation. It is virtually essential that the surface of the section be protected by an electrodeposit for this purpose (p. 300). Protection by an electrodeposit is particularly necessary in taper sectioning (p. 137) — a sectioning technique that can often be applied with profit where thin surface

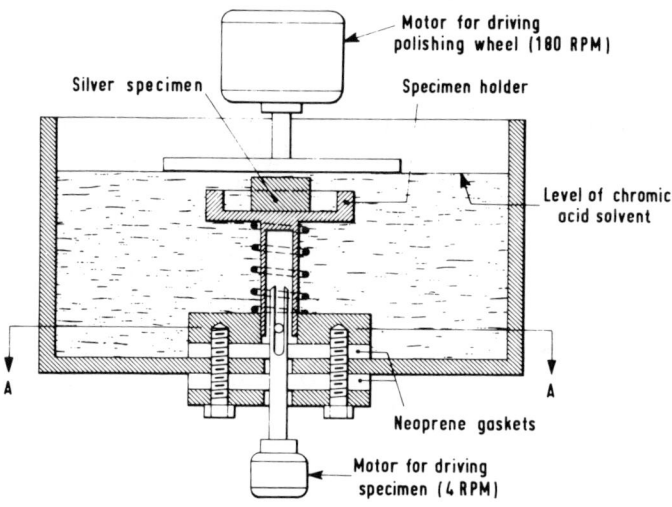

FIG. 11.9. Sketch of apparatus for chemical polishing of silver. *(Gilpin and Worzala, Ref 4)*

The lower surface of the upper driven disk is covered with an acid-resistant napless cloth. The specimen holder is spring loaded so that very light contact is maintained between the specimen surface and the cloth. No abrasive is used.

layers of the type under discussion are to be examined. Standard preparation methods will then adequately retain and polish the tinned layer. It may be helpful, however, to intersperse etching and polishing treatments, using an etchant which attacks the base metal and the electrodeposit but not the tinned layer. The intermediate polishing treatments are continued for just long enough to remove the etch.

TITANIUM

Artifacts

Dilute alloys are susceptible to abrasion artifacts in the form of both massive twins and strain-induced transformation products. Polishing rates are low. A rough-polishing stage of maximum cutting efficiency consequently is highly desirable, and care must be taken to ensure that polishing is continued for an adequate period of time.

Some alloys contain unstable beta phases, and these alloys are susceptible to structural alteration if significant surface heating occurs during machining or abrasion. The precautions discussed on p. 128 then need to be adopted.

Some alloys which have abnormally high hydrogen contents contain precipitates of a hydride phase. These hydrides may dissolve and reprecipitate, and their distribution may be altered, if the alloy is hot mounted or otherwise heated.

Final Polishing

An adequate finish can be obtained by standard techniques using 0.1-μm gamma aluminum oxide, particularly with harder alloys. Use of one of the etch-attack techniques listed in Table 9.1 (p. 274) is often advantageous. The chromic acid solution given in this table is described as being suitable for rough polishing, but can also be used for final polishing

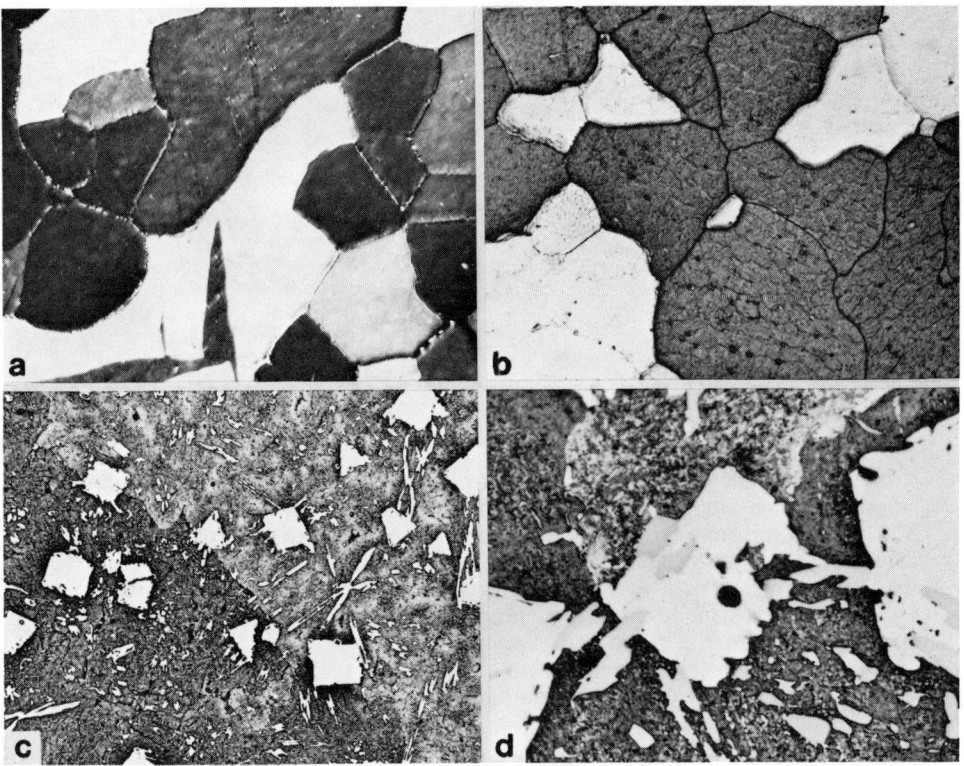

FIG. 11.10. Tin and tin alloys finish polished by skidding and standard techniques.

(a) Pure tin finish polished by skidding on magnesium oxide. Polarized light. Magnification, 100×. (b) Pure tin finish polished by skidding on magnesium oxide. Etched in 10% ammonium persulfate solution. Bright-field illumination. Magnification, 100×. (c) A tin-base bearing metal (10% Sb, 4% Cu) finish polished by a standard method on magnesium oxide. Etched in a ferric chloride reagent. Bright-field illumination. Magnification, 100×. (d) As for (c). Magnification, 500×.

The structure in (a) and (b) is composed of equiaxed grains with some twins. The structure in (c) and (d) is composed of cubes of a Sn-Sb compound, plates of Cu_6Sn_5, and a matrix of tin-rich Sn-Sb solid solution.

FIG. 11.11. Titanium and titanium alloys finish polished on fine aluminum oxide and by an etch-attack method.

(a) Titanium alloy (6% Al, 4% V) prepared by standard procedures and finish polished using fine aluminum oxide. Magnification, 100×. (b) As for (a). Magnification, 500×. (c) Commercially pure titanium prepared by standard procedures but finish polished by an etch-attack method using a chromic oxide solution. Magnification, 100×. (d) Titanium alloy (6% Al, 4% V) prepared by standard procedures but finish polished by an etch-attack method using a chromic oxide solution. Magnification, 250×. All specimens etched in a mixed-acid reagent.

The structures in (a), (b) and (d) are composed of plates of alpha phase with interleaving areas of beta phase. The structure in (c) is composed entirely of equiaxed grains of alpha phase.

with a fine aluminum oxide abrasive. These techniques are likely to introduce some relief between grains and phases, but this can be kept within acceptable limits. They are certainly easier to control if they are applied to surfaces that have already been reasonably well polished on diamond abrasives than if they are used for rough polishing as well. Examples of the standard of results that can be obtained are given in Fig. 11.11.

TUNGSTEN

See the section "Precious Metals and Refractory Metals" in Chapter 9, p. 326.

Artifacts

Artifacts are likely to originate only from the abrasion fragmented layer. However, tungsten causes rapid deterioration of abrasive papers, and this must be kept in mind during the abrasion stages of preparation. Moreover, polishing rates are low. Hence a rough-polishing stage of maximum cutting efficiency should be used. An electromechanical technique is available (Table 9.2, p. 277).

Final Polishing

Adequate finishes usually can be obtained by standard methods with fine diamond abrasives (Fig. 9.41c and d, p. 329). Etch-attack techniques are also available (Table 9.1, p. 274), although they tend to give barely satisfactory results (Fig. 9.5a, p. 273). Electromechanical methods give much more satisfactory results (Fig. 9.5b).

URANIUM

Artifacts

Uranium and its alloys are susceptible to abrasion artifacts in the form of massive twins. Abrasion and polishing rates are relatively high. Little difficulty is encountered in removing the abrasion artifacts with an efficient rough-polishing process.

Final Polishing

The most satisfactory method of revealing the structures of uranium and uranium alloys is examination under polarized light (see p. 330). Results satisfactory for this purpose can be attained by skidding on magnesium oxide, although this is difficult. Etch-attack techniques which are simpler to apply are available (Table 9.1, p. 274). The most satisfactory of these etch-attack techniques probably is the one using hydrogen peroxide. The hydrogen peroxide oxidizes the surface and the abrasive continuously removes the oxide layer at a rate sufficient for the reaction to proceed.[5] It is essential that the oxide buildup should not outpace the removal. Polarization contrast may be further improved by immersing the specimen in hydrogen peroxide for 5 min after polishing; this forms a film of reaction product which, if need be, can subsequently be removed by dipping in dilute sulfuric acid.

Note on Safe Handling

Uranium is mildly radioactive. Moreover, a toxic dust may be produced during its preparation. Uranium should be prepared only by personnel who have been trained in its safe handling.

ZINC

Artifacts

Zinc is susceptible to abrasion artifacts of the recrystallized grain and massive twin types. These artifacts and their origins have been illustrated (p. 125 and 226). Abrasion and polishing rates are high, so that it is not too difficult to eliminate the abrasion artifacts. However, an efficient-cutting rough-polishing stage and comparatively long polishing times are necessary. Smaller twin artifacts may be produced during polishing, features which were illustrated on p. 240 and 332.

The most common zinc alloy is that used for die casting. Difficulties involving artifacts do not arise with this alloy.

Final Polishing

Pure zinc and dilute zinc alloys may be examined in bright-field illumination, perhaps after etching, in which case any good standard final polishing method may suffice. They may, however, also be examined under polarized light; the problems that then arise were discussed on p. 330 and 332. Advanced polishing methods with significant chemical components are needed to achieve the best results, such as those illustrated in Fig. 4.5 (p. 95), 8.20(c) (p. 240) and 9.42(a) (p. 331). Spurious etch pitting may be developed in the polished surfaces unless precautions are taken to eliminate electrochemical effects (p. 271).

More concentrated alloys of zinc, such as those used for die-casting alloys, are simple to finish. Good standard finish-polishing methods suffice, the standard of results to be expected being illustrated by Fig. 11.6(c) and (d).

Zinc is commonly used to coat metals — particularly steels — for corrosion protection. Zinc coatings can be prepared by standard methods provided that precautions are taken to prevent electrolytic etching effects between the coating and the base metal (see p. 247). An example of the standard of results to be expected is given in Fig. 8.23(b) (p. 247).

ZIRCONIUM

Artifacts

Zirconium and zirconium-rich solid solutions are extremely susceptible to abrasion artifacts of the massive twin type. Abrasion and polishing rates are low. A rough-polishing process of maximum cutting efficiency must therefore be used and the treatment at each preparation stage must be of adequate duration.

Deformation artifacts may also be introduced by any mounting or other process which applies a pressure exceeding 1.8 MPa (250 psi).

Final Polishing

The use of the etch-attack technique listed in Table 9.1 (p. 274) is desirable, particularly if the specimen is to be examined in polarized light. A 0.5% hydrofluoric acid solution can be used as a first try.[6] This etchant forms an insoluble film on the surface, and the rate of formation of the film is greater than its rate of removal by the abrasive. Consequently, a visible surface film forms during polishing. This film can be removed at the end of the main polishing sequence by adding distilled water to the polishing slurry to dilute the etchant; polishing is then continued for as long as necessary to remove the film. Chemical polishing methods are available (Table 9.4, p. 286) which might be considered for brief final polishing after rough polishing by mechanical methods.

REFERENCES

1. "Metals Handbook" (8th Edition), Vol. 8, "Metallography, Structures and Phase Diagrams", American Society for Metals, Metals Park, OH, 1973.
2. T. Piotrowski and D. J. Accinno, *Metallography*, 1977, *10*, 243.
3. L. E. Samuels, "Optical Microscopy of Carbon Steels", American Society for Metals, Metals Park, OH, 1979.
4. C. B. Gilpin and F. J. Worzala, *Rev. Sci. Inst.*, 1964, 35, 229.
5. J. F. R. Ambler and G. F. Slattery, *J. Nuclear Materials*, 1961, *4*, 90.
6. J. F. R. Ambler, *Int. Metallographic Soc. Proc.*, 1970, 1.

APPENDIX:
Glossary of Terms Used in Metallography

Abrasion Artifact. A false structure introduced during an abrasion stage of a surface-preparation sequence.

Abrasion Fluid. A liquid added to an abrasion system. The liquid may act as a lubricant, as a coolant, or as a means of flushing abrasion debris away from the abrasion track.

Abrasion Process. An abrasive machining process in which the surface of the workpiece is rubbed against a two-dimensional array of abrasive particles under approximately constant load.

Abrasion Rate. The rate at which material is removed from a surface during an abrasion process. Usually expressed in terms of the thickness removed per unit of either time or distance traversed.

Abrasive. A substance which is capable of removing material from another substance in a machining, abrasion or polishing process. It usually takes the form of a number of small, irregularly shaped particles of a hard material.

Abrasive Cloth. A coated abrasive product in which a cloth is used as the backing sheet.

Abrasive Device. Any arrangement of abrasive particles used to abrade or grind a surface.

Abrasive Machining. A machining process in which the points of abrasive particles are used as machining tools. Grinding is a typical abrasive machining process.

Abrasive Paper. A coated abrasive product in which a paper is used as the backing sheet.

Alternate Etch and Polish Technique. A polishing technique in which the polished surface is etched at intervals during polishing, each intervening polishing stage ideally being continued for just long enough to remove the effects of the preceding etch. This technique is commonly used to increase the material removal rate, an objective for which it is not very appropriate. It may also be employed to reduce the relief developed between phases during polishing, by use of an etchant which attacks preferentially the slower-polishing phase. This use is an appropriate one.

Artifact. A false structure introduced during the preparation of a surface. See also *Abrasion Artifact*, *Mounting Artifact* and *Polishing Artifact*.

Beilby Layer. A layer that was once thought to cover mechanically polished surfaces. Smoothing of the surfaces during polishing was thought to occur by the layer being spread over the surface to fill irregularities, and the layer was thought to have an amorphous, or at least an amorphous-like, structure. Now known not to exist.

Beilby Smearing. Lateral spreading of a surface layer during mechanical polishing by a mechanism implied by the Beilby theory of polishing. See *Beilby Layer*.

Belt Surfacing. An abrasion process in which the specimen is held against a coated abrasive product in the form of a moving endless belt.

Burnishing. Smoothing a surface by rubbing with a smooth tool of a harder material.

Capping (of abrasive particles). A mechanism of deterioration of abrasive points in which the points become covered by caps of adherent abrasion debris.

Chemical Machining. See *Etch Machining*.

Chemical Polishing. A process which produces a polished surface by the action of a chemical etching solution. The etching solution is compounded in such a way that peaks in the topography of the surface are dissolved preferentially.

Chip. A fragment of material removed from a surface during a machining process. See *Fracture Chip* and *Machining Chip*.

Clearance Angle (of a machining tool). The angle between the clearance face of a machining tool and the new surface that is being produced.

Cleavage Crack. A crack that extends along a plane of easy cleavage in a crystalline material.

Clogging (of an abrasive device). A mechanism of deterioration of an abrasive device by which the spaces between the active particles of the abrasive device become filled by a packed mass of abrasion debris.

Cloth. Material woven or felted from either natural or artificial fibers.

Coated Abrasive Product. A two-body abrasion device in which a backing of paper or cloth is coated with a layer of abrasive grits, the abrasive grits being cemented to the backing.

Comet Tails (on a polished surface). A group of comparatively deep unidirectional scratches formed adjacent to a microstructural discontinuity during mechanical polishing. They have the general shape of a comet tail. Comet tails form only when a unidirectional motion is maintained between the surface being polished and the polishing cloth.

Cone Crack. A ring crack formed at the surface of a brittle material which has extended into the material approximately along the surface of an expanding cone.

Corundum. A naturally occurring and impure form of α aluminum oxide. A purer form of the oxide than emery.

Critical Rake Angle. The rake angle at which the action of a vee-point tool changes from cutting to plowing.

Cutting. A machining process in which a layer of material is separated from the surface of a workpiece. The separated layer is known as a machining chip.

Damaged Layer. A layer produced during machining of a surface in which any characteristic of the base metal has been changed. The implication of "damaged" is that the changes have been detrimental, but this is not necessarily so.

Deformed Layer. A plastically deformed surface layer produced during machining. Constitutes one form of damaged layer. See also *Fragmented Layer* and *Significant Deformation*.

Depth of Field. The depth in the subject over which features can be seen to be acceptably in focus in the final image produced by a microscope.

Diffuse Reflection. The condition wherein all of the incident light is reflected at random angles with respect to the angle of incidence. The reflecting surface then has a dull or matte appearance to the naked eye.

Edge-Trailing Technique. The use of a unidirectional motion perpendicular to and towards one edge of the specimen during abrasion or polishing. The aim is to improve the retention of the edge concerned.

Electric Discharge Machining. Controlled removal of material achieved by establishing high-energy sparks between the workpiece and an electrode of appropriate shape. Each spark melts a small volume of material on the workpiece surface, some of this material being ejected when the arc breaks.

Electrolytic Polishing. A chemical polishing process in which the metal to be polished forms the anode in an electrolytic cell. Conditions are arranged so that peaks in the surface topography are removed preferentially so that the surface becomes specularly reflecting.

Electromechanical Polishing. An etch-attack polishing method in which the chemical action of the polishing fluid is enhanced or controlled by the application of an electric current between the specimen and the polishing wheel.

Embedded Abrasive. Fragments of abrasive particles forced into the surface of a workpiece during a grinding, abrasion or polishing process.

Emery. A naturally occurring and impure form of α aluminum oxide. A less pure form of the oxide than corundum.

Epitaxy. A term used to describe the phenomenon in which a crystalline deposit grows on a crystalline substrate with the same crystal orientation as that of the substrate.

Etch-Attack Polishing. A mechanical polishing method in which the polishing fluid is chemically active with respect to the material being polished, but only to the extent that it supplements the mechanical action of the abrasive.

Etch Cutting. See *Etch Machining*.

Etch Machining. Removal of material in a controlled manner by chemical dissolution. An etching reagent is applied continuously to, but only to, that region of the surface that is to be removed.

Felt Cloth. A cloth made by aggregating irregularly aligned fibers.

Fiber. An individual filament of a natural or artificial product. Threads are composed of a number of fibers that have been twisted (or spun) together. Threads are woven into cloth.

Final-Polishing Process. A polishing process the primary objective of which is to produce a final surface suitable for microscopical examination.

Fracture Chip. A fragment of material broken out of the surface of a workpiece by the action of a machining tool.

Fragmented Layer. The outermost portion of the deformed layer produced by machining processes in which the plastic strains are particularly high. The structure of the layer is distorted correspondingly.

Graded Abrasive. An abrasive powder in which the sizes of the individual particles are confined within certain agreed-upon limits.

Grinding. An abrasive machining process employing a vitreous-bonded abrasive device. Relative speeds between abrasive and workpiece characteristically are high in comparison to those encountered during abrasion.

Grit. Individual particles of abrasive.

Heat-Distortion Temperature (of a plastic). The temperature at which a plastic begins to deform under load. The temperature must be related to an acceptable strain rate at a given applied stress.

Lap. A term that may be applied to any abrasive device that produces a fine finish. Frequently, it is a device in which the abrasive particles are supported by a flat plate of metal, glass or plastic.

Levigation. A process by which a powder is separated into a fraction with a restricted range of particle sizes.

Machining Chip. A ribbon of material separated from the surface of the workpiece during a machining process.

Machining Process. A process in which material is removed from a surface by a tool of a harder material, the tool preferably being of a specific geometry.

Mechanical Polishing. A process which produces a specularly reflecting surface entirely by the action of machining tools, usually the points of abrasive particles.

Mount. A device by which a specimen may be held during preparation of a section surface.

Mounting Artifact. A false structure introduced during the mounting stages of a surface-preparation sequence.

Napped Cloth. A woven cloth in which some fibers are aligned approximately normal to one of its surfaces.

Oriented Overgrowth. Alternative term for *Epitaxy*.

Paper. A thin sheet made of compactly interlacing fibers.

Perpendicular Section. A section cut perpendicular to a surface of interest in a specimen. Compare with *Taper Section*.

Plastic Smearing. Lateral spreading of the surface layers of a specimen by normal plastic deformation.

Plowing. A machining process in which a groove is produced in a surface but material is not separated. Material is displaced plastically ahead of, and to the side of, the machining tool.

Polished Surface. A surface which reflects a large proportion of the incident light in a specular manner.

Polishing Artifact. A false structure introduced during a polishing stage of a surface-preparation sequence.

Polishing Rate. The rate at which material is removed from a surface during a polishing process. Usually expressed in terms of the thickness removed per unit of either time or distance traversed.

Rake Angle (of a cutting tool). The angle between the rake face of a cutting tool (the face over which the separating chip passes) and the normal to the workpiece surface. The angle is reckoned to be positive when the rake face slopes away from the direction of relative motion of the tool, and negative when the rake face slopes toward the direction of relative motion.

Regular Reflection. Alternative term for *Specular Reflection*.

Ring Crack. A crack formed at the surface of a brittle solid when an indenter is forced into the surface, the crack extending around the line of contact between the indenter and the surface.

Rough-Polishing Process. A polishing process the primary objective of which is to remove the layer of significant damage produced during earlier machining and abrasion stages of a preparation sequence. A secondary objective is to produce a finish of such quality that a final polish can be produced subsequently with minimum difficulty.

Scratch. A groove produced in a surface by an abrasive point.

Scratch Trace. A line of etch markings produced on a surface at the site of a pre-existing scratch, the physical groove of the scratch having been removed. The scratch trace develops when the ray of deformed material extending beneath the scratch has not been removed with the scratch groove, and when the residual deformed material is attacked preferentially during etching.

Serial Sectioning. A technique in which an identified area on a section surface is observed repeatedly after successive layers of known thickness have been removed from the surface. Thereby a picture of the three-dimensional morphology of structural features can be constructed.

Shales. Abrasive particles of platelike shape. The term is applied particularly to diamond abrasives.

Shelling. A mechanism of deterioration of coated abrasive products in which entire abrasive grains are removed from the cement coating that held the abrasive to the backing layer of the product.

Significant Deformation. That deformation introduced into the surface layers during the preparation of a surface which is likely to affect significantly the observations that are to be made on the surface.

Skid-Polishing Process. A mechanical polishing process in which the surface to be polished is made to skid across a layer of paste comprising the abrasive and the polishing fluid without contacting the fibers of the polishing cloth.

Slivers. Abrasive particles of rodlike shape, the rods having an aspect ratio greater than three. The term is applied particularly to diamond abrasives.

Specular Reflection. The condition wherein all of the incident light is reflected at the same angle as the angle of the incident light with respect to the normal at the point of incidence. The reflecting surface then appears to be bright, or mirrorlike, when viewed with the naked eye.

Strain Markings. Manifestations of prior plastic deformation that are visible after etching of a metallographic section. These markings may be referred to as "slip strain markings", "twin strain markings", etc., to indicate the specific deformation mechanism of which they are a manifestation.

Taper Section. A section cut at an acute angle to a surface. Layers parallel to the surface are then geometrically magnified by a factor which depends on the acuteness of the section angle.

Tempered Layer. A surface or subsurface layer in a steel specimen which has been tempered by heating during some stage of the preparation sequence. When observed in a section after etching, the layer appears darker than the base material.

Thread. Twisted filaments of fiber.

Three-Body Abrasion Process. An abrasion process in which the individual abrasive particles can roll between the workpiece surface and another hard surface. The abrasive powder is usually spread thinly over the hard surface and may be pressed into the surface with a roller.

Two-Body Abrasion Process. An abrasion process in which the individual abrasive particles are fixed in space, presenting a two-dimensional array of points to the surface of the workpiece. The workpiece is pressed against and translated past the array of points.

Vent Crack. A subsurface crack formed adjacent to an indentation in a brittle material. A *median* vent crack forms beneath the point of the indentation and extends into the workpiece. A *lateral* vent crack forms at the sides of the indentation and extends from the base of the indentation toward the workpiece surface.

Vibratory Polishing Process. A mechanical polishing process in which the specimen is made to move around the polishing cloth by imparting a suitable vibratory motion to the polishing system.

White-Etching Layer. A surface layer in a steel which, as viewed in a section after etching, appears whiter than the base metal. The presence of the layer may be due to a number of causes, including plastic deformation induced by machining or surface rubbing, heating during a preparation stage to such an extent that the layer is austenitized and then hardened during cooling, and diffusion of extraneous elements into the surface.

Woven Cloth. A cloth formed by interlacing long threads running in two perpendicular directions.

Index

NOTE: Materials (principally metals and alloys) are indexed when there is substantial discussion in text or data presented in tables. Materials are not indexed when they are cited only for purposes of illustrating a generally applicable principle or practice. The symbol (T) following a page number indicates that information on the subject is presented in a table. —M.H.

Abrasion artifacts. *See* Artifacts, abrasion
Abrasion fluids, 373. *See also* Polishing fluids
   influence on abrasion rate, 54, 59, 62, 63, 65
   use with cut-off devices, 249, 250
Abrasion processes, 373. *See also* Abrasive machining; Abrasive polishing
   hand, 251-252, 253-254, 292
   metallographic, 251-253
   metallographic polishing, 253-260
Abrasion rates, 373. *See also* Polishing rates
   aluminum oxide nonwaterproof papers, 58, 70, 71
   deterioration curves and materials groups, 55-70
   diamond laps, 71-76(T), 77
   effects on edge retention, 289, 290-291
   influencing factors, 48-55, 57-69
      abrasion fluids, 54, 59(T), 62, 63, 65
      abrasive grade, 58, 60, 63-64, 69(T)
      abrasive hardness and fracture characteristics, 52-54, 60-62, 65-69
      abrasive type, 57-58(T), 63
      applied pressure, 57, 63
      batch and brand of paper, 59-60, 65
      material hardness, 48-50, 52-53, 55, 57(T), 62-68, 73-77(T), 177(T), 187-189, 208, 316-318
      rake angle, 50-52, 53
      soaking papers in water, 59, 65
      specimen material, 48, 57(T), 58-59, 60-61, 65
      work hardening, 48-49
   mathematical model for, 46-48, 206-208

   measurement, 55, 80-81
   of mounting plastics, 16, 17-18, 289, 290(T), 298, 299, 300
   of precious and refractory metals, 326, 327(T)
   silicon carbide and aluminum oxide abrasive papers, 55-71
Abrasive belt surfacers, 250
Abrasive hardness, relationship to abrasive machining and polishing rates, 52-53
Abrasive machining, 373. *See also* Abrasion processes
   artifacts, 216-233
   coated papers for, 40-46, 55-71
   depth of deformed layer, 110-114(T), 123(T), 124(T), 125, 216
   equipment and methods, 251-253, 259-260
   for edge retention, 291-292
   mechanisms, 32-40, 46-48, 51-52
      brittle materials, 203-209
   models for, 32-40, 46-48, 206-208
   preliminary, 250-251
   principles, 31-85
   surface deformation by, 87-140
Abrasive paper, 40, 373. *See also* Coated abrasive papers
Abrasive polishing. *See also* Abrasion processes; Polishing, final; Polishing, rough
   artifacts, 233-241
   depth of deformed layer, 199(T), 200, 201(T), 215, 216
   equipment and methods, 253-260, 297
   mechanisms, 142-161
      brittle materials, 203-209

Abrasive polishing *(continued)*
  optimum conditions, 190-191
  principles, 141-194
  surface deformation by, 195-202
  variables in, 161-169
    abrasive types, 162-167(T)
    polishing cloth types, 167-169, 170
Abrasive/workpiece interaction in abrasive machining, 32-40, 47
Abrasives. *See also* Diamond abrasives; Embedded abrasives
  effect on abrasion rate, 52-54, 57-58, 60-62, 65-69
  effect on edge retention, 290-291
  effect on polishing rate, 162, 165(T)
  for final polishing, 244-247, 267-269, 270, 271
  for graphite retention in cast irons, 76, 308, 309
  for hard materials, 317-318
  for retention of nonmetallic inclusions, 75, 76, 306-307
  for rough polishing, 162-167
  for skidding techniques, 267-269, 270, 271
  for soft materials, 316
  for vibratory polishing, 282-283
  grain size, 80, 81(T)
  shapes and geometry of cutting points, 32-34, 39-40, 44-46, 47, 54
  types and grades, 31-32, 80, 81(T)
Acrylic plastic specimen mounts, 13(T), 14, 19, 22, 23
  abrasion and polishing rates, 290(T)
Allyl plastic specimen mounts, 13(T), 14, 21
  abrasion and polishing rates, 290(T), 291
Alternate etch and polish technique, 285, 288, 373
Alumina. *See* Aluminum oxide
Aluminum and aluminum alloys
  abrasion rates, 56, 57-62(T)
  artifacts, 230, 271, 352
  embedding of abrasives in, 132, 133
  etch pitting, 269, 271, 353, 354, 355
  polishing rates, 177(T)
  polishing techniques, 270, 271, 279(T), 281, 283, 284, 285, 286(T), 352-354, 355
Aluminum oxide
  abrasive grain size, 80, 81(T)
  abrasives for polishing, 162-163, 165(T), 244-247
    grading method, 191-192
    particle sizes, 192
    polishing rates, 246-247
    preparation for final polishing, 245-246
    skidding techniques, 268, 269
    vibratory polishing, 282-283
  crystal structure and hardness, 31
  filler for mounting plastics, 15, 290(T), 298, 300

Aluminum oxide abrasive papers
  abrasion rates with, 56, 57(T), 58, 60, 64, 65
  characteristics, 46, 55-71
  fractured points in, 65-68
  nonwaterproof types, 58, 70, 71
  used in preparation sequences, 242(T)
Ammonium hydroxide – hydrogen peroxide etchant, 91, 134-135
Amorphous layer in Beilby polishing theory, 141, 143, 154
Arc cutting. *See* Spark cutting and machining
Artifacts, 373
  abrasion, 216-233, 373
    crack type, 228, 230
    in deformed layer, 220-233
    in fragmented layer, 216-220
    strain-induced, 225
    twin-type, 225-227
  in common metals and alloys, 351-371
  in hard materials, 318, 319
  in oxide layers, 324, 325
  in precious and refractory metals, 327, 328
  polishing, 233-241, 376
    scratch traces, 237-241
  preparation, 87, 141, 217
    removal methods, 217-218
  tempering, 227, 229, 362
  time required for removal, 223-225(T), 227

Banded structures. *See also* Strain markings
  as indication of artifacts, 227, 228
  optical microscopy of, 87-95
Beilby layer, 141, 142-145, 146, 147, 149, 150, 151, 198, 373
Beilby theory, 142-145, 146, 147, 149-154
Belt surfacing, 373
  artifacts, 354-356
  polishing techniques, 274(T), 277(T), 356-357
  toxicity, 357
Brashear process for silvering, 301
Brass. *See also* Copper and copper alloys
  abrasive and conventional machining processes for, 112(T), 113(T), 114
  deformed layers in, 89-91, 92, 94-95, 96, 100-105, 115, 195-197, 199(T), 221, 222, 224(T)
  polishing rates and abrasives, 165(T)
Brittle materials. *See also* Nonmetallic inclusions
  abrasion rate, 208
  cracked layer after abrasion, 209-213, 228, 229, 230, 321
  mechanisms of abrasion and polishing, 203-209, 320, 321
  special preparation techniques, 319-322
  surface damage by abrasion, 209-213, 321
  surface scales and oxides, 325

Brittle phases in fragmented layer, 118, 119
Brittle-to-ductile transition in brittle materials, 208

Cadmium and cadmium alloys
    abrasion rates, 57(T)
    polishing techniques, 286(T)
Carrier pastes. *See* Diamond abrasives, carrier pastes
Cast iron, graphite flakes and cavities in, 118-121, 307-313
Cavities
    abrasion artifacts, 219-220
    caused by creep straining, 303-306
    comet tails, 255, 374
    in fragmented zone, 118-119, 121, 219-220, 303-304
    polishing methods for, 255
    special techniques for, 303-305, 306, 307-308, 309
Ceramics and cermets, special preparation techniques, 317
Chemical analysis of alloys, surface preparation effects, 231-232(T)
Chemical-mechanical polishing, 159-161, 270-285
Chemical polishing, 161, 286-288(T), 316, 317, 374
    apparatus for, 355, 356
Chip-cutting mechanism of material removal, 34-36, 38
    brittle materials, 208, 209, 320-321
    efficiency of, 51-52, 53
    in polishing, 154-159
Chip formation
    in brittle materials, 208, 320-321
    in polishing, 155-158, 317-318
    shear strains in, 98, 99, 100
    translations parallel to surface, 107-110
Chips. *See* Fracture chips; Machining chips
Chromium and chromium alloys
    abrasion rates, 56, 57(T)-62, 76, 327(T)
    artifacts, 357
    polishing rates, 177(T), 327
    polishing techniques, 274(T), 277(T), 327(T), 328, 357
    structure and properties, 327(T)
Clamps for specimen mounts, 9-10
Cleaning and drying of specimens, 260-262
Clogging of abrasive papers and cloths, 62, 63, 65, 70, 113, 115, 374
    diamond laps, 72, 73
    diamond polishing cloths, 182-184
    fine laps, 77
Cloths. *See* Polishing cloths
Coated abrasive papers and cloths, 374
    abrasion rates, 55-70

characteristics, 40-46(T), 55-71, 253
deterioration of contacting points, 44-46
diamond laps, 71-76(T), 77
embedding of abrasives from, 133-134
fine fixed-abrasive laps, 76-79
silicon carbide and aluminum oxide, 55-71, 78
use in preparation sequence, 242(T)
vitreous-bonded wheels and laps, 79-80
wear of, 111, 114
Color contrast, in etching, 235-237
Comet tails, 255, 374
Complexing agents for vibratory polishing, 279-281
Composite specimens, edge retention at interface, 302
Conventional machining
    depth of deformed layer, 113(T), 114, 123, 125
    preliminary, 250-251
Copper and copper alloys. *See also* Brass
    abrasion rates, 56, 57-62(T)
    artifacts, 222, 237, 239, 358-359
    deformed layers in, 87-89, 106, 128
    effect of preparation methods on chemical analysis, 231-232
    polishing rates, 177(T)
    polishing techniques, 236, 268, 269, 274(T), 278, 279(T), 282, 283, 286(T), 358-359, 360
Corrosion of polishing equipment, 272-273, 276
Corundum, 31, 374
Cracked layer in brittle materials, 209-213, 228, 229, 230, 321
Cracks
    abrasion artifacts, 228-230
    cone, 205, 206, 374
    in brittle materials, 203-206, 210-213
    lateral vent, 204-206, 211, 212, 377
    median vent, 203-206, 211, 212, 377
    revealed by spark cutting, 242, 243
    ring, 205, 206, 376
Crystal structure and orientation
    effects on surface deformation, 87-91
    polarized light studies, 330-333
Cubic metals, plastic deformation structures, 87-93
Cupric ammonium chloride etchant, 91, 135
Cut-off wheels, vitreous-bonded, 79-80, 248-250
Cutting. *See* Cut-off wheels; Etch cutting; Sectioning equipment; Spark cutting
Cutting mechanisms in abrasion and polishing, 154-159. *See also* Chip-cutting mechanism
    brittle materials, 209, 320-321

Damaged layer in abrasive machining, 111, 374. *See also* Deformed layer
Deformation. *See* Deformed layer; Plastic deformation

Deformation-band strain markings, 94-95, 96, 103, 105, 123-124
Deformed layer, 101-102, 374.
   artifacts in, 220-233
   depth
      maximum, 111
      produced by abrasive machining, 107(T), 110-114(T), 123, 125(T), 215, 216
      produced by abrasive polishing, 199(T), 200, 201(T), 215, 216, 220-221
      produced by grinding, 215, 216
      produced by microtoming, 347-349
      significant, 110-113(T), 123(T), 199, 215, 220-221, 224(T), 237, 238, 376
      stages in preparation sequence, 215, 216
   in brittle materials, 203-206, 209-213, 321
   isostrain boundaries, 102, 104
   structure produced by abrasive machining, 99-110, 121-126
      general pattern of deformation, 99-103
      modifications of, 121-131
      origin, 98
      surface heating effects, 126-131
   structure produced by abrasive polishing, 195-201, 220-222
      modifications of, 199-201
   time required for removal, 223-225(T), 227
Diamond
   crystal structure and hardness, 31, 162, 164, 165(T), 175
   polishing of, 209
Diamond abrasives
   carrier pastes, 164, 166, 171-172, 173, 175-179
      preparation, 192-193
      standard color system, 193
   particle size and shape, 162-166, 172-175
   polishing mechanisms, 157-159, 160
   polishing rates, 162, 165, 172-191
   polycrystalline vs monocrystalline, 162, 164, 165, 175, 176, 177, 189-190
   properties, 166
   skidding techniques with, 267, 268
   slivers and shales, 164, 166, 376
   types and grades for rough polishing, 162-166, 172-177, 242
   used in preparation sequences, 242(T)
Diamond laps
   characteristics and abrasion rates, 71-76(T)
   for hard materials, 317-319
   plastic-abrasive mixtures, 82-83
Diamond tools for microtome cutting, 345, 346
Dislocations in plastic deformation of surfaces, 87, 89, 92, 94, 211, 212
Drill (cotton) polishing cloths, 169, 170, 180, 181, 183
Drying of specimens. See Cleaning and drying
Ductile cutting mode in brittle materials, 208, 209, 211, 320-321

Edge retention, 288-303
   at metal/scale interface, 323-325
   effects of abrasion stages, 289-293(T)
   effects of polishing stages, 294-298
   electrodeposition techniques for, 137, 139, 296, 300-302, 323-324, 333-335
   fine laps for, 76, 78
   mounting plastics for, 289, 290-296, 298-300
   specimen mounting techniques for, 289, 298-303
Edge trailing technique, 292-293, 297, 374
Elastic-plastic boundary in deformed surfaces, 98, 99, 101, 103
Electric discharge machining, 340-344, 374
Electrical conductivity provided in plastic mounts, 18-19, 28-29
Electrochemical polishing. See Electrolytic polishing; Electromechanical polishing
Electrodeposited metals for edge protection, internal stresses in, 302(T), 324
Electroless coating with metals for edge retention, 300, 334
Electrolytic etching, 136, 137
Electrolytic polishing, 137, 161, 277, 278, 288, 305, 306, 374
Electromechanical polishing, 273, 276-278(T), 328-330, 374
Electron diffraction patterns on polished surfaces, 150-154, 155
Electroplating for edge retention, 137, 139, 296, 300-302, 323-324, 333-335
Embedded abrasives, 131-134, 201, 313-314, 315, 375
   in polishing cloths, 157-159, 160
Emery, 31, 113, 375
Epitaxy, 149-151, 330, 375
Epoxy plastic specimen mounts, 13(T), 14-20, 21, 23-25, 28-30
   abrasion and polishing rates, 290(T)
   filled, 15-16, 290, 298, 299-300
Equipment. See Preparation systems, equipment and devices
Etch and polish. See Alternate etch and polish technique
Etch-attack techniques of polishing, 272-273, 274-275(T), 328, 330. See also Vibratory polishing, etch-attack method
   for edge retention, 295
Etch cutting and machining, 339, 340, 375
Etch faceting. See Faceting
Etch pits in aluminum alloys, 269, 271, 353, 354, 355
Etchants, 134-137
   for etch-attack polishing, 273, 274-275(T)
   for revealing abrasion and polishing artifacts, 220-221, 222, 237-239
   in vibratory polishing, 279(T), 284

Etching. *See also* Alternate etch and polish technique
  in chemical-mechanical polishing methods, 159-161, 270-275, 284
  polishing effects on color contrast, 235-237
  polishing scratches shown by, 234-237
  scratch trace development by, 237-241
Etching methods, 134-137
  for showing plastic deformation, 101

Face-centered cubic metals. *See* Cubic metals
Faceting, polarization contrast effects, 330-332
False structures. *See* Artifacts
Felt polishing cloths, 167, 168, 375
Ferric chloride etchant, 135
Filled plastics. *See* Mounting plastics, filled
Fine fixed-abrasive laps, 76-79, 83-84
  use in preparation sequence, 242
Fine polishing. *See* Polishing, final
Form factor for abrasive points, 38-40, 42(T), 44
Formvar plastic specimen mounts, 13(T), 14, 18, 21, 22
  abrasion and polishing rates, 290(T), 291
Fraction of contacting points that cut in abrasive machining, 38(T), 41-44(T)
Fracture chips, 34, 205-206, 209, 317-319, 320, 375
Fracture of abrasive cutting points, 44-46, 54, 60-62, 65-69
Fragmented layer, 100, 375
  artifacts in, 216-220
  depth, 107, 110-111, 114, 199(T), 216
  metallographic characteristics, 102, 114-121
  on abraded surfaces, 103-107, 128-129, 130, 216-220, 221
  on microtomed surfaces, 349
  on polished surfaces, 195-197, 199, 237, 238
  transformed layer, 115, 117
  white-etching, 115, 116, 117, 129, 130, 377

Gadolinium, polishing techniques, 286
Germanium
  artifacts, 228, 229
  polishing techniques, 286(T)
Glasses, polishing, 208-209
Glassy layer. *See* Beilby layer
Gold and gold alloys
  abrasion rates, 56, 57(T)-62, 327
  artifacts, 359-360
  polishing rates, 177(T), 327(T)
  polishing techniques, 274(T), 277(T), 352, 360
  structure and properties, 327(T)
Grain color contrast. *See* Color contrast
Graphite
  flakes in fragmented layer of gray cast iron, 118-121

retention methods for ferrous alloys, 307-313, 314
Gray cast iron. *See* Cast iron
Grinding, 375
  depth of deformed layer, 215, 216
  temperature effects on deformed layer, 127-128
Grinding wheels
  for preliminary machining, 250
  vitreous-bonded, 79
Grits, abrasive, 31-32, 375
Grooves. *See also* Scratch grooves; Scratches
  contours in chip cutting and plowing, 51-52, 53, 155

Hand abrasion. *See* Abrasion processes, hand
Hard materials. *See also* Material hardness
  abrasive cutting mechanisms, 38-39
  specimen preparation methods, 316-319, 320
Hardness. *See* Abrasive hardness; Material hardness
Hardness tests, abrasion work hardening effects, 230
Hot spots. *See* Transient temperatures

Inclusions. *See* Nonmetallic inclusions
Indium, polishing, 286(T), 360
Interface retention between metal and surface oxides or scales, 323, 325
Intermetallic phases, polishing techniques, 353-354, 355
Internal stresses. *See* Stresses
Iron, polishing techniques, 286(T)
Iron-base alloys. *See also* Case iron; Steels
  austenitic
    abrasion rates, 56, 57(T)-62
    artifacts, 225, 361
    final polishing, 361
    strain-induced transformations in, 91-93, 121-123
    deformed layers in, 96-98, 108-110, 115-118, 121-123
    surface temperature effects, 126-131
  ferritic
    abrasion rates, 62-69
    artifacts, 218-219, 362
    polishing, 218-219, 362
  martensitic
    artifacts, 362
    final polishing, 362
Iron-base alloys, graphite containing. *See* Cast iron

Kerosine as polishing fluid, 160, 184-187, 188, 201
Kink bands, misorientation under polarized light, 330, 331

Lamellar structures
  fragmented layer in, 117, 118, 216-217
  plastic deformation effects, 95-98
Laps, 36, 375. *See also* Diamond laps; Fine fixed-abrasive laps
Lead and lead alloys
  abrasion rates, 57(T)
  polishing and abrasion techniques, 286-287(T), 313-316, 317, 362, 363
  polishing rates, 177(T)
Levigation, 375
  aluminum oxide polishing abrasives, 191-192
Lubricants. *See* Abrasion fluids; Polishing fluids
Lüders bands, effect of surface preparation on macrostructure, 222-223

Machining. *See* Abrasive machining; Conventional machining; Microtome cutting and machining
Machining chips, 34-35, 98, 99, 100, 155, 156, 205, 208, 209, 375
Macrostructures, effect of surface preparation and artifact removal, 232-233
Magnesia. *See* Magnesium oxide
Magnesium and magnesium alloys artifacts, 356, 362
  polishing techniques, 274(T), 279(T), 356, 362, 364
Magnesium oxide
  abrasives for final polishing, 242, 244-245
  skidding techniques, 268, 269, 270, 271
  vibratory polishing, 282-283
Material hardness. *See also* Hard materials
  relationship to abrasive machining and polishing rates, 48-50, 52-53, 55, 57(T), 62-68, 73-77(T), 177(T), 187-189, 208, 316-318
Material removal
  during polishing, 154, 155
    chemical-mechanical mechanisms, 159-161
  principles, 31-85
    abrasive/workpiece interaction, 32-40, 47
    for brittle materials, 205-208, 209, 320-321
  spark cutting, 340-344
Material removal efficiency, 34, 38, 51-52, 206
Material removal rates. *See also* Abrasion rates; Polishing rates
  for grinding, abrasive machining and polishing, 215, 216, 223-224(T)
Mathematical models for abrasive machining, 32-40, 46-48, 206-208
Melting point, correlation with polishing theory and polishing rates, 147-148, 188-189
Metastable alloys
  phase transformation by plastic deformation, 91-93, 121-123
Microbands in deformed surfaces, 88-89, 197

Micromachining, polishing mechanism, 208, 209
Microscope objectives, depth of field, 288, 289
Microtome cutting and machining, 344-349
  of soft materials, 315, 345
Mineragraphic specimens, preparation methods, 321-322
Mineral-filled plastics. *See* Mounting plastics, filled
Models. *See* Mathematical models
Molding methods for plastic mounts, 16, 17, 21-25
Molds for plastic mounts, manufacture, 29-30
Molybdenum and molybdenum alloys
  artifacts, 364
  polishing techniques, 277(T), 364
Mounting of specimens, 9-30, 375
  clamping, 9-10
  defects caused by molding, 22-23
  electrical conductivity devices, 18-19, 28-29
  fissuring in, 11, 14-17
  for effective edge retention, 289, 298-303
  identification marking, 28
  mount dimensions, 26, 27
  soft materials, 313
  thin shapes, 27-28
Mounting plastics
  abrasion and polishing rates, 16, 17-18, 289, 290(T), 294-296, 298, 299, 300
  applications and fields of usefulness, 20-21
  casting, 23-26, 299
    vacuum impregnation of porous specimens, 25-26
    vacuum treatment of, 24-25
  chemical resistance, 12-14
  effects on edge retention, 289, 290(T)-296, 298-300
  filled, 15-16, 290(T), 298-300
  heat distortion, 12, 13(T), 25, 375
  molding conditions for, 13(T)
  molding techniques, 16, 17, 21-25, 28-30, 299
  properties, 12-20(T)
  reflectivity, 19-20
  removal from specimen, 28
  requirements of, 11-12
  temperature effects, 12, 13(T), 25
  thermal expansion coefficients, 13(T), 15-16
  transparency, 13(T), 19

Napped (piled) polishing cloths, 169, 170, 180, 181, 183, 375
Nickel, deposited on diamond laps, 71, 72
Nickel and nickel alloys
  abrasion rates, 56, 57(T)-62
  polishing rates, 177(T)
  polishing techniques, 287(T), 364
Niobium and niobium alloys, polishing techniques, 277(T)

Nital etchant, 135
Nonmetallic inclusions
  effect on plastic deformation of steel, 118, 119
  special techniques for retention, 77, 78, 305-307
Nylon polishing cloths, 168, 169, 180, 181, 182, 183, 184

Optical-working techniques for brittle specimens, 321-322
Oriented overgrowth of crystals on abraded surfaces, 149-151, 375
Oxalic acid etchant, 136
Oxides, surface, specimen preparation methods, 322-326

Papers. *See* Coated abrasive papers; Polishing papers
Pearlite. *See* Lamellar structures
Phase transformation strain markings, 91-93, 121-123
Phase transformations
  in surface layers caused by heating, 126-131
  strain-induced, 91-93, 121-123
Phenolic-diamond molded laps, 82-83
Phenolic plastic specimen mounts, 13(T), 18, 20
  abrasion and polishing rates, 290(T), 291
Picral etchant, 136
Plastic deformation. *See also* Deformed layer
  by abrasive machining, 87-140
  detection by optical microscopy, 87-98
Plastic-diamond laps, 82-83
Plastic flow, translations parallel to surface, 107-110, 198
Plastic smearing, 199, 375
Plastics. *See* Mounting plastics
Platinum and platinum alloys
  abrasion rates, 327(T)
  artifacts, 364
  polishing rates, 177(T), 327(T)
  polishing techniques, 277(T), 326-329(T), 364
  structure and properties, 327(T)
Plowing mechanism of material displacement, 34-36, 376
  efficiency of material removal, 51-52, 53
  in polishing, 154-159
Polarization contrast, 330-333
Polarized light
  for examination of crystal orientations, facets and epitaxial films, 330
  reflectivity of plastic mounts, 20
Polished surface, 376
  reflectance, 141-142
  strains at, 195-197
Polishing. *See also* Abrasive polishing
  final, 243-248, 267-288, 375
    abrasives for, 244-247, 267-269, 270, 271
    advanced and special preparation methods, 267-288
    basic preparation stages, 243
    chemical-mechanical, 271
    common metals and alloys, 351-371
    equipment and methods, 255, 259-260, 269-270
    etch-attack techniques, 274-275(T)
    fluids, 244, 246, 247, 267-269
    polishing cloths for, 248
    polishing rate, 239-240
    vibratory, 278, 279(T), 285
    vs rough, 161, 224-225, 233
  load application, 187, 255, 279, 281-282
  rough, 376
    basic preparation stages, 242-243
    chemical-mechanical, 270
    equipment and methods, 255, 259-260
    etch-attack techniques, 274-275(T)
    vs final, 161, 224-225, 233
  standard metallographic processes, 141-194
Polishing cloths
  accumulation of polishing debris, 182-184
  effects on edge retention, 296-297
  embedding of abrasives in, 157-159, 160
  for final polishing, 248
  for graphite retention in cast irons, 308-311
  for hard materials, 318-319
  for vibratory polishing, 265
  types, 167-169, 170, 179-184
  washing and recharging, 182
Polishing fluids. *See also* Abrasion fluids
  effect on embedded abrasive, 201
  effect on polishing rates, 184-187, 188
  for final polishing, 244, 246, 247, 267-269
  for vibratory polishing, 279(T)-284, 333
Polishing mechanisms
  Beilby theory, 142-145, 146, 147, 149-154
  Bowden-Hughes theory, 143, 145, 147-148, 154
  chemical and electrochemical, 161, 273, 276-278
  chemical-mechanical, 159-161, 270-272
  mechanical cutting, 154-159
  Samuels and Sanders experiments, 152-154, 155
Polishing papers, 167-168, 180, 181
Polishing rates, 376. *See also* Abrasion rates
  aluminum oxide and magnesium oxide abrasives, 244, 245, 246
  determination, 80-82, 169-171
  effect on edge retention, 294-296
  for typical abrasive types, 162, 165(T)
  influencing factors, 171-191
    agglomeration of polishing debris, 182-184
    applied load, 187, 190
    diamond particle size, 162-166(T), 172-175

Polishing rates *(continued)*
   influencing factors *(continued)*
      diamond types and grades, 172-177(T)
      method of adding diamond abrasive, 171-172, 173, 190
      polishing cloth types, 176, 179-184, 190-191
      polishing fluid, 184-187, 188, 190-191
      quantity of diamond abrasive, 175-179
      specimen material, 177, 187-190
      specimen melting point, 188, 189
   of mounting plastics, 16, 17-18, 289, 290(T), 294-295, 298, 299, 300
   of precious and refractory metals, 177(T), 326, 327(T)
   with skidding techniques, 267, 269
Polishing techniques
   alternate etch and polish, 285, 288
   chemical, 161, 286-288(T), 316, 317
   chemical-mechanical, 159-161, 270-285
   electrolytic, 137, 161, 277, 278, 288, 305, 306
   electromechanical, 273, 276-278(T)
   etch-attack, 272-273, 274-275(T), 295, 328
   skidding, 267-270, 271
   vibratory, 258-259, 262-265, 278, 279(T), 285
Polishing wheels. *See* Rotating disks or wheels
Polymethyl methacrylate. *See* Acrylic plastic specimen mounts
Polythene-diamond molded laps, 83
Polyvinyl chloride specimen mounts, 13(T), 14, 18, 21, 24, 28, 29
   abrasion and polishing rates, 290(T)
Polyvinyl formal. *See* Formvar
Porous materials, vacuum impregnation with plastics, 25-26
Precious metals
   abrasion and polishing rates, 326-327(T)
   specimen preparation methods, 326-330(T)
   structures and properties, 327(T)
Precipitate particles obscured by polishing artifacts, 241
Preparation systems
   advanced and special methods, 267-337
   equipment and devices, 248-265
   nonabrasive techniques, 339-350
   practical procedures, 241-262
      basic sequences and stages, 241-243(T)
   removal of artifacts, 215-241
      precautions in, 223-224
   stages of, 215, 216, 223-225
Propylene glycol for polishing slurries, 244, 279(T), 280, 333

Rake angles, 376
   critical, 36-38(T), 50-52, 53, 374
   in abrasive machining model, 36-40(T), 41, 43-44, 50, 51, 52, 67, 68

Recrystallization
   by plastic deformation, 94, 95, 105-106, 126
   in deformed layer, 125, 126, 200-201
Reflectance. *See* Specular reflection
Refractory metals
   abrasion and polishing rates, 326-327(T)
   specimen preparation methods, 326-330(T)
   structures and properties, 327(T)
Retention of edges. *See* Edge retention
Rhenium, polishing techniques, 277(T)
Ribbon formation in polishing, similarity to chip formation, 155-156
Rotating disks or wheels. *See also* Cut-off wheels
   corrosion by etchants, 272-273
   for abrasive machining, 251, 252-253
   for determining abrasion rates, 80-81
   for polishing, 256-259, 269-270, 276, 365, 366
   specimen-handling devices, 80-81, 252-253, 254, 256-257
Ruthenium, polishing techniques, 277(T), 327(T), 328

Scales, surface, specimen preparation methods, 322-326
Scratch grooves produced by abrasive points, 36-39, 43, 46-48
Scratch traces, 376
   development by etching, 237-241
Scratches, 376
   abrasion
      depth, 111, 112(T), 114, 199
      nature of, 100, 102, 103, 197
      on polished surfaces, 144-145, 146, 147
      removal, 221, 223-224
   effect on polarization contrast, 332-333
   polishing, 144-145, 146, 147, 156-157, 195, 197
      depth, 199(T)
      detection, 233-235
      etching effect, 234-237
Sectioning equipment and methods, 79, 248-250, 376
   etch cutting, 339, 340
   taper sectioning, 100, 101, 137-139
Selvyt polishing cloths, 169, 170, 180
Shear bands in deformed surfaces, 88, 89, 90, 92, 197
Shear zone in surface deformed layer, 98, 99, 100
Shrinkage cavities. *See* Cavities
Significant deformation. *See* Deformed layer, depth, significant
Silicon carbide
   abrasive grain size, 80, 81(T)
   abrasives for polishing, 162-163, 165(T)
   crystal structure and hardness, 31
Silicon carbide abrasive papers
   abrasion rates with, 57(T), 58, 59(T), 64, 65, 69(T)

characteristics, 40-46(T), 55-70
  deterioration of cutting points, 45, 46
  used in preparation sequences, 242(T)
Silicone mold-release compound as abrasion fluid, 59, 65
Silicone rubber, molds for plastic mounts, 24, 29-30
Silver and silver alloys
  abrasion rates, 56, 57(T)-62, 327(T)
  artifacts, 365
  polishing rates, 177(T), 327(T)
  polishing techniques, 277(T), 287(T), 352, 365, 366
  structure and properties, 327(T)
Silvering, for edge retention by electrodeposition, 301, 323
Skidding techniques of fine polishing, 267-270, 271, 376
Sledge microtomes, 344, 345
Slip strain markings in deformed surfaces, 89-91, 94, 96, 102, 195, 196, 242, 243
Smeared layer. *See* Beilby layer; Plastic smearing
Sodium bisulfite etchant, 136
Sodium thiosulfate etchant, 89, 91, 136-137
Soft material
  abrasive cutting mechanisms, 36-38, 51-52, 53
  embedding of abrasives in, 132-133
  etch cutting, 339
  microtome cutting and machining, 315, 345
  polishing techniques, 352-353
  specimen preparation methods, 313-316, 317, 352-353
Sorby, Henry Clifton
  biography, 3-7
  specimens prepared by, 5, 6
Spark cutting and machining, 340-344(T)
Specimen hardness. *See* Material hardness
Specimen mounts. *See* Mounting of specimens; Mounting plastics
Specimen preparation. *See* Preparation systems
Specular reflection of polished surface, 141, 142, 376
Stacking-fault energies, effect on plastic deformation detection, 87-91
Stains, prevention and removal, 261, 262, 312-313
Steels. *See also* Iron-base alloys
  corrosion resistant
    vibratory polishing, 279(T)
  effect of hardness on abrasion rates, 49, 62, 63-64, 65, 66, 68, 70, 71, 73-76
  low-alloy
    vibratory polishing, 279(T)
  polishing rates, 177(T)
  structural changes induced by abrasion, 126-131
  tempering artifacts, 227, 229, 362

Strain-induced phase transformation, 91-93, 121-123, 225
Strain markings, 376
  deformation-band, 94-95, 96, 103, 105, 123-124
  in abrasive-machined surfaces, 87-95
  in abrasive-polished surfaces, 195-197
  microband, 88-89, 197
  shear band, 88, 89, 90, 92, 197
  slip, 89-91, 94, 96, 102, 195, 196, 242, 243
  transformation, 91-93, 121-123
  twin, 90, 91, 92, 94, 95, 102, 103, 105, 123-126, 348-349
Stresses, in electrodeposited metals used for edge retention, 302(T), 324
Suede (synthetic) polishing cloths, 169, 170, 180, 181, 183
Surface deformation. *See* Deformed layer; Plastic deformation; Plastic flow
Surface finish, characterization by mean depth of scratch grooves, 46-47
Surface layers, nonmetallic. *See* Scales
Surface temperature, microstructural effects in abrasive machining, 126-131
Surfaces. *See* Oxides, surface; Polished surfaces; Scales

Taper sectioning, 100, 101, 137-139, 376-377
Temperature. *See* Surface temperature; Transient temperature
Tempering artifacts, 227, 229, 362
Terms used in metallography, 373-377
Thermal expansion of plastics, 13(T), 15-16
Thin specimens, mounting, 27-28
Thorium, polishing techniques, 274(T)
Three-body abrasion, 35, 206-207, 209, 210, 377
Tin and tin alloys
  abrasion rates, 56, 57(T)-62,
  artifacts, 365, 367
  coatings, 365-366
  final polishing, 365, 367
  polishing rates, 177(T)
Titanium and titanium alloys
  abrasion rates, 56, 57(T)-62
  artifacts, 366-367
  polishing rates, 177(T)
  polishing techniques, 274-275(T), 367-368
Trailing the edge. *See* Edge trailing technique
Transient temperatures in abrasive machining and polishing, 126-128, 143, 145-148
Transformations. *See* Phase transformations
Translations parallel to surface. *See* Plastic flow
Tungsten and tungsten alloys
  abrasion rates, 327(T)
  artifacts, 369
  polishing rates, 177(T), 327(T)
  polishing techniques, 273, 275(T), 277(T), 326, 328, 329, 369

Tungsten and tungsten alloys *(continued)*
  structure and properties, 77, 327(T)
Tungsten carbide
  abrasion rates, 76(T)
  polishing rates, 177(T)
  special preparation techniques, 316-319, 320
Twin-band deformation mode, 90, 91, 92, 123-126
Twin strain markings in deformed surfaces, 90, 91, 92, 94, 95, 102, 103, 105, 123-126, 349
Twinning
  in deformed layer, 123-126, 197, 199-200, 201(T), 225-227, 238-239
  massive, 94, 95, 123-126, 225, 330, 331, 349
Two-body abrasion, 36, 206-207, 209, 210, 377

Ultrasonic cleaning of specimens, 260-261
Uranium and uranium alloys
  artifacts, 369
  polishing techniques, 275(T), 369
  safe handling, 369

Vacuum impregnation of porous specimens, 25-26
Vacuum treatment of mounting plastics, 24-25
Vanadium and vanadium alloys, polishing techniques, 277(T)
Velvet polishing cloths, 169, 170
Vibratory amplitude, frequency relationships and measurement, 258-259, 262-264
Vibratory polishing, 258-259, 262-265, 278, 279(T)-285, 377
  etch-attack method, 278, 279(T)-285, 316, 317
  limitations, 284-285
Vitreous-bonded wheels and laps, 79-80, 248-249

Water as polishing fluid, 160, 184-186, 201
Wax-filled abrasive laps, 72, 73, 78, 79, 83-84, 133, 134
Wheels. *See* Cut-off wheels; Grinding wheels; Rotating disks or wheels
White-etching layer. *See* Fragmented layer, white-etching
Wire cutting of specimens, 339, 340
Work hardening
  effect on abrasion and polishing rates, 48-49
  introduction of abrasion artifacts in hardness testing, 230
Woven cloths for polishing, 167-169, 170, 377

X-ray diffraction, artifact patterns, 233
X-ray fluorescence analysis, surface preparation effects, 231-232(T)

Zinc and zinc alloys
  abrasion rates, 57(T)
  artifacts, 226, 238, 240, 370
  coatings, 370
  deformed layers in, 94, 95, 124-126(T)
  polishing rates, 177(T)
  polishing techniques, 287(T), 330, 331, 360, 370
Zirconium and zirconium alloys
  artifacts, 370
  polishing techniques, 275, 287(T), 370-371